THE PROTOZOA

THE PROTOZOA

BY

GARY N. CALKINS, PH.D.

INSTRUCTOR IN ZOÖLOGY, COLUMBIA UNIVERSITY

"Lies dieses Buch, und lern dabey,
Wie gros Gott auch im Kleinem sey."

D. G. L. Huth: Rösel von Rosenhof.

MJP PUBLISHERS

CHENNAI NEW DELHI TIRUNELVELI

First published: 1901
MJP's first reprint: 2012

ISBN 978-81-8094-137-5
Printed and bound in India
MJP 119

MJP PUBLISHERS
New No. 5, Muthu Kalathy Street
Triplicane, Chennai 600 005

BO: 4264/3, First Floor, Ansari Road, Daryaganj, New Delhi-2. ✆98961 92710

BO: 70B, Perumal North Car Street, Tirunelveli Junction, Tirunelveli-1 ✆0462-2330248

To

EDMUND B. WILSON

IN GRATEFUL RECOGNITION OF HIS VALUABLE ADVICE
AND FRIENDLY INTEREST

PREFACE

THE Protozoa not only claim the interest of the professional turalist, but also that of a wider circle of nature students who, th the aid of the microscope, have always found here a fascinating ld for observation and research. In writing the present volume, bodying a summary of the more recent discoveries concerning se minute animals, I have aimed to keep in mind the needs of latter class of naturalists, as well as those who search more eply in the unicellular organisms for the solution of many morpho-ical problems which remain unsolved in the higher animals, or for al processes which afford a transition from the manifestations of in its simplest expression to life as seen in the lower forms invertebrates.

The subject-matter of the volume is treated from three points view: (1) The historical, to which the first chapter is devoted. The comparative, to which five chapters are given: one to the oup of Protozoa as a whole, the other four to the main classes. The general, to which three chapters are devoted. One of these given to the phenomena of old age or senile degeneration in otozoa and renewal of youth through the union of two individuals, d to the bearing of these phenomena upon sexual reproduction in neral. Another is given to the special structures of nuclei and ntrosomes of the Protozoa; this, the most technical chapter in the ok, is introduced because of the growing importance which the otozoa have in the problems of cellular biology, especially with ose dealing with the origin of the division-centre and its accom-nying structures in the cells of the Metazoa. The last chapter devoted to a consideration of the physiology of the Protozoa, with pecial reference to the Protozoa as *organisms* endowed with the wers of coördination and of adaptation, which up to the present ne have eluded physical and chemical analysis.

Every one who works with the Protozoa is mindful of the debt owe to Professor Otto Bütschli, whose indefatigable labors of many years, embodied in his masterly treatise upon this grou of organisms, will remain the standard for years to come. My ow obligation is manifested by the repeated references in these pages In common with other students of the group I am also indebted t Professors R. Hertwig, Maupas, Klebs, Labbé, Verworn, Schaudinn Siedlecki, and a host of others whose labors have thrown so muc light upon the problems connected with the Protozoa. I regret tha I could not make use of Professor Lang's valuable addition to th literature of the Protozoa, as it appeared after the present volum had gone to press.

Many of the above investigators have given admirable figures o the Protozoa which I have freely used, in addition to numerou original drawings, in illustrating the present work. The majorit of the figures copied from other works have been made by my wife whose skilful penwork and unfailing help in preparing the manu script have been of the greatest assistance.

In conclusion I wish to express my deep obligation to Professo Henry F. Osborn for many valuable suggestions and for his gener ous assistance which has made the present volume possible. To Pro fessor Edmund B. Wilson finally, to whom I take the greatest pleasur in dedicating this book, I owe a debt of gratitude not easily ex pressed. I wish to thank him, amongst many other things, for th use of the electrotypes of several figures, for the generous interes and friendly advice throughout, and, above all, for his invaluabl criticisms and suggestions based upon the careful reading of my manuscript.

G. N. C.

TABLE OF CONTENTS

PAGE

List of Figures xiii

INTRODUCTION AND CHAPTER I

Introduction 1
A. Historical Review 5
B. Modern Classification of the Protozoa 15
C. Animals and Plants 22
D. Generation *de novo* 26

CHAPTER II

General Sketch

A. General Morphology 34
   1. The Endoplasm 34
   2. The Ectoplasm 38
   3. Nuclei 40
   4. Organs of Locomotion 43
B. General Physiology 46
   1. Nutrition 48
   2. Excretion 52
   3. Reproduction 54
   4. Irritability 60
C. Some Economic Aspects of the Protozoa 62

CHAPTER III

The Sarcodina

A. Shells and Tests 71
B. Pseudopodia 79
C. The Nucleus 86
D. The Contractile Vacuole 87
E. Encystment 90
F. Nutrition 90
G. Reproduction 92
H. Inter-relationships of the Sarcodina 99
Classification 105

CHAPTER IV

THE MASTIGOPHORA

PAGE

A. Protoplasmic Structure 11
B. The Flagella 11
C. The Nucleus 12
D. Food-taking 12
E. Vacuoles 12
F. Reproduction 12
G. Inter-relationships of the Mastigophora 13
Classification 13

CHAPTER V

THE SPOROZOA

A. Protoplasmic Structure 14
B. The Nucleus 14
C. Food-taking 14
D. Motion 14
E. Reproduction 14
F. Inter-relationships of the Sporozoa 16
Classification 16

CHAPTER VI

THE INFUSORIA

I. The Ciliata 17
A. Protoplasmic Structure 17
B. Contractile Vacuoles 18
C. The Nucleus 18
D. Encystment 19
E. Reproduction 19
II. The Suctoria 19
III. Inter-relations of the Infusoria 19
Classification 20

CHAPTER VII

SEXUAL PHENOMENA IN THE PROTOZOA

A. Phenomena of Conjugation 21
1. The Permanent or Temporary Union of Similar Adult Individuals (Isogamy) 21
2. The Union of Similar but Different-sized Individuals (Anisogamy) . . . 22
3. The Union of Swarm-spores (Isogamy and Anisogamy) 22
4. The Union of Eggs and Spermatozoa 22
B. The So-called Maturation-phenomena in Protozoa 23
C. General Considerations 24

CHAPTER VIII

Special Morphology of the Protozoan Nucleus

PAGE

A. The Nuclear Membrane 249
B. The Linin Network 252
C. The Nucleolus 253
D. The Chromatin 253
E. Kinetic Structures 258
  1. Intra-nuclear Division-centres 259
  2. Extra-nuclear Division-centres 265
  3. The Relation of Extra-nuclear to Intra-nuclear Division-centres . . . 268
  4. The So-called "Centrosomes" in Protozoa 272
F. General Considerations 273

CHAPTER IX

Some Problems in the Physiology of the Protozoa

A. Intra-cellular Digestion in Protozoa 280
B. Respiration 288
C. Secretion and Excretion 290
D. Irritability 294
E. General Considerations 301

Bibliography 311

Index of Authors 329

Index of Subjects 335

LIST OF FIGURES

INTRODUCTION AND CHAPTER I

FIG. PAGE

1. Types of Protozoa 3
2. *Actinophrys sol* Ehr. 16
3. A radiolarian, *Actissa princeps* Haeck. 17
4. Types of Flagellidia 19
5. Dinoflagellidia 20
6. Coccidiida in epithelial cells 20
7. *Leptotheca agilis*, a Myxospore 21
8. *Plasmodium malariæ* in human blood 22
9. A sphæroidal colony, *Uroglena americana* 25

CHAPTER II

10. Protoplasmic structure in different Protozoa 36
11. Flagellates with stigmata 37
12. Ectoplasmic modifications 39
13. Shells and tests 41
14. Types of nuclei 42
15. Types of pseudopodia 44
16. Cilia and myonemes of Infusoria 45
17. Types of cysts 47
18. Food-taking 49
19. *Actinobolus radians* 51
20. Tentacles of Suctoria 52
21. *Frontonia leucas* 53
22. Division of *Euplotes* 54
23. Budding of *Euglypha alveolata* 55
24. *Microgromia socialis*, a gregaloid colony 56
25. *Uroglena americana* 56
26. *Codosiga cymosa*, an arboroid colony 57
27. *Eiermocystis polymorpha*, a catenoid colony 58
28. Exogenous budding in *Ephelota bütschliana* 58
29. Degeneration in *Onychodromus grandis* 59
30. Conjugation of *Onychodromus grandis* 60
31. Internal parasites 63

CHAPTER III

32. *Actinophrys sol* 68
33. The protoplasmic regions of a radiolarian 69
34. Central capsules of Radiolaria 70

FIG. PAG

35. Types of marine rhizopod shells . . . 7
36. Polythalamous shell types schematized . . . 7
37. A complex polythalamous shell . . . 7
38. Megalospheric and microspheric shells . . . 7
39. *Clathrulina elegans* . . . 7
40. Types of spicules in Heliozoa . . . 7
41. Skeleton formation . . . 7
42. *Lichnaspis giltochii* . . . 7
43. *Plagiocarpa procortina* . . . 7
44. Types of pseudopodia . . . 8
45. *Camptonema nutans* . . . 8
46. *Ciliophrys*, pseudopodia and axial filaments . . . 8
47. *Amœba proteus* pseudopodium, diagram . . . 8
48. *Amœba verrucosa*, ectoplasm and vacuoles . . . 8
49. Contractile vacuole of *Amœba proteus* . . . 8
50. *Gromia oviformis* . . . 9
51. *Microgromia socialis* in division . . . 9
52. *Paramœba eilhardi* . . . 9
53. Spore-formation in Heliozoa . . . 9
54. Conjugation of *Actinophrys sol* . . . 9
55. *Dimorpha nutans* . . . 10
56. *Nuclearia delicatula* . . . 10
57. Protozoa with pseudopodia and flagella . . . 10
58. Membrane in *Actinosphærium eichhornii* . . . 10

CHAPTER IV

59. *Proterospongia haeckeli* . . . 11
60. *Codonœca costata* . . . 11
61. *Dinobryon sertularia* . . . 11
62. *Phacotus lenticularis* . . . 11
63. *Distephanus speculum* . . . 11
64. *Gymnodinium ovum* and *Peridinium divergens* . . . 11
65. *Synura uvella* . . . 11
66. Primitive forms of Dinoflagellidia . . . 12
67. *Phalansterium digitatum* . . . 12
68. Types of collars in Choanoflagellida . . . 12
69. A Choanoflagellate type . . . 12
70. Nuclear division in *Noctiluca miliaris* . . . 12
71. *Megastoma entericum* . . . 12
72. *Gonium pectorale* . . . 12
73. *Gonium pectorale* in division . . . 12
74. Budding in *Noctiluca miliaris* . . . 13
75. *Cercomonas crassicauda*, division and spore-formation . . . 13
76. *Pandorina morum*, conjugation . . . 13

CHAPTER V

77. Life history of a gregarine; schematic . . . 14
78. *Pfeifferia tritonis* in *Triton* cells . . . 14
79. *Leptotheca agilis*, a Myxospore . . . 14

FIG.		PAGE
80.	*Clepsidrina munieri*, myonemes	145
81.	*Lymphosporidium Truttæ*	147
82.	Movement of a gregarine	148
83.	Types of spores	150
84.	Sporulation in a gregarine, schematic	152
85.	*Gamocystis tenax*, sporoducts	153
86.	*Myxobolus*, capsule formation	155
87.	*Monocystis ascidiæ*, conjugation	157
88.	Life history of a *Coccidium*	160
89.	*Klossia helicina*, conjugation	161
90.	Life history of *Plasmodium malariæ*	163

CHAPTER VI

91.	Types of Ciliata	172
92.	*Dileptus anser*	173
93.	*Coleps hirtus*	176
94.	*Lacrymaria coronata*	177
95.	Supposed shifting of the mouth of Ciliata	178
96.	*Zoothamnium arbuscula*, myonemes	179
97.	Myonemes and cilia	180
98.	Schematic hypotrichous ciliate	182
99.	*Urocentrum turbo*	183
00.	*Actinobolus radians*	184
01.	Buccal apparatus of ciliates	186
02.	Anterior end of *Ophrydium eichhornii*	187
03.	Types of macronuclei	189
04.	*Loxophyllum meleagris*, nucleus	190
05.	*Spirochona* and *Paramæcium*, mitosis	191
06.	*Stentor rœselii*, division	193
07.	*Epistylis umbellaria*, conjugation	194
08.	Tentacles of Suctoria	196
09.	*Dendrosoma radians*	197
10.	*Ephelota bütschliana*, exogenous budding	198
11.	Endogenous budding in Suctoria	199
12.	*Multicilia lacustris*	200
13.	Illustrating Bütschli's origin of Hypotrichida	201
14.	Illustrating Bütschli's origin of the Vorticellidæ	202
15.	Ciliates with tentacles	205

CHAPTER VII

16.	Conjugation in *Cercomonas*	215
17.	Conjugation in *Tetramitus rostratus*	216
18.	Conjugation in Rhizopoda	217
19.	Conjugation in *Arcella vulgaris*	218
20.	Conjugation in *Polytoma uvella*	222
21.	Conjugation in *Lagenophrys ampulla*	223
22.	Conjugation in *Epistylis umbellaria*	224
23.	Conjugation in *Chlorogonium euchlorum*	225

FIG.
124. Conjugation of *Monocystis ascidiæ*
125. Division of *Gonium pectorale*
126. Conjugation of *Pandorina morum*
127. Formation of microgametes in *Klossia*
128. Life history of a *Coccidium*
129. Conjugation of *Euglypha alveolata*
130. Conjugation of *Actinophrys sol*
131. Conjugation of *Paramœcium caudatum*
132. *Onychodromus grandis*, degeneration

CHAPTER VIII

133. Diagram of a cell
134. Types of protozoan nuclei
135. Nucleus and karyosome in *Klossia*
136. Mitosis in *Euglena*
137. Division in *Amœba crystalligera*
138. Various nuclei in division
139. Mitotic division in the Infusoria
140. Nuclear division in *Actinosphærium*
141. Mitosis in *Noctiluca miliaris*
142. *Paramœba eilhardi*, sporulation and division
143. Mitosis in *Tetramitus chilomonas*
144. Nuclear division and spore-formation in Heliozoa

CHAPTER IX

145. Starch grains in Ciliata after partial digestion
146. Digestion in Reticulariida
147. Digestion in *Carchesium*
148. Excretory granules in *Paramœcium*
149. Shell-formation in *Gromia fluviatilis*
150. Phosphorescence in *Noctiluca miliaris*
151. Motor response in *Paramœcium*
152. Isolated nucleus of *Thalassicolla nucleata*
153. Reactions of *Amœba verrucosa* and of fluids

THE PROTOZOA

INTRODUCTION

"In the clearest waters and in muddy pools, in acid as well as alkaline waters, in brooks, lakes, rivers and seas, often, also, in the interior fluids of living plants and animals, abundantly in living men, and periodically borne on the dusts and vapors of our atmosphere, there exists a world unknown to the ordinary senses of man, of minute, peculiar forms of life." — C. G. EHRENBERG, 1838.

BEYOND the ordinary range of unaided vision there exists a world of minute animal organisms, technically known as the *Protozoa*.[1] They abound in the dust of the air, in the sea, in freshet and ditch, in brackish and potable waters — wherever, in short, there is air and moisture, while even air is apparently superfluous for the vast majority of parasitic forms which make their homes in the living bodies of higher plants and animals. Their beauty, their varied modes of life, the suddenness of their appearance and disappearance, the simplicity of their structure, and modes of reproduction, combine to make them, even to the superficial observer, a fascinating group. Apart from their superficial attraction, however, the Protozoa have a deeper significance to the student of zoölogy. As the name Protozoa indicates, they are primitive animals, and in the scale of living things they are not far removed from the colorless bacteria on the one hand, and the primitive green plants on the other. Their chief significance, however, and the main feature which distinguishes them from the higher animals or *Metazoa*, centres in the fact that they consist of but a single cell within the confines of which are carried on all of the essential vital functions which characterize the highest many-celled animals.

In their main characteristics these cells do not differ from those which make up the tissues and the body of higher animals. Like a tissue-cell the protozoön consists of protoplasm differentiated into *nucleus* and cell-body or *cytoplasm*, both parts being variously modified in the several types (Fig. 1). Unlike tissue-cells, however, the Protozoa are not specialized for the performance of any one function. They invite attention, therefore, from both the morphological or structural and the physiological or functional points of view. Morphologically they are equivalent to the isolated epithelial, muscle- or nerve-cell; physiologically, they are equivalent not merely to the muscle- or nerve-cell, but to the entire group of cells which collec-

[1] The term Protozoa was first used in its modern sense by von Siebold ('45).

tively constitute a higher animal. The Protozoa are, in short, complete, but unicellular organisms, and are to be regarded as the most generalized of single cells.

Considered as a complete animal, the protozoön cell at once arouses the inquiry as to the nature of the organs by means of which the vital functions are carried on. Lending themselves readily to the experimental method of investigation, the Protozoa have already contributed not a little to knowledge of the localization of function in the cell. The importance of the nucleus in the economy of cell-life, which Barry and Goodsir early pointed out in animal tissues, has been fully confirmed by the researches of Gruber, Balbiani, Hofer, Verworn, and others upon the Protozoa. From the structural point of view, the protozoön nucleus with its accompanying structures must ultimately throw considerable light on the vexed questions connected with the finer structures of metazoan cells. As the sequel will show, considerable advance has already been made in this direction through the efforts of Bütschli, Schaudinn, Balbiani, R. Hertwig, and many others. Here, the generalized structures, especially those elements concerned in cell-division, although difficult of analysis, must, when more thoroughly studied, aid the interpretation of the more specialized structures in Metazoa which are now involved in some of the most deeply-lying problems of biology.

Physiology likewise has been and is still to be greatly enriched by the study of unicellular animals. Bichat's theory of tissues, propounded at the very outset of the last century (1801), formed the basis of Virchow's development of the cell-theory along physiological lines ('58). It was Virchow who put on a working basis Schwann's conception that the vital activities of an animal are the sum of all of its parts, and that each part, a cell, or, as Brücke suggested, an "elementary organism," performs all of the characteristic activities of life. Thus while the older physiologists were satisfied with the knowledge that the function of the kidney is to secrete urine containing the waste matters of living activity, the modern problems, as Virchow intimated, centre more especially in the inquiry as to the activity of the kidney cells as such. Again, the modern physiological problem of contractility or of nervous action is concerned with the muscle- and ganglion-cell, and is therefore a cell-problem. For investigations upon cellular physiology there are obvious advantages in studying the unicellular organisms, which, says Verworn, "seem to have been created by nature for the physiologists, for, besides their great capacity for resistance, of all living things they have the invaluable advantage of standing nearest to the first and the simplest forms of life."[1]

[1] Lee ('98), p. 50.

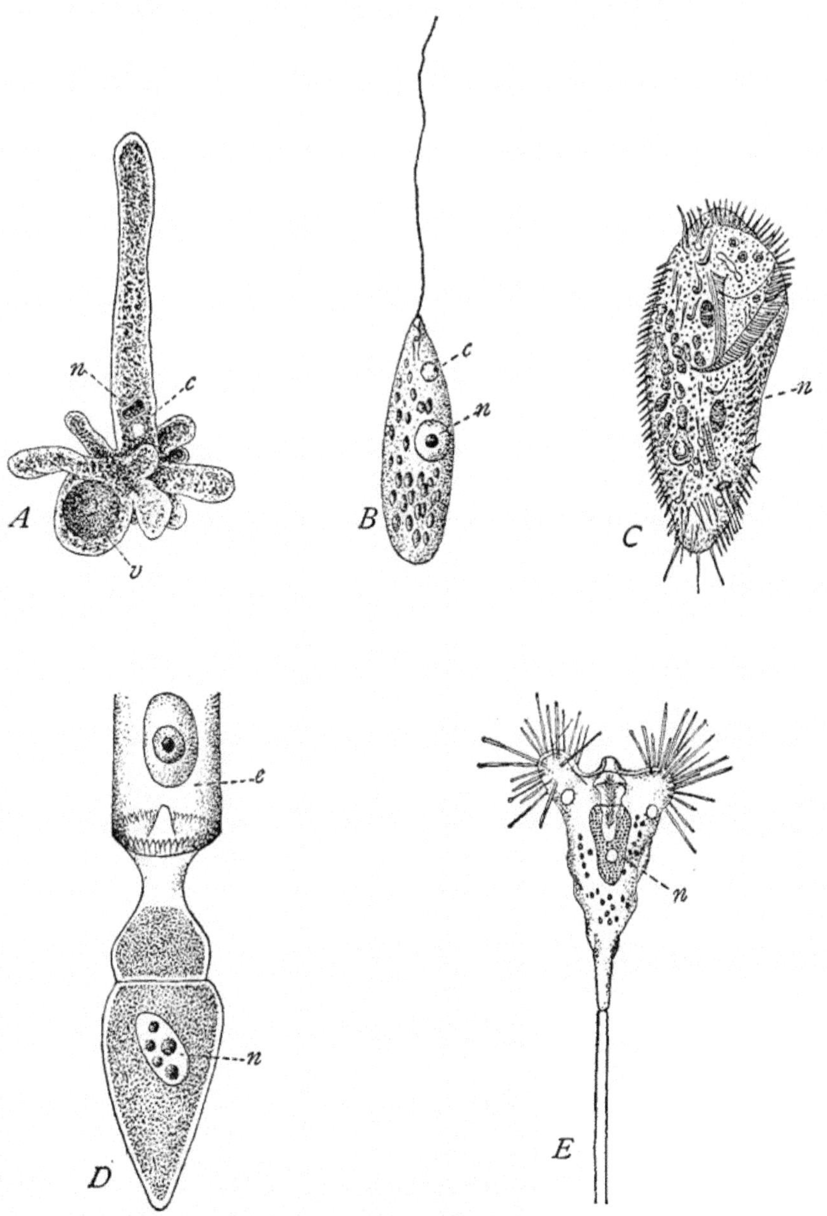

Fig. 1. — Types of Protozoa.

A. Amœba proteus, a rhizopod. *B. Peranema trichophorum*, a flagellate. [BÜTSCHLI.] *C. Stylonychia mytilus*, a ciliate. [BÜTSCHLI.] *D. Pyxinia sp.* a sporozoön. [WASIELEWSKY.] *E. Tokophrya quadripartita*, a suctorian. [BÜTSCHLI.] *c.* contractile vacuole; *e.* epithelial host-cell; *n.* nucleus; *v.* food or gastric vacuole.

Although, as Virchow pointed out, each cell of a tissue is a complete organism performing all of the functions of living matter, some one of these functions predominates over the others and gives to the cell and to the tissues of which it forms a part its special characteristics. In these specialized cells the secondary functions, *i.e.* those acting only for the good of the cell itself, fall into the background and are not readily investigated. In the Protozoa, on the other hand, no one function predominates, and despite their primitive nature, the protoplasm of which they are composed appears quite similar to that of the most highly specialized tissue-cell. In it, however, lies the secret of digestion and assimilation, of the kidney's secretion, and of muscular contraction.

The Protozoa invite attention from still another point of view As the lowest animals they show the beginnings of sex differentiation, of maturation, or the changes which the germ-cells undergo before fertilization, and of fertilization, while their union into cell-aggregates or *colonies* with incipient division of labor among the constituent cells, points the way toward the Metazoa, and makes them significant in the light of evolution.

As complete primitive organisms, therefore, the Protozoa are important from many points of view: structurally, they contain in simple form the elements which in higher tissue-cells are moulded into more complicated organs of the cell; functionally, they epitomize the life activities of even the highest many-celled animals, but their vital processes are more easily observed and correlated; theoretically, they occupy a prominent place in questions of phylogeny, of sex, and of reproduction, and finally, placed as they are at the lowest limit of animal life, they must ever be closely connected with problems concerning its origin.

With this conception of the Protozoa in mind the present volume has been written. The work makes no pretence of a comprehensive description of the Protozoa or of any one group, but aims rather to give an intelligible idea of the main types, to point out the problems of biology with which the Protozoa are most closely connected, and, so far as possible in a limited space, to survey the work already accomplished.

In the present introductory chapter there is a brief historical review of the stages by which the Protozoa have come to be regarded as single cells, and at the same time as complete animal organisms. Here, too, is a short account of the interesting position which the Protozoa have held in the time-honored dispute over the limitations of the animal and plant kingdoms, and in theories of spontaneous generation. The second chapter deals with the general structures and functions of the Protozoa as a group, and introduces the four following chapters, which are devoted to the structural and functional adaptations of the organisms in each class. The last three chapters, finally, deal with the relations of the Protozoa to more general problems.

A. HISTORICAL REVIEW

The Dutch microscopist, Anton von Leeuwenhoek (1632–1723), using crude lenses of his own make, was one of the first to apply the microscope to scientific investigation. His contributions to microscopic anatomy and to physiology, inaugurating as they did the invaluable services of the microscope in biological research, marked an epoch in the history of science. An ardent follower of Harvey, he was one of the first to offer experimental evidence against the current belief that many of the lower organisms arise by spontaneous generation, and on every occasion he sought to establish the truth of Harvey's dictum *ex ovo omnia.* In 1675, while searching for evidence of spontaneous generation, Leeuwenhoek discovered "living creatures in Rain water, which had stood but four days in a new earthen pot, glased blew within."

"This invited me," he continues, "to view this water with great attention, especially those little animals appearing to me ten thousand times less than those represented by Mons. Swamerdam, and by him called *Water-fleas* or *Water-lice,* which may be perceived in the water with the naked eye. The first sort by me discover'd in the said water, I divers times observed to consist of 5, 6, 7, or 8 clear globuls, without being able to discern any film that held them together, or contained them. When these *animalcula* or living Atoms did move, they put forth two little horns, continually moving themselves. The place between these two horns was flat, though the rest of the body was roundish, sharp'ning a little towards the end, where they had a tayl, near four times the length of the whole body, of the thickness (by my Microscope) of a Spider's-web; at the end of which appear'd a globul, of the bigness of one of those which made up the body; which tayl I could not perceive, even in very clear water, to be mov'd by them. These little creatures, if they chanced to light upon the least filament or string, or other such particle, of which there are many in water, especially after it hath stood some days, they stook entangled therein, extending their body in a long round, and striving to dis-intangle their tayl; whereby it came to pass, that their whole body lept back towards the globul of the tayl, which then rolled together Serpent-like, and after the manner of Copper or Iron-wire that having been wound about a stick, and unwound again, retains those windings and turnings. This motion of extension and contraction continued awhile; and I have seen several hundreds of these poor little creatures, within the space of a grain of gross sand, lye cluster'd together in a few filaments." [1]

This is the first description of a protozoön; and although the description is incomplete, it undoubtedly refers to a species of *Vorticella.* Leeuwenhoek observed several other forms at the same time, but for the most part their identity is uncertain.

[1] See *Phil. Trans.,* London, Vol. XII., 1677, p. 821.

At this period, although the term *cell* had already been used by Robert Hooke (1665), the idea of simplicity of organization, apart from minuteness of the organs, was unknown, and until the cell-theory was established in 1838, the Protozoa were regarded as complex animals having all of the parts and organs, although of microscopic size, found in Metazoa. Leeuwenhoek allowed his imagination to see what his eyes could not. "When we see," said he, "the spermatic animalcula [spermatozoa] moving by vibrations of their tails, we naturally conclude that these tails are provided with tendons, muscles, and articulations, no less than the tails of a dormouse or rat, and no one will doubt that these other animalcula which swim in stagnant waters [Protozoa], and which are no longer than the tails of the spermatic animalcula, are provided with organs similar to those of the highest animals. How marvellous must be the visceral apparatus shut up in such animalcula!"[1]

The minute size of the Protozoa made it impossible for the early investigators with their crude instruments, to follow out any life-cycle, and the prodigious numbers and the sudden appearance of certain forms in stagnating waters led to the belief already current in respect to other forms, that they arose *de novo*. Two misconceptions thus sprang up almost at the beginning of our knowledge of the Protozoa: one, that Protozoa are provided with organs like higher animals; and, two, that they arise by spontaneous generation; and one of the main tasks of research on the Protozoa down to our own times has been the correction of these early errors.

It is not strange that Leeuwenhoek and his immediate followers considered Protozoa as complicated organisms. Organisms without organs were as novel to them as animals without cells would be to us, and they described only what experience had taught them to expect. With increasing knowledge of many forms and with constantly improving microscopes, the conception of simplicity of organization gradually gained ground until Dujardin, about 1840, defined the Protozoa as simple, slightly differentiated structures composed of a fundamental living substance to which he gave the name of *sarcode*.

Despite the crudity of their instruments, the early microscopists obtained wonderful results. Leeuwenhoek himself, although studying these low forms only incidentally, gave recognizable descriptions of twenty-eight species, and in addition, noted the rapid increase of some of the larger forms, saw conjugation or the temporary union of two individuals, and discovered so-called embryos. The early literature soon became crowded with notices of new and interesting forms, found in all sorts of hitherto unthought-of localities. Forms with

[1] Quoted from Dujardin ('41), pp. 21, 22.

whip-like appendages, or *flagella;* with *cilia,* or motile appendages, similar in general form to eyelashes; with changeable processes, or *pseudopodia;* or with no motile apparatus whatsoever; forms of the most diverse size and shape, including many higher Metazoa, such as worms, rotifers, ctenophores, crinoids, and crustacean larvæ, as well as many plants, were all described as "*animalcula.*" Descriptions of internal organs soon began to accompany the descriptions of types. The *contractile vacuole,* a characteristic pulsating vesicle of the Protozoa, was discovered by Joblot (1754-'55), who also showed that cilia on Infusoria have a definite arrangement in different species, and that many forms are provided with cuticular stripings. All of these forms, sometimes called insects and sometimes fish, were still generally supposed to be microscopic reproductions of higher animals. Dujardin's criticism of Joblot's work might well be applied to that of many others of this early period: "Several of the figures which he gives," says Dujardin, "bear the impression of a too lively imagination for scientific purposes, and are frequently so bizarre and fantastic as to discredit the use of the microscope."[1] It is easy to understand this criticism when we think of Joblot's picture of the worm *Anguillula* with a serpent's head, or the flagellated protozoön *Euglena* with a broad mouth, flagellum, and well-developed mammalian eyes. "For his picture of the ciliate *Paramœcium aurelia,*" says Dujardin, "he apparently used his own slipper as a model."

The life history of a protozoön was first made out by Trembley (1744-'47), who saw the microgonidia or young spore-forms of certain Vorticellidæ leave the parent-colony and begin the formation of new colonies by longitudinal division.

The discoveries made by means of the microscope were regarded with complete scepticism by Linnæus in his earlier scientific works, and the very existence of Leeuwenhoek's animalcula was at first denied by him, but in the later editions of his *Systema Naturæ* they were grudgingly admitted under the significant generic name of *Chaos* [*Chaos proteus* (*Amœba*), *Chaos redivivum,* etc.]. The organized nature of *Volvox globator,* a form which had been discovered and fairly well described by Leeuwenhoek, was admitted at this time, and finally, in the twelfth edition (1767), the animal nature of a *Vorticella.*

Many of the early investigators studied the Animalcula from the physiological standpoint, and attempted to ascertain the functions of many of the so-called organs. Their efforts were often strikingly successful and have been confirmed by later observations. Corti (1774), Spallanzani (1776), and Gleichen (1778) are the most familiar names in this line of research. Corti and Gleichen compared the

[1] Dujardin ('41), p. 7.

contractile vacuole of *Vorticella*, with its regular pulsations, to a beating heart, while Spallanzani, distinguishing the vacuole from its canals, assigned to it the function of respiration. The mouth was found in a number of forms, by Gleichen, who first used the now common experiment of feeding the Protozoa with minute particles of colored substances, such as carmine, indigo, etc. A considerable knowledge of reproduction was also obtained. Longitudinal division, discovered by Trembley (1744), was confirmed by Spallanzani, who, in addition, observed transverse division in no less than fourteen species, while his friend Saussure followed out for the first time the division of an encysted *Colpoda;* an observation confirmed by Corti and Gleichen as well as by Spallanzani himself, who saw a *Colpoda* slip out of its cyst, which he not unnaturally mistook for an egg-case.

These early discoveries were, in most cases, so bound up with fantastic speculations that their zoölogical value was greatly impaired. Many of these early inaccuracies were, however, weeded out by Otto Friedrich Müller (1786), to whom we are also indebted for the scientific naming of the Animalcula, which up to his time had been called by long descriptive names given according to the fancy of each observer, and often based on far-fetched resemblances. Müller, adopting the Linnæan binomial nomenclature, described and named some 378 species, of which about 150 are retained to-day as Protozoa. His classification was the first successful attempt to bring order out of the heterogeneous collection of forms included under the name Animalcula. He used Ledenmüller's (1760–'63) term *Infusoria*, for the name of the entire group, which he placed as a class of the worms.[1] While he eliminated the inaccuracies, he confirmed the substantial observations of the earlier observers, extending many of them to all groups of the Protozoa. He ascertained the presence of an anus, showed that many Infusoria are carnivorous, and observed the process of *conjugation*, his description of the latter being more accurate than that of any of his predecessors or followers until the time of Balbiani in 1858–'59.

Like his predecessors, Müller included among the Protozoa many other organisms; placing here diatoms, nematode worms, *Distomum* larvæ, and larval forms of cœlenterates and molluscs, as well as the rotifers. The majority of these miscellaneous forms were, however, properly classified before 1840. The larvæ of molluscs and cœlenterates, and the worms were the first to be removed from the "animalcula," while finally spermatozoa (discovered by Ludwig Hamm, who is said to have been a pupil of Leeuwenhoek), which had been universally regarded as Protozoa inhabiting the seminal fluid, were withdrawn during the present century.

[1] Cf. Bütschli ('83), p. 1129.

Unlike his predecessors, Müller did not regard the Protozoa as complicated animals, but considered them as the simplest of all living things, composed of a homogeneous gelatinous substance, a view in which he was followed by a majority of the "Nature-philosophers" (Lamarck, Schweigger, Treviranus, Oken), most of whom gave little or no study to the Protozoa, but, accepting Müller's work as final, based many of their speculations upon it.

It is difficult to understand why, after Müller's work, the next great authority, C. G. Ehrenberg (1795–1876), the renowned Berlin microscopist, using much finer achromatic lenses, should have returned to the crude view of Leeuwenhoek, assigning to the Protozoa a system of minute but complete organs. His conclusions on Protozoa were brought together in one great work, the title of which alone shows his point of view: "The Infusoria as Complete Organisms" (*Die Infusionsthierchen als vollkommene Organismen*). He was primarily a student of their finer structure, and the details of organization, although erroneously interpreted, were clearly described. In working out the internal structures he made use of Gleichen's experiments in feeding. The animals were seen to ingest the powdered carmine, so that the boundaries of the internal gastric vacuoles were clearly marked. He followed these particles as they passed from the mouth into the œsophagus and thence into one of the many digestive or *gastric vacuoles* found in the inner plasm of nearly all Protozoa. He saw that the particles followed clearly defined paths which might be straight or curvilinear, or otherwise varied in different forms, but which always ended in a more or less clearly marked anal opening. He saw also that the parts of the supposed tract nearest the mouth fill first; that they become globular, and that successive reservoirs become filled, down to the posterior end of the body. He inferred from this the existence of a digestive tract, concluding that the parts thus filled were stomachs. As soon as the first was filled, the overflow of food passed on into another stomach. From the supposed possession of many stomachs Ehrenberg gave to this group the name *Polygastrica* or *Magenthiere*, making it a sharply defined class in the animal kingdom. To all forms in which he could find no stomachs, but in which he supposed that mouth and anus were the same opening, he gave the name of *Anentera* (gutless), while to forms with many stomachs he gave the name *Enterodela* (gut-bearing).

The red pigment spots of many forms were interpreted as true eyes, but as eyes could not be conceived without an accompanying nervous system, he sought for nerve-ganglia in different organisms, and supposed he found what he was looking for in a species of *Astasia*. He described the eye in this form, as seated upon a "spherical glandular mass," which he considered equivalent to the supra-pharyngeal ganglion of the rotifers (cf. Fig. 11). He discovered the *myonemes* or muscular elements in the stalks of *Vorticella*, in *Stentor*, and in certain other Ciliata, and interpreted them as muscles. He discovered that the flagellum of the flagellates is the motile organ, but explained its vibrations as due to the action of exquisitely fine muscle-fibres. Pigment spheres and protoplasmic granules were described as ovaries, the nucleus as a testis, while the contractile vacuole was at first regarded as a respiratory organ. With this latter conclusion he could not harmonize his subsequent observations, and finally decided that the vacuoles have the same functions as in the rotifers.

Ehrenberg's strong position as an investigator of Protozoa is due to his remarkable powers of observation, especially of the finer structure of flagellates and ciliates, which in many cases he described and accurately figured, and these justify the tribute which Bütschli pays him: —

"The great service which Ehrenberg did in furthering the knowledge of these forms cannot be clearly enough recognized. After a naturally somewhat difficult comparison, I find among the species described in 1838 a few more than 100 Infusoria (in the present sense), of which five are Suctoria. Also his system of classification was much more natural than that of any of his forerunners, and formed the basis of all subsequent efforts. Many of his genera had correct limitations which hold even to-day, although many indeed cannot be sustained. . . . With astonishing assiduity he sought to collect, study, and systematically interpret everything that had been done upon the Infusoria." ('83, p. 1145.)

Ehrenberg's interpretations, however, were not as successful as his collection of data, and it is to be regretted that throughout his life he obstinately clung to his view of the *Polygastrica*, even after the period of the complete establishment of their unicellular nature. It was surely the irony of fate which led to the publication of his immense work on the *Polygastrica* the same year ('38) that the Dutch botanist Schleiden made the greatest advance in the conception of the cell as the unit of structure.

A formidable opponent of Ehrenberg soon appeared in France, — Felix Dujardin, — who, influenced by long study of the *Rhizopoda* (or Protozoa with changeable processes), came to the conclusion, in 1835, that the marine forms (*Foraminifera*), which, up to that time, had been classed with cephalopod molluscs, are in reality the simplest of organisms, composed of a simple, homogeneous substance which he called *sarcode*. He showed that the many stomachs, which, according to Ehrenberg, constituted the digestive tract, were mere vacuoles without definite walls, which become filled with water taken in from the outside with the food. He denied Ehrenberg's assertion that an anus terminates the digestive tract, although in some cases his generalization was made without adequate observation, for at first he denied the presence of a mouth as well as anus.

Dujardin further showed that the motile organs of the Protozoa, whether cilia, flagella, or pseudopodia, are mainly the prolongations of the outer coatings of the organism, and are in no sense similar to the hairs of higher animals, and he even suggested the transition, which has since been shown to occur, between the simplest of pseudopodia and the more complex flagella. He contradicted Ehrenberg's theory as to the function of the contractile vacuole, reverting to the interpretation given by Spallanzani. He also denied the complexity of the reproductive organs, as described by Ehrenberg, but made a singular mistake in regarding the granules inside of the body as germs.

Many besides Dujardin had begun to criticise Ehrenberg's theory. Carus ('32) insisted, on purely theoretical grounds, that animals must exist whose structure is as simple as that of an egg, since all animals begin with the simple egg structure. Two years later, he criticised Ehrenberg's theory, on the ground that inner circulation of the plasm in *Paramœcium* (discovered by Gruithuisen, '12), which resembles the circulation in the plant *Chara*, does not accord with Ehrenberg's description of the digestive apparatus. A similar objection was raised by Focke ('36), based on observations upon the streaming plasm in *Vaginicola* and *Paramœcium*.

The first suggestion that Protozoa might be single cells was made by Meyen ('39), who compared the entire infusorian body to a single plant-cell. The cell-theory, according to Bütschli, however, was first applied directly to the Protozoa by Barry ('43), who asserted that *Monas* and its allies among the Flagellidia are single cells, and that the nucleus found within them is the equivalent of the cell-nucleus of higher animal forms. At the same time Barry expressed the view that cells increase only by division, and he compared the processes of multiplication in *Volvox* and *Chlamydomonas* with the cleavage of eggs which he, with Schwann, regarded as single cells.[1]

Barry's view was accepted in part by Owen, who thought, however, that the Infusoria could not be included with the Flagellidia as single cells, because of their higher differentiation. It was von Siebold ('48), however, who finally asserted the unicellular nature of all Protozoa.

Ehrenberg's theory was not given up without a struggle, and, among others, we find Schmidt ('49) coming to his support with the fact that the *trichocysts*, or stinging threads of the Infusoria, found by Ellis (1769), and by Spallanzani (1776), and the contractile vacuoles

[1] Cf. Bütschli ('83), p. 1153.

with their canals, show the strongest similarity to the corresponding structures in flatworms. The trichocysts were particularly difficult for the opponents of Ehrenberg to explain. Stein ('56) regarded them as "taste-bodies" (*Tästkörperchen*), and we find even Leydig ('57) regarding them, together with the microsomes in the stalks of *Vorticella*, as the nuclei of very minute cells.

Kölliker ('48, '49), following von Siebold, but at first almost alone, strenuously maintained that all Protozoa are single-celled animals, in spite of severe criticism, especially by the ardent and brilliant young naturalists, Claparéde and Lachmann, who, not able to make out the cell-membranes and nuclei in many cases, placed the Protozoa with the Hydroida, making them a subdivision of the Cœlenterata. It should be noted, in justice to Claparéde, that later he admitted his error.

The opponents of Kölliker were, however, gradually convinced. Max Schultze ('63) showed the identity of Dujardin's sarcode with protoplasm; Stein ('67) vigorously assailed the objections of Leydig and Haeckel, while the latter ('73) brought back his Infusoria from the Articulata, to which he had consigned them in 1866, and became an ardent advocate of Kölliker's theory. The next few years saw the remarkable researches of Bütschli, Engelmann, and Hertwig, upon the physiology and finer organization of the Protozoa, and their work, with that of the hosts of others since them, aided by modern technique, has fully demonstrated the unicellular nature of all Protozoa.[1]

The theory of *alternation of generations* (the alternation of a sexual with an asexual method of reproduction) also became curiously involved in the foregoing controversy. Steenstrup ('42) applied his discovery of alternation of generations to the Protozoa, regarding the parasitic Infusoria which he found in certain molluscs, as the larval form of the liver-fluke, *Distomum*. The same view was found in various forms in the works of Claparéde and Lachmann, Perty, and Kölliker, and finally as the "Acineta-theory" in the works of Fr. Stein. This theory was based upon the supposed metamorphosis of one form of Protozoa into another. The first suggestion of such a metamorphosis seems to have been given by Pineau ('45), who ob-

[1] L. Agassiz ('57), adopting a point of view which has appeared sporadically since von Siebold announced that the Protozoa are single-celled animals, advocated the abandonment of the Protozoa as a group, placing some divisions with the lower plants, others with the larvæ of worms, and still others with the Bryozoa. His most curious error was in placing *Trichodina pediculata*, one of the higher forms of Protozoa, as the medusa-generation of the fresh-water polyp *Hydra*. "I have seen for instance," says Agassiz, "a *Planaria* lay eggs, out of which *Paramæcium* were born, which underwent all of the changes these animals are known to undergo up to the time of their contraction into a chrysalis state; while the *Opalina* is hatched from *Distomum*-eggs" ('57, p. 182). Similar views were held by Alder ('51), Burnett ('54), and even recently by Lameere ('91), and by Villot ('91).

served that decaying flesh apparently breaks into minute granules (bacteria). Influenced no doubt by the teachings of the nature-philosophers, Buffon and Oken, he further thought that he had observed the formation of larger forms of life from these minute granules, and among them, some *Podophryas*, which changed into *Vorticellas*, and these again into *Oxytrichinas*.

Stein's famous Acineta-theory was first brought out in his work of 1849, in which he described the many division phases of *Vorticella*, and gave a very good account of the process of encystment. The encysted animal, he thought, breaks down into a great number of minute particles, having at first the form of certain flagellates, which develop into young *Vorticellas*. Later he adopted a second hypothesis, equally untenable, *viz.* that *Acineta* is derived from the cysts of *Vorticella*. This conclusion was based upon the fact that he had seen the preparatory stages of encystment of the Suctorian *Podophrya*, which he thought were transition phases from the encysted condition to the adult free *Podophrya*, while the cysts from which he supposed they had come he thought were formed by *Vorticellas*. Generalizing from this supposed fact, and seeing supposed confirmation in many different directions, he finally regarded the entire division of the Suctoria as merely reproductive phases of the genus *Vorticella*. An apparent support for his theory was found in 1854, when he discovered the ciliated embryos in Suctoria, which closely resemble the Vorticellidæ. Not once did he follow out the development of these embryos by actual observation; it was purely hypothetical, and the discovery of *Acineta* embryos, since found to be parasites in various other ciliates, only strengthened him in this point of view.

Stein's theory soon found opponents. Perty ('52) feebly opposed it, while Johannes Müller and his pupils, Claparéde and Lachmann, and Cienkowsky ('55) traced the development of the supposed young *Vorticellas* to the adult forms of Suctoria. The theory was finally completely overturned by Lachmann ('56) and Balbiani ('60), the former showing by actual observation that neither does *Vorticella* develop into *Acineta*, nor do embryos of the latter develop into *Vorticella*, while the latter discovered that the supposed embryos are in reality parasitic Suctoria, a view in which he was ably supported by Metchnikoff ('54) and Kölliker ('64).

Balbiani's researches in the life history of the Protozoa at first led him into a curious error, a reminiscence apparently of Ehrenberg's and the older point of view. O. F. Müller had observed and correctly interpreted conjugation in different forms, but his successors down to Balbiani regarded this interpretation as incorrect, maintaining that he had seen only stages in simple division. Balbiani ('61) returned to Müller's view, and clearly stated that, in addition to simple division, another and a sexual method of reproduction occurs. His interpretation of the sexual organs of the Protozoa was given in 1858, when he maintained that the larger of the two kinds of nuclei of Infusoria, the *macronucleus*, is the ovary, and the smaller one, or *micronucleus*, the testis. He saw and pictured the striped appearance of the micronucleus prior to division, and interpreted the stripes as spermatozoa. The eggs were said to be formed in the macronucleus, to be fertilized and then deposited on the outside, where they develop into new ciliates. Stein at first opposed this assumption, but in the second volume of his work on the Infusoria, misled by his Acineta-theory, he practically adopted it, maintaining, however, that the embryos develop in the nucleus first, and only later leave the mother organism. Bütschli ('73) was apparently the first to point out Balbiani's error, and in his epoch-making work of 1876, after demonstrating the "striped" appearance of many egg-cells during division (*mitotic figure*), he concluded that the stripings which Balbiani held to be spermatozoa were no other than this striated condition of the nucleus during division. He held, therefore, that in addition to the macronucleus there is a second and a smaller nucleus in Infusoria, and to this he gave the name *Nebenkern* or micronucleus ('76). Bütschli further showed at the same time that during conjugation the macronucleus or larger nucleus disintegrates, and that the parts which Balbiani regarded as eggs are eliminated, to be replaced by one of the subdivisions of the *Nebenkern* (micronucleus). His interpretation of the process was equally happy. After observing that a continued asexual division of certain forms resulted in decreased size and a general "lowering of the life energy," he concluded that the function of conjugation is to bring about a *rejuvenescence* (*Verjungung*) of the participants. He called attention to the similarity between conjugation and fertilization of the egg in animals and plants, and, at the same time, made the classic comparison between the body of the metazoön and the chain of individuals which arise from one individual protozoön subsequent to conjugation.

In the same year Engelmann ('76) obtained very similar results. Quite independently of Bütschli he also proved the error of Balbiani's view, and came to a conclusion not far different from Bütschli's. "The conjugation of the Infusoria," he said, "does not lead to reproduction through 'eggs,' 'embryonic spheres,' or any other kind of germ, but to a peculiar developmental process of the conjugating individual, which may be designated as reorganization." (*Reorganization.*)[1]

Neither observer noted the conditions which induce conjugation or the mutual interchange of parts of the micronuclei, although both, indeed, suspected that the latter might take place. The actual inter-

[1] Engelmann ('76), p. 628.

change was first made out by Balbiani ('82), by Jickeli ('84), by Gruber ('86, '87[1, 2]), and by Hertwig ('89), and in great detail by Maupas ('88, '89), by whom the conditions leading to conjugation were for the first time made known.

B. MODERN CLASSIFICATION OF THE PROTOZOA

Although the Protozoa are the simplest forms of animal life and the most generalized of cells, it does not follow that they are simply organized and devoid of complicated structures. On the contrary, in many cases they are highly differentiated, and in the Infusoria, the highest group of the Protozoa, they become so complex, that Stein, who was never an ardent advocate of the simplicity of Protozoa, remarked: "The adult Infusoria must ever be considered doubtful single-celled organisms, for they are not simply cells which have undergone further development, but the original cell-structure has given place to an essentially different organization entirely foreign to typical cells."[1] On the other hand, there are simpler forms of Protozoa in which the undifferentiated protoplasm falls within the description of sarcode, as given by Dujardin in 1835. Between the two extremes of structure lie the vast majority of Protozoa, showing among them all gradations from extremely simple to extremely complex forms. A partial explanation of their frequent complexity of structure lies in the fact that, unlike tissue-cells, they live free and usually motile lives, and, like other independent organisms, are subject to changes of form and to intracellular modifications in response to their mode of life.

Notwithstanding the innumerable forms and the various intracellular modifications, differentiation, as a rule, has followed comparatively few general lines (Fig. 1). It thus becomes possible to arrange the Protozoa in groups or classes, with numerous divisions and subdivisions. The four classes which are now generally recognized are the *Sarcodina*, the *Mastigophora*, the *Sporozoa*, and the *Infusoria*.

The first attempt to classify the Protozoa was made by O. F. Müller in 1786. Rude and simple, and based upon the presence of visible motile organs (Bullaria), or upon their absence (Infusoria *s. str.*), this nevertheless was retained as the chief system of classification until the time of Ehrenberg, who made use of it in his own system, based upon the presence or absence of "stomachs."

Despite the progress made by Dujardin, his classification, in its main divisions, based upon unnatural differences of symmetry was equally imperfect. His two divisions were very unequal, the "asym-

[1] Stein, *Organismus*, etc., II. ('67), p. 22.

metrical" Infusoria including all but one or two known forms. The subdivisions were, however, remarkably happy, and were based upon natural lines which have never been displaced. While Ehrenberg's subdivisions were based upon gross external characters, such as the presence of hairs, the position of the mouth, etc., Dujardin's were based upon the means of locomotion, and in this early grouping we see the first use of our modern terms "*rhizopod*," "*flagellate*," and "*ciliate*." Von Siebold ('45) was the first to divide all Protozoa into

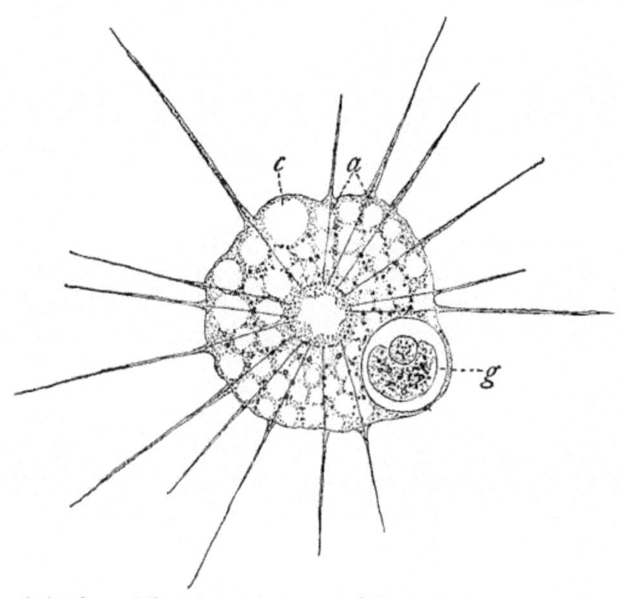

Fig. 2. — *Actinophrys sol* Ehrenberg, a heliozoön. [After GRENACHER from BÜTSCHLI.]
An individual with a large gastric vacuole (*g*), contractile vacuole (*c*), and axial filaments (*a*) in the ray-like pseudopodia.

two classes, *Rhizopoda* and *Infusoria*, a system which formed the basis of our modern classification; the Mastigophora or flagellates, regarded by von Siebold as plants, found no place in his zoölogy.

Three of the four great classes recognized to-day were thus outlined by Dujardin in 1841. The modern Rhizopoda (Sarcodina) were characterized as "animals provided with variable processes"; the Mastigophora as "animals provided with one or several flagelliform filaments, serving as motile organs"; and the Ciliata as "ciliated animals" (Fig. 1). The further subdivisions, which, little by little, have been developed along the lines laid down by Dujardin, have brought order out of this heterogeneous group of organisms, which at the present time includes nearly sixteen hundred genera and many thousands of species. As the name of a class, the term Rhizopoda, as used by von Siebold, may be replaced by the more comprehensive term Sarcodina, given by Bütschli ('83), while the term Rhizopoda may be applied to one group (subclass), characterized by an amœboid or changeable adult condition, and with lobose or reticulate motile processes or *pseudopodia*.[1] Two other subclasses are now universally included in the Sarcodina, the *Heliozoa* (Haeckel), or "sun animal-

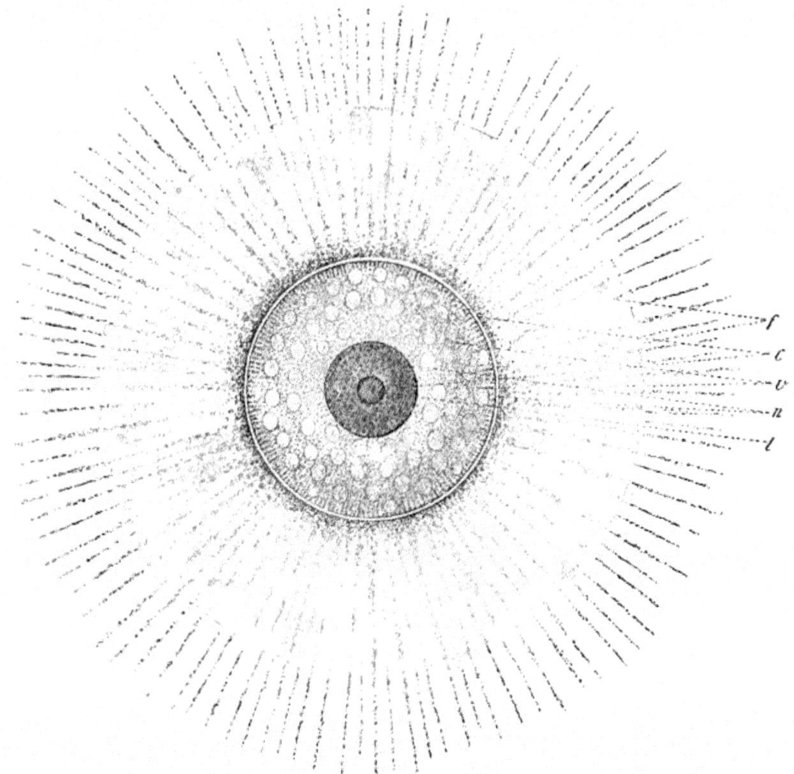

Fig. 3. — A radiolarian, *Actissa princeps* Haeck. [HAECKEL.]
The central capsule (*c*) separates the inner protoplasm (*v*) containing the nucleus (*n*) with its nucleolus (*l*), from the outer protoplasm which gives rise to the pseudopodia (*f*).

cula" (Fig. 2), characterized by fine ray-like pseudopodia, which often contain a central axial thread of stiffened protoplasm, and the *Radiolaria* (Haeckel), characterized, in addition to the ray-like pseudopodia, by the possession of a central portion of protoplasm, which is surrounded by a perforated membrane, the "*central capsule*" (Fig. 3).

Before the modern system of classification was established, many of

[1] The term "pseudopodia" was given by von Siebold to replace Dujardin's more descriptive phrase "changeable processes" (*expansions variables*).

the forms which are now recognized as Heliozoa or Radiolaria, were variously interpreted. Ehrenberg did great service in describing the skeletons of many Radiolaria, especially of the fossil forms, but he had no conception of their organization, and placed them with the Bryozoa, Rotifera and Echinodermata as a special class (*Tubulata*). Under the name, *Actinophryens*, Dujardin grouped the Heliozoa, together with a modern subdivision of the Infusoria (*Suctoria*), as "forms with slowly contractile appendages." The structure of the Radiolaria was first made out by Huxley ('51), who recognized them as Protozoa, and correctly compared *Thalassicolla* with the heliozoön *Actinosphærium*. The pseudopodia, however, were not recognized, and he was inclined to regard these forms as higher in organization than a single cell, and placed them between the Protozoa and the Sponges. Johannes Müller ('55–'58) first saw the resemblance between the fine ray-like pseudopodia of the Radiolaria and of the Heliozoa, and his pupils, Claparéde and Lachmann ('58), discovered the same granule-streaming in their pseudopodia that Schultze had observed in some of the Rhizopoda. With these data, Müller included the Radiolaria and the Heliozoa in the class Rhizopoda of von Siebold, under the name *Rhizopoda radiaria*, which was modified into its modern form, Radiolaria, by another of his pupils, Ernst Haeckel ('62). Four years later Haeckel ('66) separated the Radiolaria from the similar fresh-water forms, to which he gave the name Heliozoa.

The further subdivisions of the subclass Rhizopoda have been made upon two bases having almost equal value. In one system they are divided according to the nature of the pseudopodia into the orders *Lobosa* (*Amœbæa* of Ehrenberg) and the *Reticulariida* (*Reticularia* of Carpenter, '62). In the other they are subdivided according to the absence or presence of a shell, into the orders *Amœbida* (Ehbg.) and *Testacea* (M. Schultze, '54).[1] The former system is adopted by Delage and Hérouard, by Lankester and the English zoölogists generally; the latter by Bütschli. A third order under the name *Mycetozoida*, is usually included with the Amœbida and the Reticulariida. Although generally recognized in part at least, by zoölogists as Protozoa, the taxonomic position of the organisms included in this order is in dispute. Under the name *Myxomycetes* they are included with the fungi by most botanists, while by the zoölogists they are usually placed as a class of the Rhizopoda under the name *Mycetozoa* (de Bary, '59). The relation to the fungi is claimed on account of their saprophytic mode of life (terrestrial forms), and their mode of spore-formation in sporangia which are often complicated by the presence of stalks, columellæ, and other plant-like

[1] *Thalamophora*, R. Hertwig ('74); *Foraminifera*, d'Orbigny ('26).

structures such as elastic capillitia for dispersion of the spores. The relation to the Protozoa, on the other hand, is claimed on account of the unicellular nature, development of the swarm-spores, and occasional holozoic mode of nutrition. The spores leave the sporangium as amœboid or flagellated organisms and may increase by simple division during the swarming stage. If flagellated, the spore after a time loses the flagellum and becomes amœboid, in which condition division may again occur; finally numerous amœboid indi-

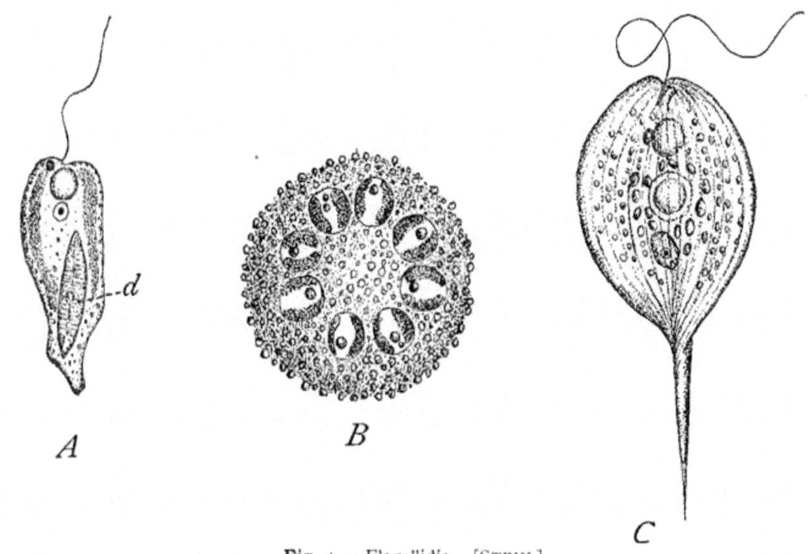

Fig. 4. — Flagellidia. [STEIN.]

A. Chrysomonas (*Chromulina*) *flavicans*, Ehr. with chromatophores and an engulfed diatom (*d*). *B.* The same encysted. *C. Phacus longicaudus* Duj.

viduals group themselves together, forming a colony or *plasmodium*. In some cases the fusion is complete, in others the outlines of the individual Amœbæ persist.

In view of the questionable position which these forms occupy, there is some danger of their being neglected altogether, the botanists refusing them because of their animal characteristics, the zoölogists because of their plant-like features. No harm can be done by including them in both kingdoms, for on purely *a priori* grounds it is to be expected that some organisms should be on the boundary line between artificial groups such as the unicellular animals and plants. The present group and the *Phytoflagellida* among the Mastigophora appear to occupy such a position, and it is advisable to include them as provisional groups of the organisms with which they show the greatest number of common points. With our present knowledge, the majority of Mycetozoa undoubtedly resemble fungi more than they do Protozoa, and will not be further considered in the present work; the Phytoflagellina have, on the other hand, so many obvious connections with the animal flagellates that they cannot well be omitted.

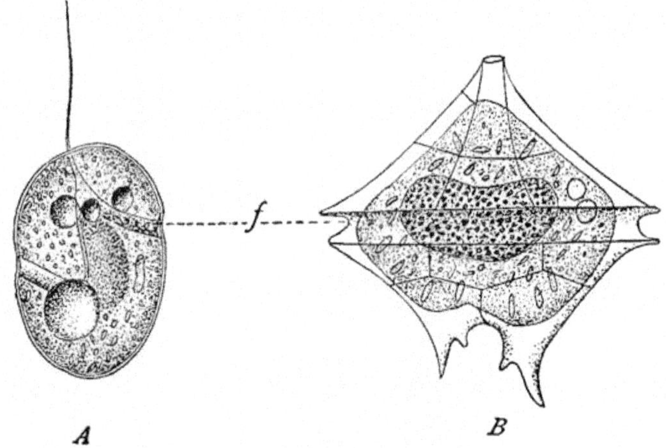

Fig. 5. — Dinoflagellidia. [SCHÜTT.]
A. Gymnodinium ovum, Schütt. *B. Peridinium divergens* Ehr. *f*, the transverse furrow.

The flagellated organisms now included under Diesing's name, *Mastigophora*, fall naturally into three subclasses: (1) the *Flagellidia* (Fig. 4) (flagellates in a strict sense), recognized by Dujardin and

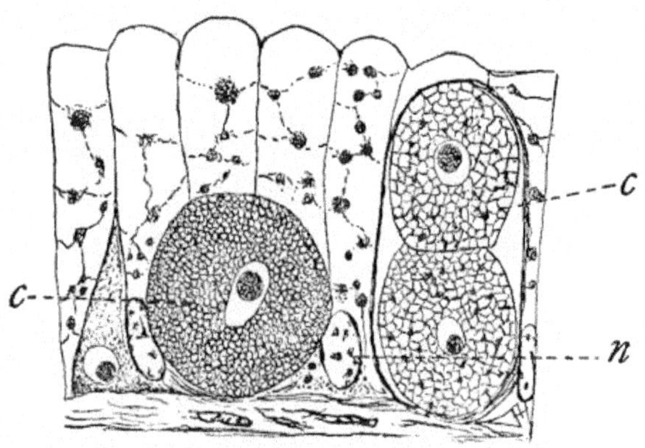

Fig. 6. — Coccidiida in epithelial cells. [LABBÉ.]
The coccidium, a species of the genus *Myxinia*, is supposed to have divided in one case (to the right). *c*, the sporozoön; *n*, the nucleus of an epithelial cell.

named by Cohn ('53); (2) *Dinoflagellidia* (Bütschli) (Fig. 5), which were first seen by O. F. Müller (1773) and later fairly well described by Ehrenberg, but curiously misinterpreted as ciliated forms (a mistake rectified only during the last twenty years), which led Claparède and Lachmann ('58), R. S. Bergh ('84), and Saville Kent ('81) to regard these organisms, under the name *Cilio-flagellata*, as intermediate forms between the Ciliata and the Mastigophora; (3) *Cysto-flagellidia* (Haeckel), including two genera, *Noctiluca* and *Leptodiscus*, the former observed during the eighteenth century, the latter discovered by R. Hertwig ('77).

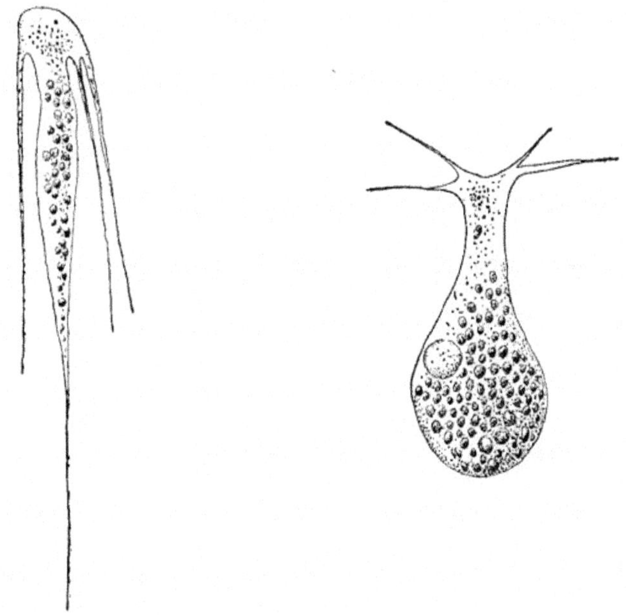

Fig. 7. — Two forms assumed by *Leptotheca agilis*, a myxospore. [DOFLEIN.]

The history of the Sporozoa as a class dates from Kölliker's ('45–'48) and Stein's ('48) works, although the name *Gregarine* now used as the title of an order (*Gregarinida*) goes back to Leon Dufour ('28), and the first observation to Redi in the seventeenth century.[1] The different kinds of Sporozoa were first grouped together by Leuckart ('79) under the present name, and he subdivided the group into the *Gregarinida* (Fig. 1, *D*) and the *Coccidiida* (Fig. 6), the former dwelling in cavities of various invertebrate hosts, the latter inside epithelial cells in, chiefly, vertebrate hosts. Under the term *psorosperms* (Joh. Müller, '41), a number of fish parasites belonging to the Sporozoa were known early in the century, and these were grouped together by Bütschli under the term *Myxospo-*

[1] Cf. Diesing, p. 183.

ridia (Fig. 7), and in the present classification form a fourth order of the Sporozoa. A third order under the name *Hæmosporidiida* (Labbé, '94), includes the sporozoan parasites dwelling in blood-cells and plasm of different vertebrates (Fig. 8).

The Infusoria finally have been variously classified since von Siebold restricted the term as given by Ledenmüller (1760–'63) to its modern significance. Until their unicellular structure was definitely established, the various types were placed among the higher animals, sometimes with the worms, sometimes with the Cœlenterata. Even Perty ('52), who was the first to bring the ciliated forms together under their present name Ciliata, believed them to be combinations of

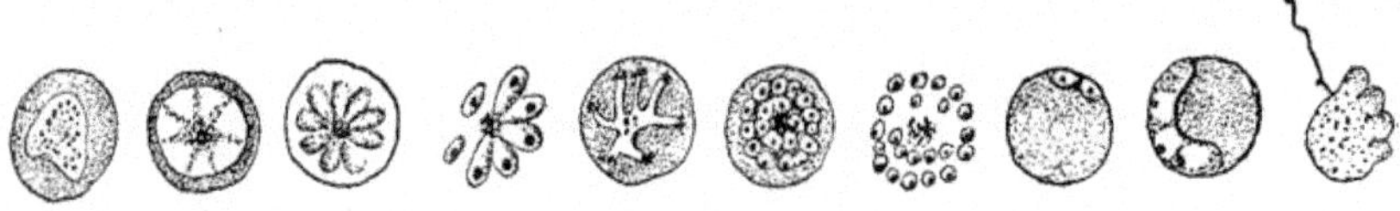

Fig. 8. — A blood parasite or Hæmospore, *Plasmodium malariæ*. Amœboid, spore-forming and sexual phases are shown. [WASIELEWSKY.]

cells. He included the Suctoria and the Ciliata as subdivisions of the Infusoria. Stein ('57), who put the classification of the Ciliata upon its final modern basis, had previously confused the relations of Suctoria and Ciliata in his Acineta-theory. Claparède and Lachmann ('58–'61), after showing that Stein's interpretation was incorrect, raised those Infusoria which are provided with suctorial tentacles and are ciliated only during the embryonic phases, to the grade of a separate subclass to which they gave the modern name *Suctoria* (Fig. 1, *C*, *E*).[1]

Since Stein's work there have been but few important changes in the classification of the Infusoria. Bütschli ('83–'88) divided the Ciliata into two unequal groups distinguished by the nature of the mouth parts, the *Gymnostomata* and the *Trichostomata*. Stein's subdivisions based upon the arrangement of the cilia are more simple, however, and the advantage of Bütschli's division is somewhat questionable.

C. ANIMALS AND PLANTS

In determining the boundaries of the subkingdom Protozoa, two very interesting controversies have arisen, one relating to the boundary between animals and plants, the other to the relations of Protozoa to Metazoa. The modern attitude toward the first of these problems is well expressed by Delage ('96), who says: "The question is not so important as it appears. From one point of view, and on purely

[1] Cf. Stein, II ('67), p. 142.

theoretical grounds, it does not exist, while from another standpoint it is insoluble. If one be asked to divide living things into two distinct groups of which the one contains only animals, the other only plants, the question is meaningless, for plants and animals are concepts which have no objective reality, and in nature there are only individuals. If, in considering those forms which we regard as true animals and plants, we look for their phylogenetic history, and decide to place all of their allies in one or the other group, we are sure to reach no result; such attempts have always been fruitless." [1]

No one at the present time denies the extremely close relation which Huxley ('76) has so clearly pointed out between the lower algæ and some of the flagellates, and it is the general opinion that no one feature separates the lowest plants from the lowest animals, and the difficulty — in many cases the impossibility — of distinguishing between them is clearly recognized. Curiously enough, this modern idea was early expressed by Buffon at the time when Aristotle's view of the plant-like nature of some animals (Zoöphyta) was still accepted in regard to the Cœlenterata. Buffon wrote as follows in 1749: "From this investigation we are led to conclude that there is no absolute and essential distinction between the animal and vegetable kingdoms; but that Nature proceeds from the most perfect to the most imperfect animal, and from that to the vegetable." This statement might have been written in 1899, but Buffon unfortunately goes on to say: "Hence the fresh-water polypus (*Hydra*) may be regarded as the last of animals and the first of plants." [2]

Ehrenberg included a large number of plant forms among his Infusoria, most of which Dujardin threw out, restricting the group, practically, to the Protozoa as known to-day. But the discovery of flagellated swarm-spores of algæ cast doubt on the animal nature of the organisms which Dujardin had described as flagellates. Von Siebold ('45) was thus led to retain only the families Astasiidæ and the Peridinidæ in his zoölogy, removing the Mastigophora, as a group, to the botanical side. In this he was followed by Bergmann and Leuckart ('56), while Cienkowsky ('65) placed them as an intermediate group between animals and plants. Others went to the opposite extreme and actually excluded the algal swarm-spores from the plants, on the ground that they were merely flagellated parasites living on the plant-cells (Diesing, '65). Still others, noting that some of the flagellates are animal and some vegetable in their nature, undertook the impossible task of finding a single distinguishing character. The presence of green coloring matter or *chlorophyl*, upheld by Cohn ('76) and others as a characteristic vegetable feature, seemed to be a good

[1] ('96), p. 518. [2] Edition 1812, p. 357.

test; Oersted ('73), however, showed that in the lower plants there are forms differing only in the presence or absence of chlorophyl. These forms may be arranged in a series as follows:[1]—

With chlorophyl	Without chlorophyl	With chlorophyl	Without chlorophyl
Oscillaria.	Beggiatoa.	Spirulina.	Spirochæta.
Leptothrix.	Leptomitus.	Palmellaceæ.	Chroöcoccaceæ.
Chlamydomonas.	Chlamydomonas hyalina.	Synedra.	Synedra putrida.

A similar series can be arranged among the Protozoa, including forms which cannot be genetically separated, though some contain chlorophyl, and some are colorless. In the first of these, nutrition is *holophytic* or of the green plant type, in the second *saprophytic* or of the fungus type. The chlorophyl differential, if used here, would separate closely allied and in other respects identical forms, always to be found among the Mastigophora, and would lead to confusion. Furthermore, the chlorophyl differential would cause confusion in the classification of the fungi, where colorless representatives of several families of the Phycomycetes reproduce by colorless swarm-spores. Again, some of the Mastigophora with chlorophyl are not dependent upon this substance for their nutriment, but may combine t ꞏ plant type with the animal type of food-getting (*e.g. Chromulina* and some Dinoflagellidia, Fig. 4).

Stein sought a differential in the presence of contractile vacuoles and of nuclei, which, he maintained, are not found in vegetable swarm-spores, but are characteristic of all animal cells. This view has not been supported by later discoveries, for not only have vegetable spores been found to possess nuclei, but many of them are also provided with contractile vacuoles.

Haeckel bases the classification of animals and plants upon nutrition, which differs but little from the earlier chlorophyl differential. All forms with the power of absorbing carbon dioxide, water, and nitrogen compounds, and of combining them into proteids, he calls plants, those without this power, animals, but he considers that this division, though logical, is at best only artificial, and gives no clue to the actual phylogenetic relations of Protozoa and Protophyta. As a single differential, however, the method of nutrition is probably as satisfactory as any, for there are only a few forms which combine the two modes of food-getting. If rigorously applied, however, it cannot fail to shock the prejudices of both botanists and zoölogists in claiming for the animal kingdom forms which have usually been identified with the vegetable kingdom, and *vice versa*. Although Haeckel states that the dividing line is purely arbitrary and does not represent genetic affinity

[1] See Entz ('88).

in the least, animal forms being derived from plants in a polyphyletic series, he does not hesitate to rank certain of the fungi, together with the Sporozoa and bacteria, as animal forms; the majority of chlorophyl-bearing Protozoa, on the other hand, are placed with the plants.

Another differential, which, perhaps, has been the most widely accepted, is the power of spontaneous motion. It is supported to-day

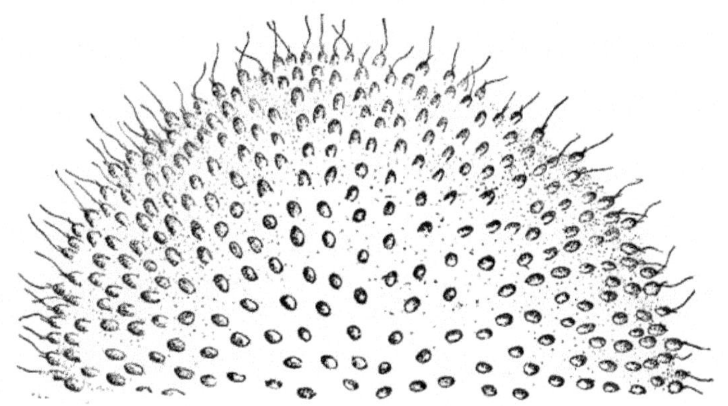

Fig. 9.—A sphæroidal colony, *Uroglena americana* Calkins, consisting of monads embedded in a gelatinous matrix.

as the most universal of the arbitrary differentials by Bütschli, Bergh, and Delage. Briefly stated, all forms, which are freely-motile in their adult life, are animals, while stationary forms are plants. This distinction is applied only to the lower forms, and not to the higher groups, but even as thus limited, this differential would necessitate some striking changes in existing schemes of classification. The freely-moving diatoms, which, since the time of Nitsch ('38) have been classed with the unicellular plants, would be included among the Protozoa, while the majority of Sporozoa, which are almost devoid of motion, would be excluded.

The point of view which demands the strict separation of animals and plants has, however, little utility save perhaps to determine the limits of a text-book or monograph. Many observers, recognizing this truth, have included all forms in which the transition from plants to animals is shown, in a special group of the Protozoa, and usually with some heading which gives a clue to their position. This is first seen in Aristotle's *Zoöphyta* (Cœlenterata); again in a more modern form in Perty's *Phytomastigoda*, and in the *Phytoflagellida* of Delage. Haeckel ('66) made a group of equivocal forms large enough to include all of the Protozoa, and, under the name *Protista*, vainly attempted to establish a third kingdom between the animal and the plant.

The arbitrary dividing line between the Metazoa and the Protozoa can be much more sharply drawn than that between animals and plants. The Protozoa are usually defined as single-celled animals, the Metazoa as many-celled; but this definition is not strictly accurate, for many forms of Protozoa live in aggregates, or colonies in which specialization and division of labor have progressed to a considerable degree (*Volvox, Uroglena, Magosphæra*, etc.; Fig. 9). As a rule, however, colonies do not form a distinct tissue of cells as in the blastula stage of Metazoa, while a still stronger point is that they never form a diblastic embryo.[1]

D. GENERATION DE NOVO

Leeuwenhoek's discovery of the Protozoa had a marked effect upon current thought, some speculative writers seeing in these minute organisms the hypothetical units of organic structure, which, from the time of Democritus to that of Descartes, had been a subject of philosophical discussion. The rapid and incomprehensible increase of Protozoa in standing water could apparently best be explained by a theory of spontaneous generation; Leeuwenhoek, nevertheless, was convinced that their origin would be found in minute eggs or germs which are carried through the air as dust, or brought from place to place by birds, etc., thus showing his firm belief in Harvey's axiom, *ex ovo omnia*. He was supported in this view by Joblot (1718), whose experiments led him to the conclusion that the lower stratum of the air is filled with the germs of various kinds of animalcula, while Réaumur (1738) asserted that the dust of the air also contains disease germs, which are the cause of epidemics. These men were, however, in the minority, and until the last fifty years only an occasional observer opposed the theory of spontaneous generation, as applied to these minute organisms.

Even in Leeuwenhoek's time it was well known that dead organic matter of any kind, when left exposed in water, gradually decomposes, while the water, at first clear, becomes murky, and minute organisms of various kinds develop in it. Adopting the view that higher organisms are composed of organic units, speculative writers inferred that the small animals discovered by Leeuwenhoek were the units which had again become freed from the aggregated condition. This is the key-note of Buffon's (1749) famous theory of generation, which, in one form or other, persisted well into the 19th century. Briefly stated, Buffon believed that all organisms are composed of an infinite number of organic particles. The In-

[1] See Saville-Kent ('81) for the obsolete theory that Sponges are colonial Protozoa.

fusoria he believed to be nothing but these particles become free. "The destruction of organized bodies is only a separation of the organic particles of which they are composed. These particles continue separate till they be again united by some active power. When, however, a man's body has nearly attained its full size, he does not require the same quantity of organic particles; the surplus is, therefore, sent from all parts into reservoirs destined for their reception. These reservoirs are the testes and seminal reservoirs" (page 397). "The different parts of the body are, however, built up of different kinds of organic units, so that upon disintegration there are different forms of animalcula, which are in no respect different from the spermatozoa of the same animal. The freed units are therefore neither animals nor plants, but the formative elements of both. Arising as the disintegrated parts of dead organisms, or rather as elements which never die, they are organisms which pass from one living state into another." This view was carried further by Needham (1748), and as the Buffon-Needham hypothesis, was generally accepted. Thus the early advocates of the theory of spontaneous generation did not maintain that living things arise from not-living substances, but that all organisms are derived from parts of those living before, — a sort of transmigration. Spallanzani, however, to whom so much credit is due for our early knowledge of the Protozoa, adhered to the view of Leeuwenhoek that the Infusoria are not the units which constitute higher organisms, but distinct forms of life which, like other organisms, are derived from definite germs. Furthermore, he vigorously upheld their animal nature against Buffon and his school, basing his arguments upon their voluntary movements, changes of direction when moving, food taking, and upon their relations to moisture and dryness, warmth and cold, to which they reacted like higher animals. He found that Infusoria do not develop in a vacuum, and must, therefore, come from germs contained in the air. Seeing a *Colpoda* emerge from its cyst, he concluded that the cysts were eggs, mistaking the cyst-case for the egg-membrane. He separated the large from the small forms of Infusoria, a separation which was the first attempt to distinguish the Protozoa from bacteria, and which was destined to have great effect upon the theory of spontaneous generation, for it is a significant fact that the forms which have been supposed to arise by spontaneous generation have always been those approaching the limits of vision.[1]

[1] Spallanzani's work has hardly been sufficiently recognized by later writers. Never carried away by enthusiasm, but describing only what he saw, he placed himself outside the current of popular favor by opposing the tempting hypothesis of the nature-philosophers. He seems to have combined his power of observation with a remarkable breadth of view, which in some cases gave rise to daring conceptions. Thus, in 1776, he wrote: "*Pour des*

The distinction which Spallanzani drew between large and small forms was also adopted by O. F. Müller in his classification of the Protozoa. The latter maintained, however, that the lower of his two groups (Infusoria) were formed according to Buffon's view, from the disintegrated parts of higher organisms. These parts, after the disintegration, collect, forming a slimy scum on the surface of the infusion (*Zoöglœa*), which formed a most important adjunct in all subsequent theories of spontaneous generation. Minute vesicles arise later from this scum, and these remain as living organisms in the form of "Infusoria," which included bacteria, spermatozoa, and the smallest forms of flagellates. This mode of origin he limited to those Protozoa which have no visible organs or means of locomotion. The other group (Bullaria), which included the larger of the animalcula (worms, ciliated Protozoa, rotifers, etc.), he maintained were formed, as in the higher animals, from eggs. Müller also held that these units mix with the inorganic particles to form the solid and fluid portions of the body of higher forms, while alone, and without contact with foreign matter, they form the nerves and "soul."

Oken (1805), another advocate of the theory of spontaneous generation, held that all Protozoa arose in a similar manner. Since all plants and animals were built up of these Infusoria, he named the latter *Urthiere*, although, like Buffon, he held that they were neither plants nor animals. Infusions demonstrated to him that all plants and animals could disintegrate into Infusoria. Small Infusoria at first joined together to form larger ones, and out of their union arose the polyps and higher forms. While the outline of Oken's view would seem to indicate a prophecy of the cell-theory, it is quite evident from his book on creation that he had little real conception of what we now regard as the essence of that theory.[1]

The majority of contemporary naturalists followed Buffon and Oken, either absolutely or with slight modifications. Among these Treviranus ('03), Goldfuss ('20), and Carus ('23) accepted Oken's views, while Lamarck ('15), Blainville ('22), and Bory de St. Vincent ('24) followed Müller in restricting spontaneous generation to the simpler and smaller forms.

Dallinger and Drysdale ('73–'75), taking turns over the microscope by day and night, followed out the life-histories of many of these simpler forms through the process of division and spore-formation, thus showing that the monads arise, as do the higher Protozoa, from

animaux inférieurs, le changement de demeure, de climat, de nourriture, doit produire peu à peu dans les individus, et ensuite dans l'espèce, des modifications très considérables qui déguisent à nos yeux les formes primitives" (cited by Dujardin ('41) from Spallanzani).

[1] The name "Protozoa," given by Goldfuss ('20), meant the same as Oken's "Urthiere." It did not acquire its present significance until 1845, when von Siebold gave it a new meaning.

ancestors similar to themselves. They found that the spores which burst from the encysted forms were at first far beyond the limits of vision even with the high powers of the microscope at their command, but remained together in the form of a "glairy" mass in which minute specks soon appeared, and these specks were watched until they had become full-grown monads similar to the original form.

In later years the theory of spontaneous generation has been limited almost exclusively to the bacteria, but even here it has been energetically and successfully opposed by Pasteur, Tyndall, Milne-Edwards, Claude-Bernard, Quatrefages, and others, against a constantly decreasing number of advocates. No one is in a position to assert, however, that it does not take place in some organisms, although such a view is highly improbable; nor can it be maintained that it never has taken place in the past. Many theories of "archigony" (Haeckel), or the first origin of life by spontaneous generation, have been held by modern naturalists; but all such theories are of a purely inferential character and lack substantial foundation. Without attempting to discuss these[1] it may be pointed out that the eminent botanist Nägeli has advocated an hypothesis which suggests that of Buffon. Assuming that protoplasm consists of minute structural units or "micellæ," he suggests that such micellæ were first formed from not-living matter and secondarily united into organisms. Nägeli does not hesitate to say that the evolution of the simplest protozoön from inorganic compounds involved a far greater step than from the first organism to man, and in accordance with this idea Haeckel places the beginning of life in the oldest known geologic age and in the oldest period of that age, the Laurentian. This, again, is entirely speculative; for if we except the questionable form *Eozoön*, the rocks of the Laurentian contain no recognizable records of past life.[2]

The rocks of the period after the Laurentian, however, the Cambrian, possess a great number of well-marked types, families, and genera, thus indicating, even at this time, a considerable antiquity. Haeckel and Nägeli argue with Huxley, and the argument is of great

[1] For a discussion of this topic the reader is referred to the essays of Huxley, Tyndall, and Haeckel, and to Verworn's *Allgemeine Physiologie*, pp. 298–319. Lee's translation, pp. 297–319.

[2] The supposed genus *Eozoön* ("dawn of life") was discovered by Logan, of the Geologic Survey of Canada in 1865, and the name was given by Dawson in the same year. There has been a lengthy dispute, however, in regard to this supposed fossil, some asserting that it is the earliest known foraminiferon, others that it is entirely inorganic. The former opinion was held by Carpenter ('65, '66, etc.) and Dawson ('65, '75, etc.), the latter by the majority of geologists and petrologists, beginning with King and Rowney ('66) and followed by Möbius and others. Bütschli, while admitting that Dawson and Carpenter had a certain amount of evidence, inclines to the opposite view, while petrologists maintain that the same structure as that of Eozoön has been frequently observed in minerals forming parts of rocks of undoubted igneous origin.

weight, that since organized living bodies are composed of the same materials as unorganized or lifeless bodies, and after death are again resolved into those same lifeless materials, it may be logically assumed that in the beginning and under certain conditions, simple materials were combined into new compounds having the properties which we know to-day as life. Haeckel at first held that these complex compounds were primitive organisms which he called *Monera* or organisms consisting of homogeneous plasm without differentiations of any kind, since differentiation follows the localization of function and can originate only as a result of living activity. In his later work, however, Haeckel ('96) follows Nägeli in postulating simple structural units as the primitive forms of life instead of the homogeneous and formless lumps of proteid.

Nägeli maintained that two stages must be distinguished between inorganic matter and the lowest organisms known to us. The first consisted of the synthesis of the albumin compounds and the organization of these into micellæ which constitute primordial plasm. The second stage was the transformation of the primordial plasm into the simplest of living organisms. Haeckel's hypothetical *Monera*, if they existed, would approach most closely to these primordial forms of living matter, being described as "organisms without organs." They could be called structureless, however, only from the anatomical standpoint. Physically, the earliest organisms must have been already complex, for, chemically considered, an albumin molecule is an extremely complex substance, and every unit of plasm which Nägeli calls a micella must have had, and now has, that same complex composition.

Somewhere in the obscurity of this early period came the change from the plant to the animal mode of nutrition. The latter must have begun at an early time, although the possibility of change at any time from plant to animal nutrition is not excluded, as shown by the numerous instances among the higher plants of adaptation to a parasitic or a saprophytic mode of life. So too, among the Protozoa, the acquisition of a cannibalistic mode of life, or, as Haeckel calls it, *metasitism*, may have required, and probably did require, a long period, and there is little reason to doubt Haeckel's view that the Protozoa are polyphyletic in their origin. We possess no positive data for the conclusion as to which of the Protozoa were the most primitive. In considering this question, it must not be overlooked that, during the eras that have passed, the Protozoa may have been adapted and re-adapted many times over to changing conditions of environment, and living species have, in all probability, not come unchanged from that remote past.[1]

[1] See Lankester ('91), Klebs ('92), and *infra*, p. 99.

SPECIAL BIBLIOGRAPHY I

Bütschli, O. — Protozoa. In Bronn's Klassen und Ordnungen des Thierreichs. *Leipzig*, 1883-1888.

Dujardin, F. — Les Infusoires. *Paris*, 1841.

Entz, Geza — Protistenstudien. *Budapesth*, 1888.

Huxley, T. H. — On the Border Territory between the Animal and the Vegetable Kingdoms. *Collected Essays, D. Appleton & Co.* Vol. 8, Edition 1896.

Kent, W. Saville. — A Manual of the Infusoria. *London*, 1881.

Stein, Fr. — Organismus der Infusionsthiere. *Leipzig*. Abth. I., 1861; II., 1867; III., 1878.

CHAPTER II

GENERAL SKETCH

It is a widely accepted opinion among men of science that life originated in the sea, and here to-day are found the great majority of species of Protozoa. In the littoral regions, particularly in the superficial slime and upon submerged water-plants, are found a profusion of Rhizopoda. Farther out, Radiolaria and shelled Rhizopoda belonging to the order Reticulariida float upon the surface or at varying depths below it, while their empty shells, settling slowly to the bottom, have added little by little to the accumulations of the past, until to-day, under the names *Radiolarian ooze* and *Globigerina ooze*, they form vast areas, miles in extent, and often attaining a depth of many feet. By the agency of earthquakes or slow upheavals, these beds have become exposed from time to time, and we recognize the Barbadoes as composed in large part of the skeletons of Radiolaria, or the chalk cliffs of England as built up of the lime shells of reticulate Rhizopoda.

Apart from the Sarcodina, the majority of Protozoa leave no memorial in stone of their past existence. Pelagic forms such as Dinoflagellidia and Cystoflagellidia, living near the shores, and often drawn together into great aggregates by currents, winds, etc., become the food of whales, fishes, and other marine animals. Many Rhizopoda, Ciliata, and Suctoria are attached by mineral secretions, or by stalks, to rocks, submarine plants, etc. Others are parasites upon the outsides of fish and other animals, while still others are parasites within.

The fresh-water Protozoa, while less rich in species, are much better known than the marine forms, for their modes of life, habitats, and life histories are more easily observed and controlled. Many kinds of Rhizopoda, Heliozoa, and Ciliata are found both in fresh water and in salt; and numerous experiments by Verworn, Gruber, and others have shown that some forms can live either in salt water or in fresh; the change from one to the other usually results, however, in modifications of structure. In general, the Protozoa abound in fresh water which contains enough food material for their growth and reproduction, but the widespread belief that each drop of drinking water contains countless myriads of microscopic forms has absolutely no foundation. Protozoa cannot live in chemically pure water, and on the other hand, many of them cannot live in foul waters.

Thus, in sewage, one finds occasional ciliates and flagellates, but no such numbers as are sometimes found even in good drinking waters, where, obtaining their food as do the green plants through the agency of their green or yellow coloring matter (chlorophyl), the Protozoa sometimes become a source of annoyance. They thrive in standing waters where the accumulation of bacteria gives food for numerous ciliates and rhizopods; the decomposing organic matter dissolved in the water is taken in by saprophytic forms of flagellates, which multiply to prodigious numbers, and these in turn may form the food-supply for predaceous Infusoria and Sarcodina. Rotifers, Crustacea, molluscs, and worms prey upon all forms, and when the cycle is passed, the water becomes cleared of animal life. In nature, the pools rarely if ever become thus cleared, because new food is constantly brought in from fresh sources, and the cycle becomes continuous. In such places the superficial slime upon the bottom contains naked and shelled rhizopods, although the latter are more often found alive upon the leaves and stems of water-plants; here, too, are colonial Infusoria or single forms attached by their stalks. Suspended in the water are to be found the majority of species of Flagellidia, a few Dinoflagellidia, the majority of Heliozoa, many predatory ciliates, and a few rhizopods, especially certain shelled forms which secrete a bubble of gas to buoy them up.

Many forms of Protozoa are capable of sustaining life either as terrestrial or as parasitic organisms. The former, allied to the Mycetozoa, grow over damp wood, while a number of rhizopods are almost able to withstand dryness, for as Dujardin, Ehrenberg, Greeff, and others early pointed out, they live in damp moss and leaves of the woods. Forms which have become adapted to a parasitic mode of life may be found in all classes. Among the Rhizopoda, various species of intestinal *Amœbæ* may be found in all sorts of vertebrate and invertebrate hosts; about twenty species of Flagellidia and many more of Ciliata live as parasites, some in the blood, some in the intestinal fluids, and others in the cavities of various organs in man and other hosts. These forms, however, are only occasional parasites, and are more like commensals than parasites, having little significance when compared with the Sporozoa, a class of Protozoa which, without any exceptions, are parasitic. These infest all animal forms from Protozoa to man: one group lives in the digestive tract and the cavities of the body (Gregarinida); another in the cells of the digestive organs (Coccidiida); another in the muscle-cells and lymph surrounding them (Myxosporidiida, Sarcosporidiida); and still another in the blood-corpuscles and in the blood-plasm (Hæmosporidiida). Of all Protozoa these are the only forms which are known to menace the life of man.

A. GENERAL MORPHOLOGY

As might be expected from the wide distribution of the Protozoa and their varied modes of life, each of the several classes contains organisms of varying forms and grades of complexity. In fact, no one form is characteristic of any group, but in all cases where the body is plastic and subjected to an even pressure the form is spherical (homaxonic), readily changing, however, into an elongate or monaxonic condition. In the higher types, especially those which are inclosed in a firm membrane, the form is usually asymmetrical, and cannot be interpreted as the direct result of mechanical conditions. The homaxonic type prevails among Heliozoa, Radiolaria, and intra-cellular Sporozoa (Coccidiida), and occurs in the simpler types of all classes. The monaxonic form prevails among the Mastigophora and the lumen-dwelling Sporozoa (Gregarinida), while asymmetrical forms are dominant among the Infusoria. In all classes, when for any reason the surroundings become unsuitable, or at times as a preliminary to some methods of reproduction, the organisms secrete a thick and resistant protective coating or cyst which is usually homaxonic.

The various adaptations found in the Protozoa are confined almost entirely to the outer protoplasm or *ectoplasm*, the inner portion or *endoplasm* remaining approximately similar in structure throughout the group. The ectoplasm, being in direct contact with the surrounding medium, becomes hardened into ectoplasmic coatings of various kinds, serving as protective coverings for the inner endoplasm. It also becomes differentiated into various external organs of locomotion, of food-getting, of defence and offence, and, in the higher types, into organs of sensation.

1. *The Endoplasm.*

Examined under the low powers of the microscope, the body of a protozoön appears to be made up of a gelatinous, diaphanous substance which, under certain conditions, breaks out of the confines of the cell-membrane, forming irregular globular masses in the water. This phenomenon was early recognized, and under the term "diffluence" was regarded by Dujardin as a special property of sarcode.[1]

Examined under higher powers of the microscope (*e.g.* with a one-

[1] "Sarcode. Je propose de nommer ainsi ce que d'autres observateurs ont appelé une gelée vivante, cette substance glutineuse diaphane; insoluble dans l'eau, se contractant en masses globuleuses, s'attachant aux aiguilles de dissection et se laissant étirer comme du mucus, enfin se trouvant dans tous les animaux inférieurs interposée aux autres élémens de structure." — DUJARDIN, '35, p. 367.

twelfth inch objective), the mass of endoplasm is seen to consist of a more or less definite matrix, and if the cells be properly fixed and stained, a distinct structure is visible. This appears to be little more than a definite meshwork, the meshes of which are sometimes minute, compressed, and narrow, sometimes large and open. The substance of the mesh proper appears to differ noticeably from that within its spaces. The latter is fluid-like, and not infrequently contains granules of larger or smaller size; the former, also a fluid, appears more dense, and is made up of exceedingly minute granules (microsomes). Differential stains show that the various granules differ not only in size, but in chemical composition, and it has been determined that some are food particles in process of assimilation, and that others are waste matters. This protoplasmic structure, which Bütschli ('92) compares with a foam structure (*Schaumplasma*), is described by him as consisting of small drops of a liquid *alveolar substance* inclosed within the meshes of a continuous *inter-alveolar substance*, also liquid, but of a different composition. Each alveolus may be compared to a bubble in a foam structure; the air of the bubble corresponding to the alveolar substance, the walls to the inter-alveolar substance.

While the endoplasm of all Protozoa is alveolar in structure, there is considerable variation in density due to the relative sizes of the alveoli and to the nature of the granules contained within them (Fig. 10, A–D). They vary in size from minute vesicles in Sporozoa (C) to large vacuoles in many Heliozoa, Radiolaria, and Infusoria. In some cases, *e.g.* in the heliozoön *Actinosphærium* (D), or the cystoflagellate *Noctiluca*, the vacuoles are so large that the protoplasmic structure appears parenchymatous like a plant-cell. The granules in the walls of the alveoli are equally variable in size. In some cases they are exceedingly minute, and correspond apparently to the fine elementary granules which Altmann ('94) regarded as the basis of all protoplasm (*e.g. Amœba*, A); in others they are coarse and obviously of different kinds (*Pelomyxa*).

The various granules within the alveoli are sometimes inert and functionless and often crystalline in form.[1] In other cases they may have some function to play in the economy of the cell. Thus carbohydrates in the form of starch, sugar, or cellulose are generally present and serve as a reserve store of food, or of building material for the outer covering. Other granules which are invariably present may be food particles in various stages of digestion, assimilation, and excretion, or oil particles of various forms and sizes.

With the exception of the Sporozoa, every class of Protozoa includes

[1] Cf. Chap. IX., p. 286.

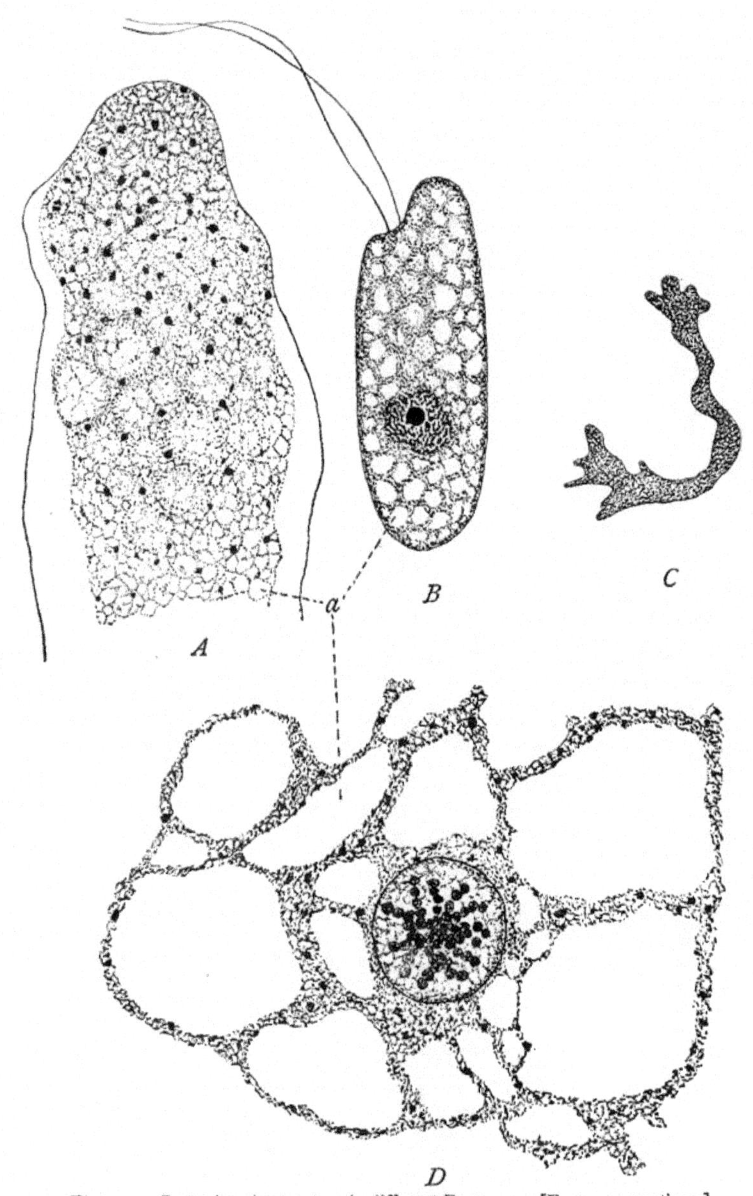

Fig. 10. — Protoplasmic structure in different Protozoa. [From preparations.]

A. Amœba proteus pseudopodium. The endoplasm has broken through the ectoplasm, ar is now in advance. *B. Chilomonas paramœcium* Ehr. a flagellate. *C. Lymphosporidium trutt* Calkins, a sporozoön. *D. Actinosphærium Eichhornii*, endoplasm and one nucleus. *a*, alveo Same magnification throughout.

some species in which colored masses of protoplasm — *chromatophores* — are present, either as living parts of the cell (Mastigophora) or as commensals or *symbionts*, the protozoön and the plant living together for mutual benefit (Infusoria, Sarcodina). The chromatophores are colored by different substances, usually green by chlorophyl (Chloromonadina, some Infusoria), or brown by diatomin (Chrysomonadidæ and Dinoflagellidia), and have a definite shape and size for each species. Brilliantly colored patches of pigment, the so-called eye-spots or *stigmata*, are frequently seen, chiefly among the Mastigoph-

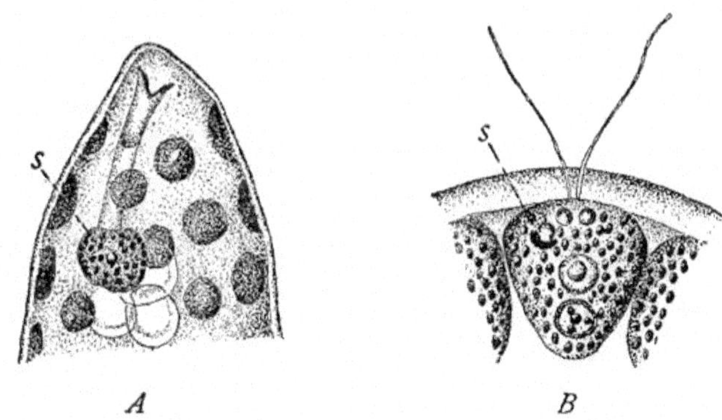

Fig. 11. — Flagellates with stigmata. [FRANCÉ.]

A. Euglena Ehrenbergii, Klebs. The color-mass (*s*) contains several concrements (lenses ?). *B. Pandorina morum*, Ehr. The color-mass (*s*) is attached to a single spherical lens.

ora, where they are situated near the base of the flagellum. These spots are supposed to have some special relation to light,[1] an unproved, though probable, view which is based chiefly upon the fact that in many of them there is a distinct lens-like body, and other structures which usually accompany eyes of primitive form in other types of invertebrates[2] (Fig. 11).

Among other inclusions occasionally found within the endoplasm are the peculiar trichocysts found in the holotrichous ciliates (*Paramœcium* and its allies). These are minute defensive or possibly offensive weapons analogous to the stinging threads of the Cœlenterata. *Nematocysts* containing a spirally wound thread, as in the Cœlenterata, are also found in two forms, one a dinoflagellate (*Polykrikos*), the other a ciliate (*Epistylis umbellaria*), while analogous thread-bearing structures are found in the spores of all Myxosporidiida among the Sporozoa (Fig. 12, *C*, *D*).

[1] Cf. Pouchet, '86.

[2] Engelmann's ('82) experiments, on the other hand, tend to show that it is not the colored body, but the colorless protoplasmic mass in front of the stigma which is particularly sensitive to light.

2. *The Ectoplasm.*

In many Protozoa, especially among the Rhizopoda, there may be no distinction between ectoplasm and endoplasm. These cases, however, are exceptions, for in the majority of forms a well-marked ectoplasm can be distinguished. In many cases the difference appears to be only in the presence or absence of granules, and their distribution depends upon the density of the fluid plasm. No great morphological importance can be attached to this regional difference, for it appears to be only an index of the physical conditions of the protoplasm. The body of the common rhizopod *Amœba*, for example, consists of a more or less fluid mass in which lie suspended the various granules, vacuoles, nuclei, crystals, and food particles, and, as Grüber ('84) pointed out, if the plasm is thin, *i.e.* more fluid, the contents can spread easily through the whole mass, while if the plasm is dense and viscous, they will be held back by the resistance, and a relatively broad ectoplasm may result. The more fluid condition is seen in rhizopods like *Protomyxa* and *Pelomyxa*, the denser in *Amœba proteus*, and the majority of fresh-water shell-bearing forms. In some of the latter and in a few Infusoria the distribution according to density is so marked that several regions can be made out. Thus Pénard ('90) described no less than four zones in the shelled rhizopod, *Euglypha*, while in many Infusoria and in some Sporozoa a membrane, ectoplasm, cortical plasm, and endoplasm, differing from one another in density, can be distinguished. It is an interesting fact that in the artificial mixtures which Bütschli has so successfully made to imitate protoplasm, a similar regional differentiation, at least as far as ectoplasm and endoplasm are concerned, may be seen.

It is perhaps to a tendency of protoplasm to stiffen while in contact with water that we can turn for an explanation, first pointed out by Grüber ('81), of the outer condensation of protoplasm resulting in the numerous membranes and tests of the Rhizopoda, and of the outer coverings of Protozoa in general. The simplest form of membrane is an almost invisible cuticle of extreme delicacy (*pellicula* of R. S. Bergh) as in the rhizopod *Amœba proteus* (Grüber, '81). In other forms of the same genus, however, the outer zone becomes greatly thickened (*A. tentaculata*, *A. actinophora*, Fig. 12, *A*), and a more or less lifeless membrane results. In these thick-skinned forms the membrane is often perforated by the pseudopodia, which form long finger-like processes, and when retracted leave minute holes in the membrane. In these cases there is usually a sharp distinction between the inner plasm and the cortical part, but in many Infusoria and Sporozoa there is a gradual increase in density from within outward, and the outside is covered by living membranes which may become complicated by the addition of muscular fibrils (*myonemes*), of sensory and tactile organs (*cirri*), or protective structures like hooks, spines, and tentacles (Fig. 12, *B–G*; see also Fig. 16).

Like many of the cells which constitute the tissues of higher ani-

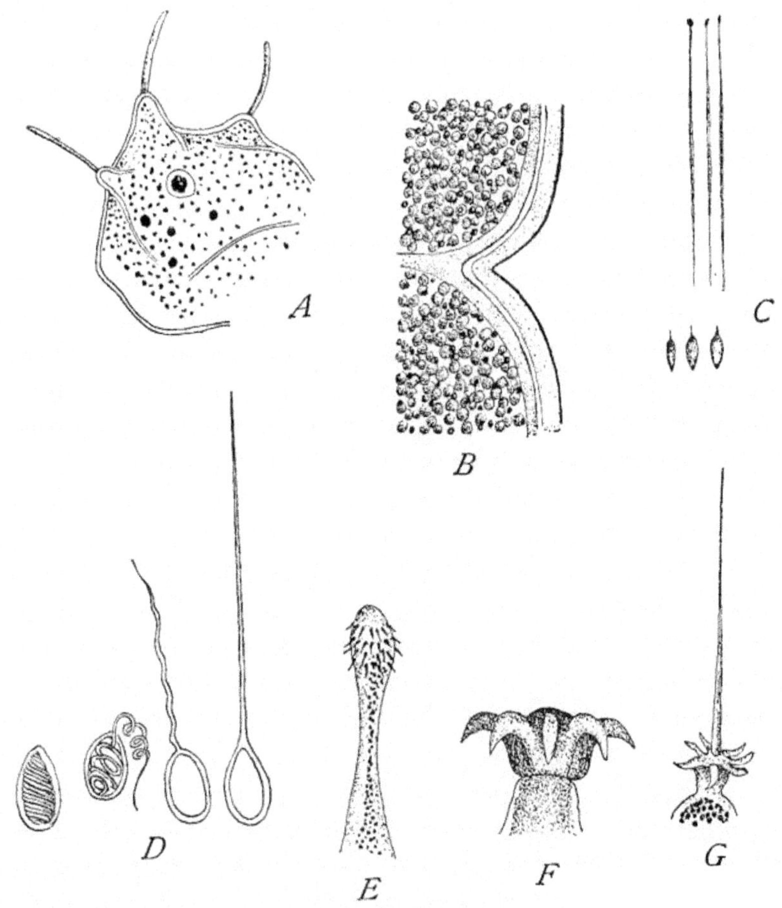

Fig. 12. — Ectoplasmic modifications.

A. Amœba tentaculata. [GRÜBER.] *B. Clepsidrina munieri.* [SCHEWIAKOFF.] *C.* Trichocysts. [SCHEWIAKOFF.] *D.* Nematocysts from the sporozoön *Myxobolus.* [BALBIANI.] *E, F, G,* attaching hooks and spines from different Gregarinida. [WASIELEWSKY.]

mals, the protozoan body has the power of forming by chemical processes over and above those which relate merely to nutrition, various products which are secreted just within the peripheral plasm, where they usually form a protective armor in the shape of shells, tests, or "houses." The materials thus formed within the cell-body may be *chitin* (composed of C, H, N, and O, and supposed to be a derivative from carbohydrates, but the exact formula is in dispute), *cellulose*

($C_6H_{10}O_5$), calcium carbonate ($CaCO_3$), and silica. The secretions may take the form of plates, of continuous deposits, or of regular skeletons which are often extremely complex (Fig. 13, *D*). In the majority of cases, the secretions are made in the ectoplasm, although in one well-authenticated case at least (*Euglypha alveolata*, Fig. 13, *A*), the plates destined to form the shell are formed in the endoplasm and in the immediate vicinity of the nucleus.[1] In other shell-bearing forms of Rhizopoda, there is usually a basis of chitin, upon which the various shell-substances are deposited, or the shell may consist of the chitin alone. In some cases it is no more than a cap covering a small portion of the body, and into which the entire protoplasmic mass could not possibly be withdrawn (*Pseudochlamys*, Fig. 13, *C*). Here the chitin which forms the shell is perfectly smooth; but in other forms it may be ornamented in various ways by pits or protuberances. Again, in many fresh-water Rhizopoda the shell-material is not secreted, but the test is composed of foreign particles, such as diatom shells, sand crystals, mud, or detritus of any kind, fused together and to a chitinous substratum by means of mucilaginous cement secreted by the organism.

3. *Nuclei.*

Haeckel's claim ('68) that there are organisms without nuclei (Monera), although it rests upon negative evidence, cannot be rejected until all of the forms considered have been shown to possess them. On purely *a priori* grounds, it is possible to conceive such organisms, although the numerous experiments which have been performed during the last decade upon nucleated and non-nucleated parts of Protozoa, show, in these cases at least, the absolute necessity of the nucleus for the life of the organism. These experiments make it probable that the so-called Monera have in reality some structure or structures which perform the functions of the nucleus, although a well-defined nucleus with membrane and other characteristic parts may be absent.

In the majority of Protozoa there is but one nucleus (many Sarcodina, Mastigophora, Sporozoa), while in some forms two nuclei are the rule (some Rhizopoda). In others, again, there may be a great number of nuclei, the number varying with the age of the organism (examples occur in all groups of the Protozoa). In many of the Protozoa, although not in all, the nucleus is provided with a membrane and contains two substances; *chromatin*, staining with certain basic dyes and consisting largely of nucleinic acid, and *achromatin*, a substance which is not stained by the chromatin dyes, in the form of a

[1] Cf. Schewiakoff ('88).

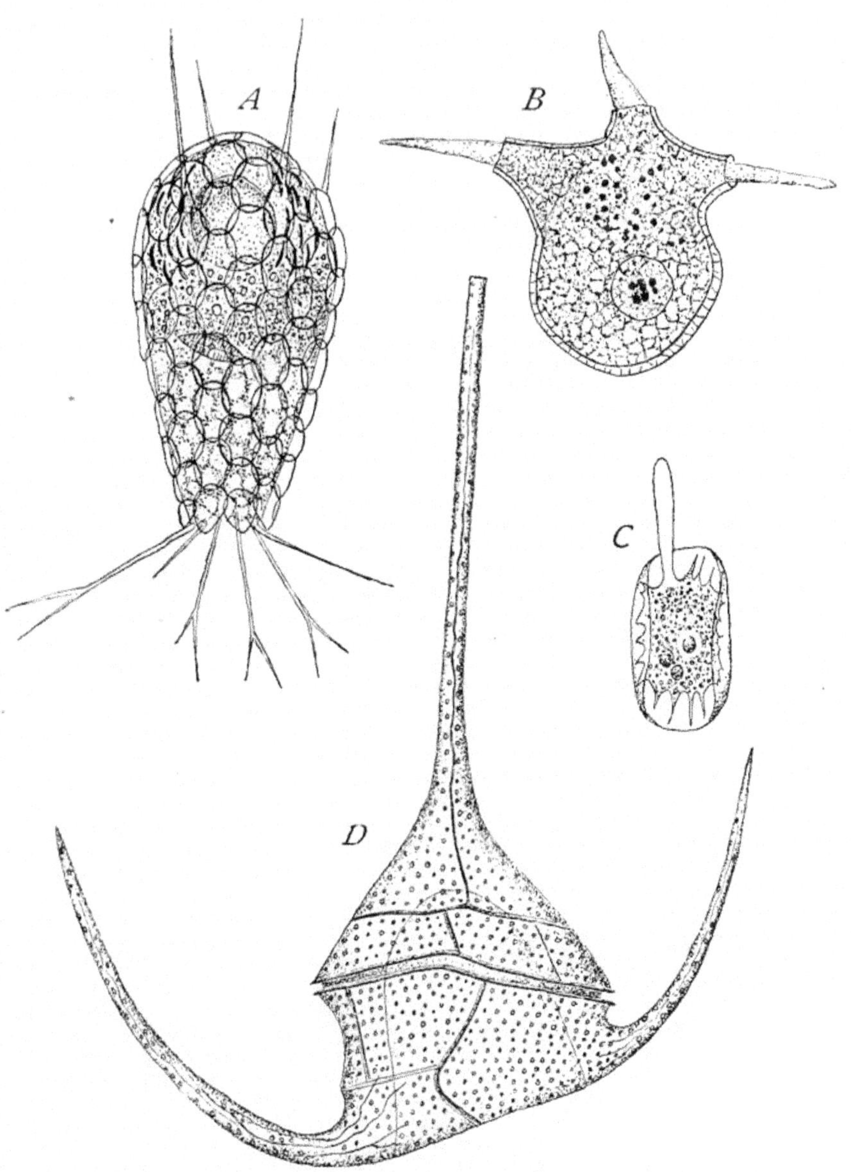

Fig. 13. — Shells and tests. [*A*, SCHEWIAKOFF; *B*, ORIGINAL; *C*, BUTSCHLI; *D*, STEIN.]

A. *Euglypha alveolata* Duj. The shell consists of oval siliceous plates glued together by a siliceous (?) cement. *B*. *Cochliopodium digitatum*, n. sp. The test is membranous and perforated for pseudopodia. *C*. *Pseudochlamys patella* Clp. and Lach. The test is membranous and shield-like. *D*. *Ceratium tripos* Nitsch. The shell consists of cellulose plates of diverse size and shape.

network, or of a homogeneous body of considerable size (*Karyosomes*). Several different types of nuclei may be distinguished; some of the most important being: (1) The *distributed nucleus*,

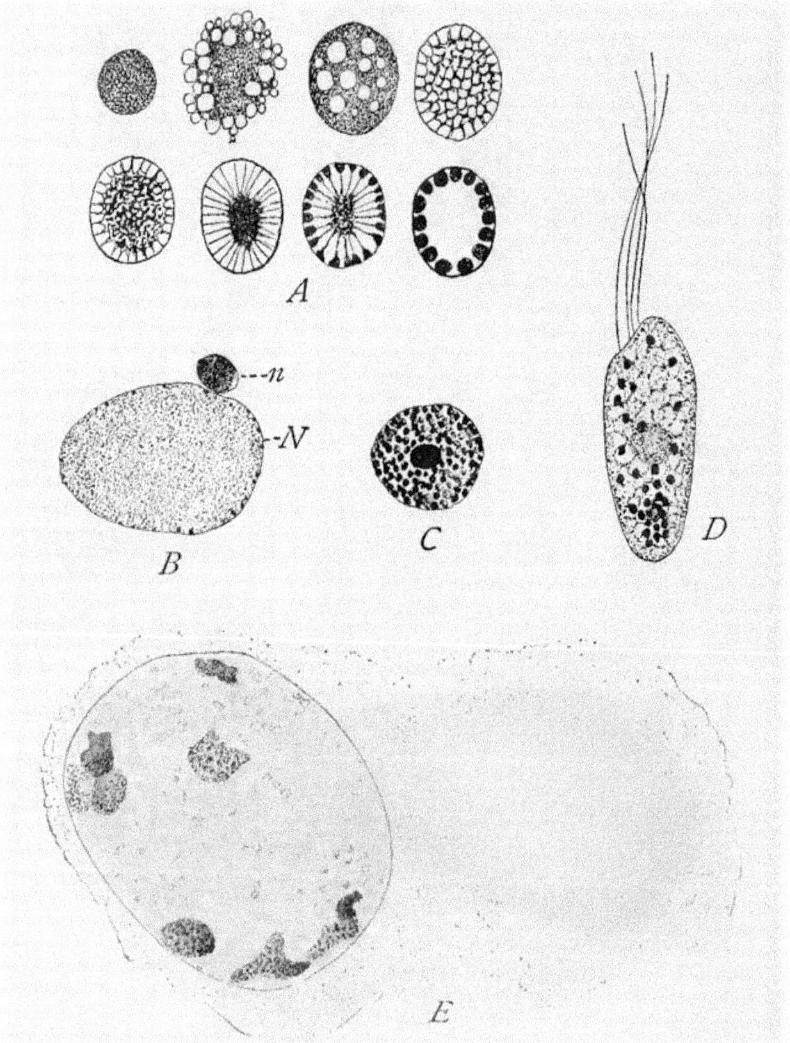

Fig. 14. — Types of nuclei. [*A. Calcituba polymorpha* Roboz, from SCHAUDINN; *B. Colpidium colpoda*, from a preparation; *C. Euglena viridis* Ehr. from a preparation; *D. Tetramitus chilomonas*, n. sp.; *E. Noctiluca miliaris* Sur., from a preparation.]

A single karyosome (*A*) becomes vesicular, and ultimately gives rise to several daughter-karyosomes (so-called "fragmentation" Schaudinn). Several karyosomes in *Noctiluca* (*E*) hold the chromatin, the rest of the nucleus is filled with "achromatic" granules. In *Tetramitus chilomonas* (*D*) the chromatin is scattered throughout the cell; the lighter-colored body in the centre of the cell is the homologue of the deeply stained central body in *Euglena* (*C*).

consisting of innumerable chromatin granules distributed throughout the cell (*Tracheloceerca*, *Chænia teres*, *Holosticha flava*, *H. scutellum*, *Tetramitus*). (2) The *homogeneous nucleus* consisting of a single mass of chromatin with a homogeneous structure throughout all stages, and with no trace of reticular substance (many Phytoflagellates). (3) *Dimorphic nuclei*, consisting of a large nucleus called the macronucleus, and a small one, the micronucleus, in the same individual. The former is generally regarded as functional chiefly in vegetation, the latter in conjugation. With the exception of *Polykrikos* among the Mastigophora, dimorphic nuclei are found only in the Infusoria (Fig. 14).

The typical form of the nucleus is spherical, although it may be discoid or ellipsoid, or, in the case of the macronucleus, drawn out into various fantastic shapes, of which the horseshoe (Vorticellidæ), the beaded (*Stentor* and *Spirostomum*, etc.), or branched (*Acineta*, *Dendrosoma*) are examples.[1]

4. *Organs of Locomotion.*

With very few exceptions, the Protozoa have the power of moving from place to place. The exceptions are found among the parasitic Sporozoa, although even here there is, in some cases, a peculiar gliding motion. In no adult sporozoön is there a special organ of locomotion, yet the Gregarinida and Hæmosporidiida actually move from place to place, although very slowly. In some cases, the motion is due to peculiar peristaltic waves of contraction; in other cases to the contraction of muscle-like fibrils, the myonemes. An analógous movement is also known in certain flagellates (Euglenidæ) and ciliates (Heterotrichida). In the majority of Protozoa, however, movement is accomplished by the activity of special motor organs, which may be either changeable processes (pseudopodia) or permanent vibratile appendages (flagella and cilia). The changeable processes or pseudopodia, found chiefly in the Sarcodina, are sometimes numerous, sometimes few; when few in number they are usually short, finger-formed, and quick to change in form and appearance by the flowing protoplasmic substance of which they are composed (Fig. 1, *A*, and Fig. 15, *A*, *B*); when numerous, they are fine-pointed, and often sticky, so that when two or more come in contact, they fuse or anastomose (Reticulariida, Fig. 15, *C*). Again, the pseudopodia may be fine and pointed, but rigid in structure and unchanging in form, a condition brought about by the presence of an axial filament of stiffened protoplasm, which runs down the centre of each pseudopodium (Heliozoa, Radiolaria, Fig. 15, *D*). Unlike pseu-

[1] Cf. Chapter VII. for further details concerning nuclei.

dopodia, the protoplasmic filaments, known as flagella and cilia, are derived solely from the ectoplasm and are constant in their position, and, save for the occasional absorption within the body, for some reason or other, they are unchangeable. Flagella-motion, characterized by energetic contractions or undulations, or by rotary motions, differs

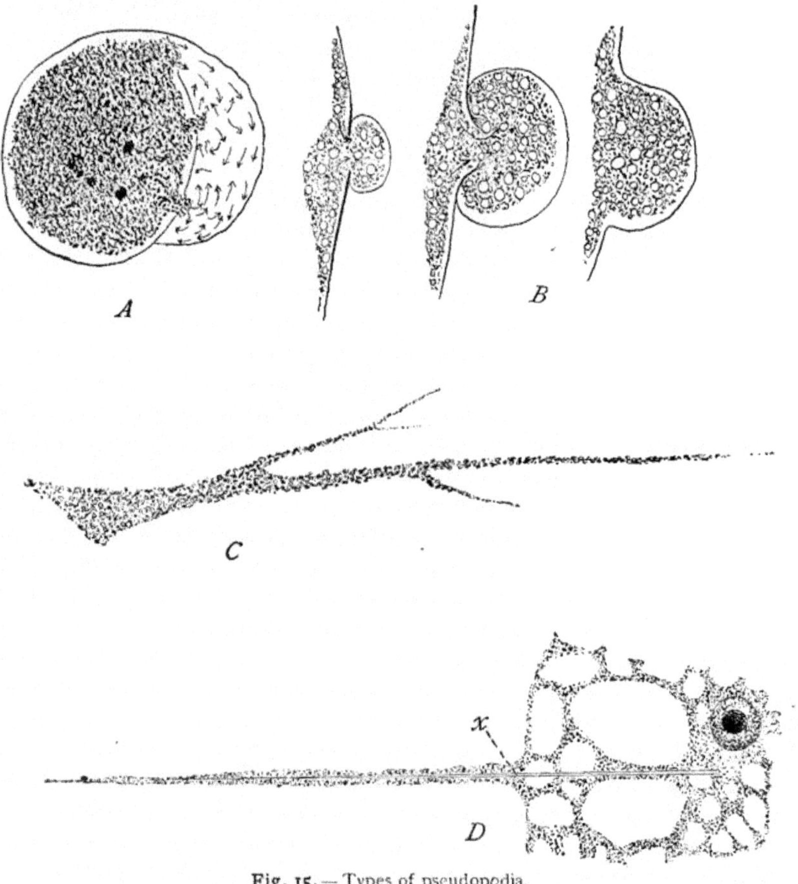

Fig. 15. — Types of pseudopodia.
A. Amœba limicola Rhmb. [RHUMBLER.] *B. Amœba blattæ* Bütsch. [BÜTSCHLI.] *C. Lieberkühnia* sp. [VERWORN.] *D. Actinosphærium Eich.* Ehr. [ORIGINAL.] *x*, the axial filament.

entirely from the slow flowing movement of pseudopodia; yet, as Dujardin first observed, in some forms pseudopodia change into flagella, and flagella into pseudopodia. In structure, flagella are long, thin, usually pointed threads of protoplasm, which, as a rule, are longer than the cell itself; they are typically single, but there may be two, three, or many. Cilia, on the contrary, are always multiple, and are never interchangeable with pseudopodia (Fig. 16). Although characteristic of various epithelial tissues in Metazoa, they are found only in a single specialized group of the Protozoa, the Infusoria. In form, they are similar to flagella, but as a rule they are shorter, never pointed, and more numerous, many of them acting in unison, with a quick regular motion like a set of oars. In some groups, the cell is completely clothed with these motile elements (Holotrichida), in others only a portion is covered either in one or more rings about the body (Peritrichida), or upon one surface only (Hypotrichida). The cilia may also become variously modified by fusion with one another,

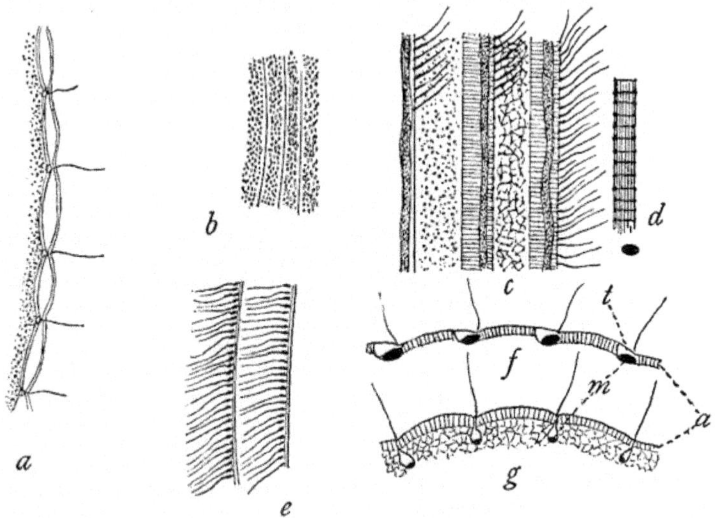

Fig. 16. — Cilia and myonemes of Infusoria. [*a*, *b*, and *e*, JOHNSON; *c*, *d*, *f*, *g*, BÜTSCHLI.]
The surface view of *Stentor cœruleus* (*c*, *e*) shows rows of cilia inserted on the borders of canal-like markings, each of which contains a myoneme (*d*). These are more clearly shown in the optical section (*f*). In *Holophrya discolor* (*g*) the canals and myonemes are inserted deeper in the cortical plasm. *a*, the membrane of *Stentor cœruleus* under pressure.

giving rise to motile organs of a more complex structure, such as the membranes, membranelles, and cirri found in different groups of the ciliated Infusoria.

In many Protozoa the adult forms have no distinct motile organs, although they may pass through embryonic stages in which such structures are present. The Suctoria, for example, are, for the most part, entirely devoid of cilia in the adult stages, although the embryos possess them. Again, many of the Rhizopoda pass through flagellated stages before assuming the amœboid condition, and certain Mastigophora pass through amœboid swarm-spore stages. Some Heliozoa and Radiolaria similarly pass through both flagellated and amœboid stages before assuming their own adult forms.

Normal movement on the part of Protozoa provided with pseudopodia consists in the simple protrusion and retraction of the changeable processes. It becomes much more definite in forms provided with flagella, where, in many cases, a steady progression with the flagellum in advance is the characteristic motion. In one group of the Mastigophora, however, the Choanoflagellida, the flagellum, like the tail of a spermatozoön, is directed backward during motion. Among the Ciliata, complex movements accompany the high organization of the cell, and the change from one form to another, apparently at the will of the organism, is extremely suggestive of conscious action. Here, in addition to the normal and constant motion of the cilia, are various forms of contractile movement varying from the simple sarcode streaming, which is characteristic of the Suctoria, to the definite contraction of distinct muscular elements in the myonemes of Heterotrichida and Peritrichida.

B. GENERAL PHYSIOLOGY

In all Protozoa, as in higher animals, the functions of nutrition, respiration, excretion, reproduction, and irritability the analogue of nerve-response, are indispensable for the life of the organism. When compared with the vital functions of the higher animals, all of these processes appear simple; yet the difference is one of degree only, and among the unicellular animals, as among the multicellular, the functions become more complicated and difficult to analyze as the cell-structures become more complex. In the simpler forms, the naked unmodified protoplasm contains the beginnings of the most complicated functions, none of which can be regarded as having a particular time and place of birth in the series of animal forms; all are characteristic of the cell, and beyond that, of living protoplasm, of which they are the distinguishing properties. The most primitive Protozoa, entirely destitute of organs, feed without mouth or digestive tract, move without appendages, react to external stimuli, excrete, and reproduce. In the higher types the cell-organism becomes differentiated into special parts for the performance of these various functions, and the relative position of the organism in the scale of Protozoa depends upon the degree of this differentiation. In no class of animals is the connection between division of physiological labor and regional differentiation so clearly marked as here. This is especially noteworthy in the outer plasm, which, directly correlated with the action of the environment, has apparently become progressively modified into external coverings, into motile organs, and into organs of sensation, while the endoplasm retains the same character throughout the group.

One function, that of *encystment*, is limited almost exclusively to the Protozoa, although occasionally seen in some Metazoa (*e.g. Macrobiotus*, or the "water bear," and some rotifers). This is a special adaptive process by which the organisms are enabled to survive when the environment is unsuitable. If a pool dries up, becomes too dense, or too foul from putrefaction or other causes,

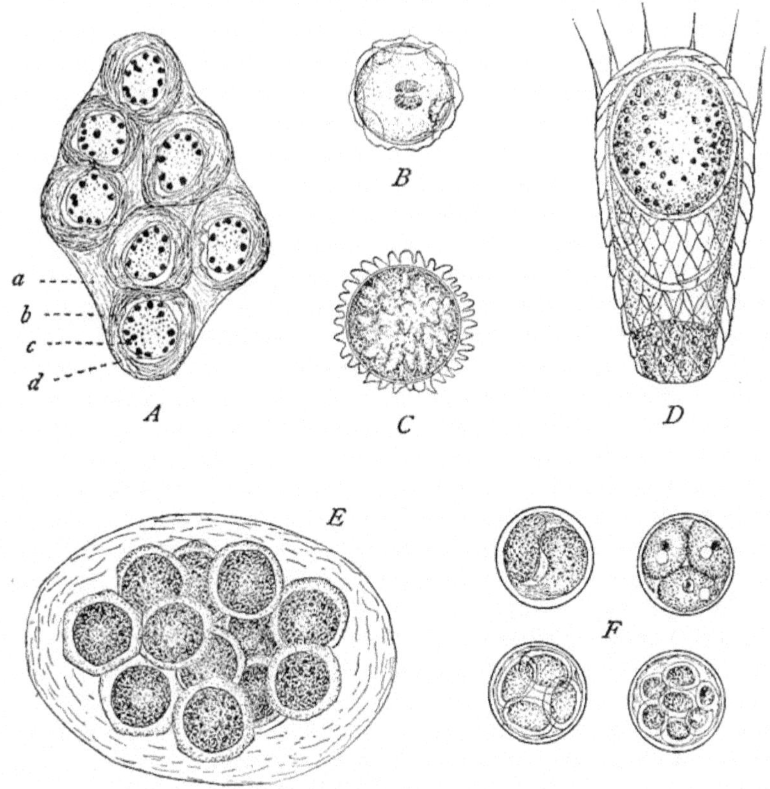

Fig. 17. — Types of cysts.

A. Amœba proteus. [SCHEEL.] *B. Stylonychia mytilus* Ehr. [BÜTSCHLI.] *C. Pleurotricha grandis* St. [BÜTSCHLI.] *D. Euglypha alveolata* Duj. [LEIDY.] *E. Actinosphærium Eich.* Ehr. [BÜTSCHLI.] *F. Colpoda Steinii.* [MAUPAS.] *a*, gelatinous matrix; *b*, outer cyst wall; *c*, middle cyst wall; *d*, inner cyst wall.

the cell draws in its appendages, rounds out into a sphere and secretes a resisting membrane, within which it can exist for a long period. When first formed, this membrane is a delicate gelatinous substance, which soon hardens and gradually acquires the peculiar characters of chitin. With the exception of the contractile vacuole, which continues to contract rhythmically for some time, all of the organs of the body are quiet at this period. The water, expelled by the vacuole, collects between the cyst and the spherical wall of the animal, the latter becoming smaller and smaller as more and more water is expelled. The nucleus appears unaltered, except for a very slight reduction in size (Fig. 17). In this condition the animal can withstand a long period of desiccation, or even extreme heat and cold, and, owing to its minute size, it may be blown hither and thither until it reaches some favorable spot where it may recommence active life. Water is then absorbed, the cyst is ruptured, and the former active life begins anew. Encystment may occur in some cases after a particularly heavy meal, or more frequently, before reproduction by spore-formation or simple division.

1. *Nutrition.*

The processes of nutrition, as in the higher animals, may be divided into three stages: 1, the capture and ingestion of food; 2, the digestion of the ingested parts; and 3, the ejection or defecation of the undigested remains.

Many of the Protozoa have no special apparatus for seizing and ingesting food, but absorb it directly from the surrounding medium. Thus many Mastigophora live, like the fungi, by absorption, through the body walls, of fluids which hold in solution the products of decomposition of other animal and plant forms. Others, as the Sporozoa and some Ciliata, live like a tapeworm and other intestinal parasites, upon the digested foods of the alimentary tract, or in nutritive fluids in other cavities of different hosts. The Phytoflagellida, also, do not ingest solid food, but, by the aid of chromatophores, they have the power of manufacturing their food in the same manner as do the green plants. The majority of Protozoa, however, take in solid food through more or less definite regions of the body. In some of the phytoflagellates there is a distinct mouth-opening, in addition to the chromatophores, and such forms may combine both animal and plant modes of nutrition.

Food-taking has been carefully examined in connection with the Ciliata, where many species living upon certain specific organisms apparently select their food. Thus Maupas ('88) distinguishes forms that are herbivorous, others that are carnivorous, and still others that are omnivorous. The food that may be thus selected consists of all sorts of lower plants, such as desmids, diatoms, zoöspores, bacteria, filamentous algæ, etc., while among the animals the Mastigophora and smaller ciliates are the most frequent victims, although rotifers and small worms are often eagerly seized. Maupas believes that the cause of these various adaptations in feeding should be sought in the modifications of the mouth. "The mouth is, in short," he says, "the dominating organ *par excellence* in the morphology and the physiology of the Ciliata. Nutrition in its manifold phases in these minute beings absorbs and completes their entire existence. This function assumes with them an intensity which, I believe, is equalled nowhere else in the animal kingdom. They are gluttons *par excellence*, absorbing and digesting night and day without repose. It results that the apparatus charged with the performance of such an intense function becomes modified, diversified, and developed to an astonishing degree, especially striking when it is remembered that these are unicellular organisms." [1]

The capture and ingestion of food, in its simplest form, occurs in

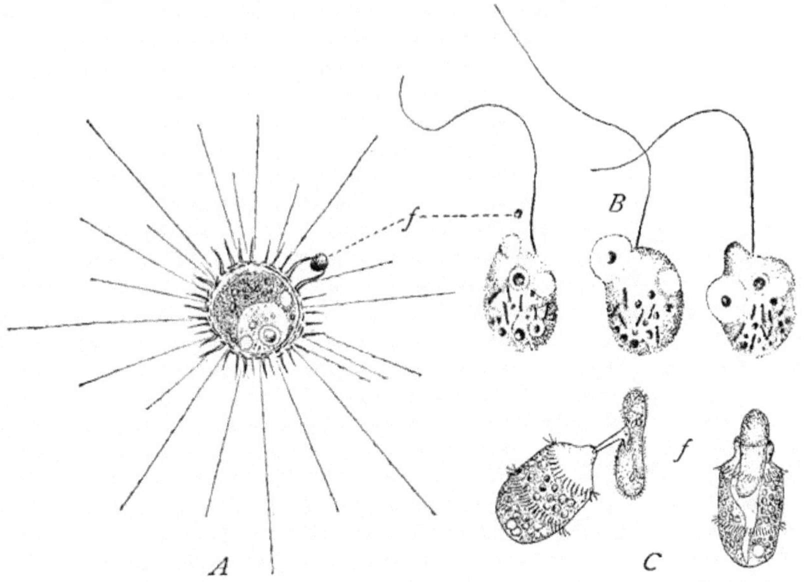

Fig. 18. — Food-taking. [*A*, PÉNARD; *B* and *C*, BÜTSCHLI.]
A. Raphidiophrys elegans Hert. and Lesser. *B. Oikomonas termo* Ehr. *C. Didinium nasutum*, O.F.M. *f*, food particles.

the group of Rhizopoda, where, as in *Amœba proteus*, any part of the body can act as a mouth. In this form pseudopodia are pushed out toward the victim (a flagellate, ciliate, minute plant form of any kind, or even a higher animal, such as a rotifer or worm) and entirely surround it, together with a certain amount of water, thus forming a gastric vacuole, or an improvised "stomach." When the rhizopod is provided with a shell, the food-taking area is limited to a mouth-opening in the shell, while in many of the shelled Heliozoa the taking of food is complicated by the presence of an unbroken coating, and a special opening must be made for each ingestion (Fig. 18). In the Reticulariida, the gastric vacuoles are on the outside of the shell, and are formed in the network produced by the anastomosis of the pseu-

[1] ('88), p. 185.

dopodia. Here the prey is digested, and the products of digestion find their way by protoplasmic streaming to all parts of the animal.

In many Mastigophora and Ciliata, the motile organs create a vortex current in the region of a well-defined mouth, which usually leads into a distinct pharynx. In some flagellates, the base of the flagellum is an area of soft plasm, through which the food particles can be readily engulfed as they strike against it, but in others there is a distinct opening which leads into the endoplasm. In other flagellates (*Noctiluca* and the Choanoflagellida), a peculiar protoplasmic funnel-shaped collar surrounds the region which answers the same purpose. In most of the Ciliata, the buccal region is surrounded by strong cilia, which are frequently fused to form membranes or membranelles; these send a powerful current of water, containing innutritious as well as nutritious particles, toward the mouth, which receives all without discrimination. In some cases, as in certain of the holotrichous Ciliata, there is a true swallowing or deglutition, by which solid food is gulped into a capacious pharynx and thence into gastric vacuoles. Many of the latter forms have offensive trichocysts, resembling the rhabdites of Turbellaria. One of these, upon the approach of its prey (usually a small ciliate or flagellate), launches its darts, which penetrate the cuticle and paralyze the prey. The victim is then swallowed, the mouth of the carnivore enlarging to accommodate it (Fig. 18, *C*). This process is strikingly illustrated by the ciliate *Actinobolus radians*, which combines the selection of food with the offensive use of trichocysts. This remarkable organism possesses a coating of cilia and protractile tentacles, which may be elongated to a length equal to three times the body-diameter, or withdrawn completely into the body. The ends of the tentacles are loaded with trichocysts (Entz, '83). When at rest (Fig. 19), the mouth is directed downward, and the tentacles are stretched out in all directions, forming a minute forest of plasmic processes, amongst which smaller ciliates, such as *Urocentrum*, *Gastrostyla*, etc., or flagellates of all kinds, may become entangled without injury to themselves and without disturbing the *Actinobolus* or drawing out the fatal darts. When, however, an *Halteria grandinella*, with its quick and jerky movements, approaches the spot, the carnivore is not so peaceful. The trichocysts are discharged with unerring aim, and the *Halteria* whirls around in a vigorous, but vain, effort to escape, then becomes quiet, with cilia outstretched, perfectly paralyzed. The tentacle, with its prey fast attached, is then slowly contracted until the victim is brought to the body, where, by action of the cilia, it is gradually worked around to the mouth and swallowed with one gulp. Within the short time of twenty minutes, I have seen an *Actinobolus* thus capture and swallow no less than ten *Halterias*.

Still another mode of food-taking is found among the Suctoria. Here there is no mouth and no motile organ to create currents, but the body is provided with distinct tentacle-like processes, through the

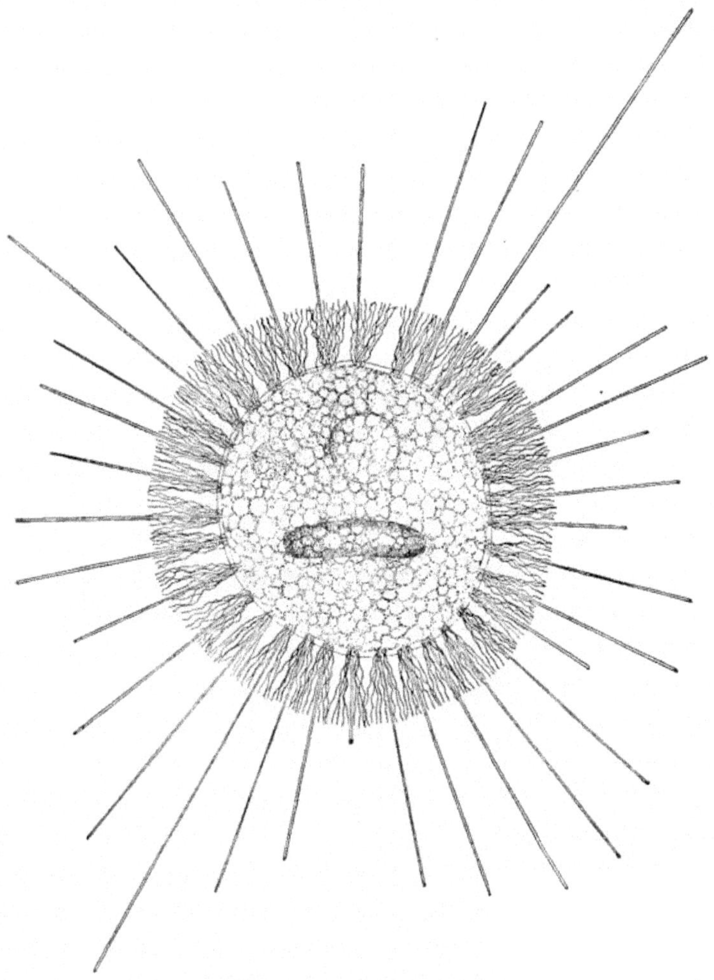

Fig. 19. — *Actinobolus radians* St.

The organism is represented at rest, with the mouth turned downward, and with the tentacles widely outstretched. At the base of each tentacle is a brush of 8 to 12 cilia which vibrate like flagella instead of striking like cilia. Within the body are represented the nucleus, contractile vacuole, and one *Halteria*.

ends of which the food substances are absorbed into the body. These tentacles are of various kinds, some sharp-pointed for piercing, others cup-shaped for attachment by suction, while others are pointed and spirally wound (Fig. 20). The cuticle of the prey is pierced by the sharper tentacles, and its fluid endoplasm passes in a current down the cavity of the tentacle and into the endoplasm of the suctorian. In other cases, the endoplasm of the tentacle passes into the body of the prey and there digests the internal substance *in situ*, the digested parts flowing back into the body of the minute carnivore. A similar mode of food-taking occurs in some Heliozoa (*Vampyrella*), where the parasite penetrates the cells of algæ and there digests the protoplasm.

In all Protozoa, digestion is accomplished within the endoplasm. The ingested proteid is contained within a gastric vacuole filled primarily with water, which is taken in with the food. The water, however, gradually changes by osmosis with the fluids of the plasm;

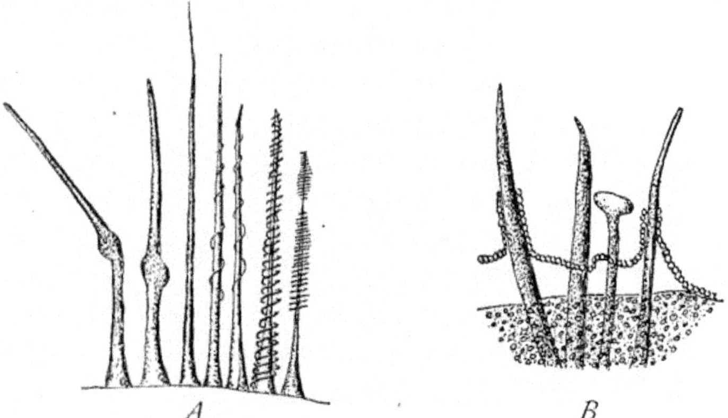

Fig. 20.—Tentacles of Suctoria. [HERTWIG.]
A. Seizing tentacles of *Ephelota*. *B.* Feeding and seizing tentacles of *Ephelota*.

among these is a digestive, acid fluid, which reduces the digestible portions of the food probably to some form of peptone. Then, again by osmosis, the digested portions are assimilated in all parts of the endoplasm. The indigestible remains of the food are excreted in various ways. Sometimes, as in the Rhizopoda, they are voided from any portion of the body, usually, however, from that part which at the time is posterior. The gastric vacuole, after the contained food has been digested as far as possible, frequently becomes a defecatory vacuole, and its contents are expelled to the outside at the posterior end of the individual. Finally, a distinct and permanent anal opening is found in the more complex ciliates.

2. *Excretion.*

Like all other animals, the protozoön uses a certain amount of protoplasm in the performance of its vital activities. In a large number of Protozoa there is no known organ by which the waste products are removed. In such forms excretion probably takes place by osmosis through the walls of the body, in the same way possibly that saprophytic forms take food. This must be the case in the Sporozoa and many marine forms as well as in certain flagellates, in which there is no specialized excretory organ. In the majority of the Protozoa, however, there are specialized structures which regularly throw to the outside of the organism a certain amount of fluid substance. These structures are the *contractile vacuoles*, which, with the exceptions of the Sporozoa and the marine forms, are found in every class of the Protozoa. In the living animal the vacuole is a clear spherical area in the endoplasm. It is formed by the slow addition of water from the endoplasm, and grows until a maximum size is reached, when it suddenly disappears, the contained water being driven to the outside. Vacuoles are frequently variable in position (Sarcodina), while the number is, to a certain extent, dependent upon the condition of the protoplasm, several observers having shown that, as the individuals lose their vitality, the protoplasm becomes more and more vacuolated. In many cases the vacuole moves about with the endoplasmic flow until, becoming heavier than the protoplasm, it remains stationary, while the rest of the endoplasm moves forward with the organism (many Rhizopoda). In this manner the vacuole, as it attains its full size, is gradually left at the posterior end of the moving organism, where it finally bursts. Again, as in some Mastigophora and Ciliata, the contractile vacuole is a stationary organ connected with the outside by a definite pore. Here, too, are numerous accessory structures in the form of canals and reservoirs, the former apparently collecting the water and waste matters from all parts of the cell, and conducting them to the contractile vesicle, the latter receiving the fluid after contraction of the vacuole, and conveying it to the outside, with which

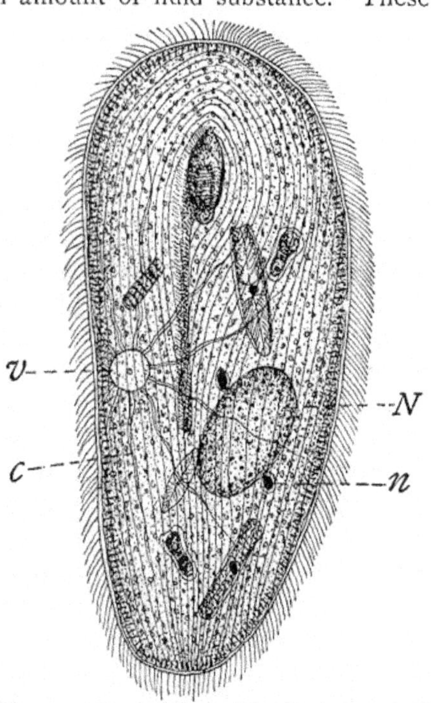

Fig. 21. — *Frontonia leucas* Ehr. [SCHEWIAKOFF.] *c*, canal; *v*, vacuole with external pore; *N*, macronucleus; *n*, micronucleus.

they are in open connection. The canal system, which some observers (*e.g.* Fabre Dumergue, '90) consider widely spread throughout the Ciliata, is often strikingly developed, as in *Frontonia*, where there is a complicated network traversing the entire cell (Fig. 21).

While the excretory function of the contractile vacuole is generally accepted, there have been only a few satisfactory experiments to demonstrate it, and the possibility of other functions is not excluded. At the present time the balance of evidence is in favor of the view that the contractile vacuole has both excretory and respiratory functions, inasmuch as it regularly empties a fluid to the outside, which carries with it the products of destructive metabolism in the form of *urea*, and probably *carbon dioxid*, although the respiratory function has never been demonstrated.[1]

Whatever may be the function of the contractile vacuole, it does not appear to be universally necessary for the life of the organism, for it is lacking in the Sporozoa and the majority of the Sarcodina (Reticulariida and Radiolaria). Furthermore, whatever the use of the vacuole, it is independent of the nucleus, non-nucleated fragments forming new vacuoles which pulsate rhythmically for some time. Hofer ('89) found that vacuoles in non-nucleated bits of *Amœba proteus* would contract for fourteen days. He also noted that whereas the regular period of pulsation was seven minutes, the periods became longer and longer, until at the end of the fourth day there was but one pulsation every two hours, and even then the contents were not completely expelled, a reaction which Pénard ('90) formulated later by the statement that the activity of the contractile vacuole is directly proportional to that of the entire individual.

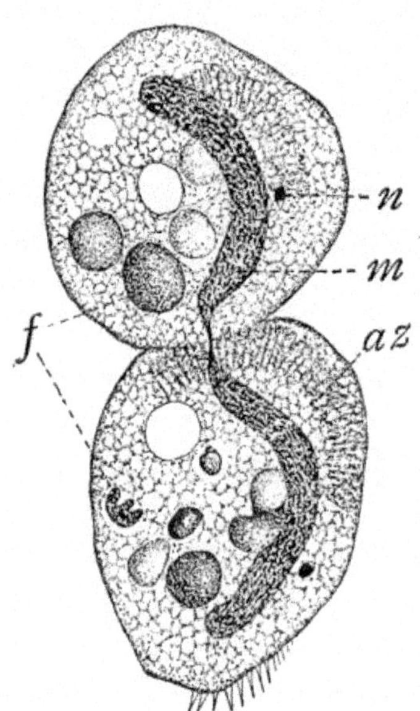

Fig. 22. — Division of *Euplotes*. [FROM A PREPARATION.]

The daughter-cells are almost ready to separate; the daughter-micronuclei (*n*) are re-formed; the macronucleus (*m*) is not quite divided; the gastric vacuoles (*f*) are equally distributed in the two daughter-cells, one of which has generated the adoral zone (*az*).

3. *Reproduction.*

With the exception of the Sporozoa, simple division, or splitting into two parts, is the characteristic mode of reproduction in all Protozoa. In the Sporozoa, and at times in most of the other Protozoa, division is replaced by

[1] Cf. Chapter IX, p. 28:.

spore-formation or the breaking up of the body into many small particles, each the germ of a new organism. While the majority of the Protozoa reproduce asexually in these ways, reproduction in some is bound up with complete sexual differentiation, and a series of forms may be selected which indicate the probable development of the sexual from the more primitive methods. In numerous cases the sexual phenomena include many of the preliminary maturation stages shown by the Metazoa, in the formation of polar bodies and reduction of the quantity of chromatin, etc.

Simple division, the most common method of reproduction, is usually a separation of the body into two equal parts either longitu-

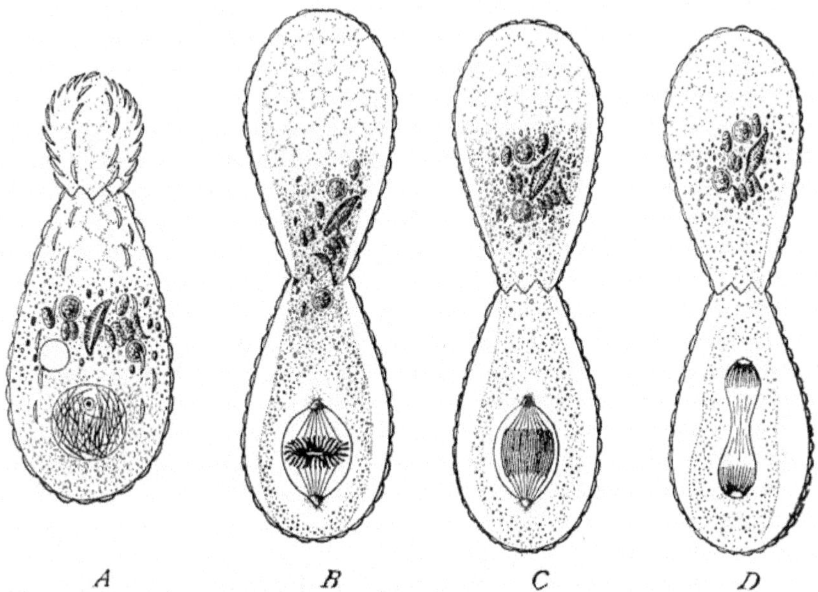

Fig. 23. — Division (budding) of *Euglypha alveolata* Duj. [SCHEWIAKOFF.]

The shell-plates which were stored in the endoplasm about the nucleus pass out with the streaming protoplasm (*A*) to form the shell of the daughter-cell. The nucleus is shown in different stages of mitosis.

dinally (Flagellidia) or transversely (Ciliata). It is invariably preceded by division of the nucleus, and is often accompanied by the equal division of certain of the internal structures of the cell, such as the chromatophores, pyrenoids, etc. It may take place either during active life or under the protection of a cyst. Ciliata in the process of division may be frequently seen swimming about actively, the connecting-strand becoming narrower and narrower, until finally only a delicate strand of protoplasm separates the daughter-cells, and this, after a few energetic contortions, gives way and the young cells are free (Fig. 22). The presence of a firm shell or coating complicates the process, especially if the shell is a secretion (Fig. 23). In many

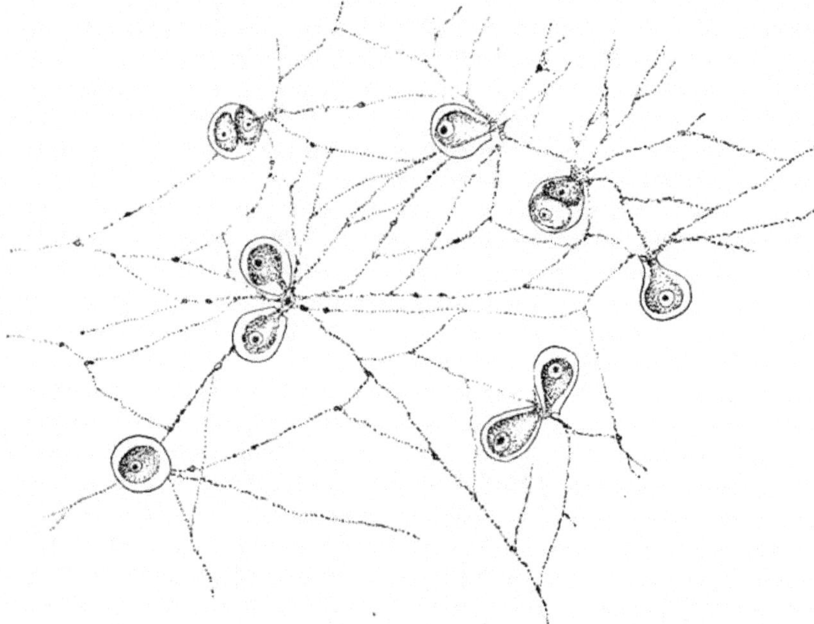

Fig. 24. — *Microgromia socialis* Hert., a gregaloid colony. [HERTWIG.]

cases the new shell-parts are secreted before the act of division begins, and when the protoplasm buds out of the original shell-mouth, they are carried with it, and the bud is covered with the newly formed

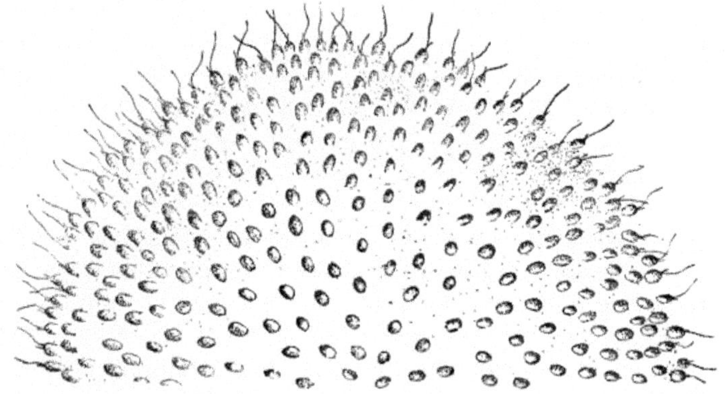

Fig. 25. — *Uroglena americana* Calkins, a sphæroid colony.

pieces, which are glued together by means of a chitinous or silicious secretion.

Simple division frequently leads to colony-formation through incomplete separation of the daughter-individuals. Four general types of these colonies are met with among the Protozoa. Adopting Haeckel's terms, they may be designated according to their general structure as (1) *gregaloid*, (2) *sphæroid*, (3) *arboroid*, and (4) *catenoid*.

A *gregaloid* colony is an aggregate of Protozoa having a round, ellipsoidal, or indefinite shape, and usually with a gelatinous basis in which the single individuals are variously distributed. The colonies may be formed by incomplete division of the individuals or by partial union of two or more adults (Fig. 24). A *sphæroid* colony is a globu-

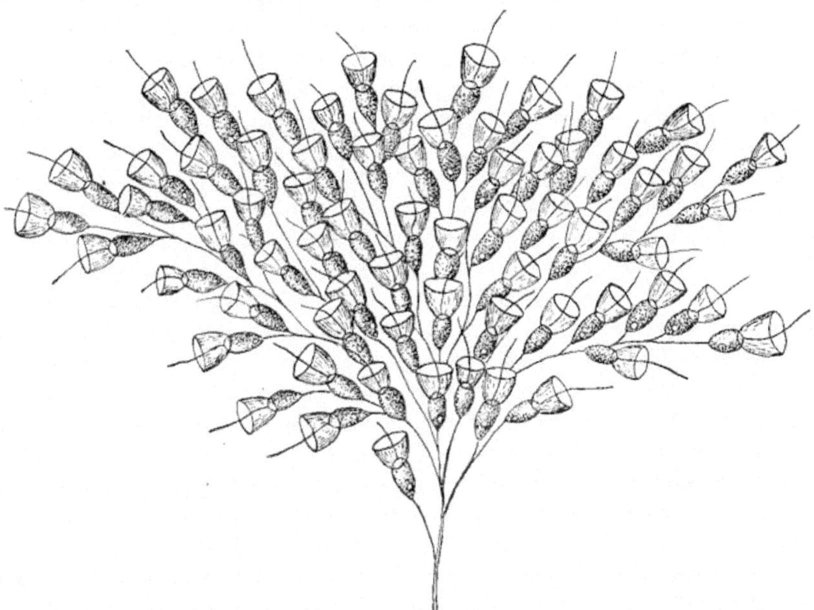

Fig. 26. — *Codosiga cymosa* Sav. K., an arboroid colony of Choanoflagellida. [KENT.]

lar, ellipsoidal, or cylindrical aggregate in which the individual cells form a superficial layer in a common gelatinous matrix. When these superficial cells are closely packed together into an almost continuous layer as in *Volvox*, *Magosphæra*, or *Uroglena*, they are extremely suggestive of certain stages in developing Metazoa (Fig. 25). An *arboroid* colony is a tree- or bush-like aggregate arising by the dendritic or dichotomous branching of a primary stalk or a gelatinous matrix. Such colonies are usually attached by the base to some foreign object and often resemble hydroids or Bryozoa (Fig. 26). They may, however, as in *Dinobryon*, be free-swimming. A *catenoid* colony is filiform or chain-shaped, arising from the union of cells end to end, or side to side, or through the continuous division of the cells in one plane (Fig. 27).

Many intermediate stages between simple binary division and spore-formation show that the latter method of reproduction was probably derived from the former. When simple division takes place within

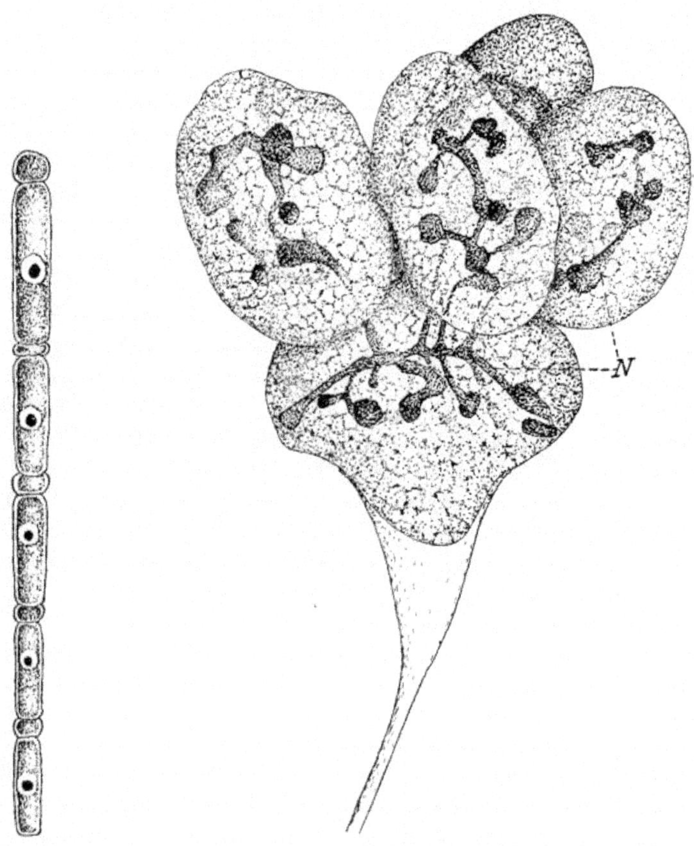

Fig. 27. — *Eiermocystis polymorpha* Lég., a catenoid colony of Gregarinida. [WASIELEWSKY.]

Fig. 28. — Exogenous budding in *Ephelota Bütschliana* Ishik. [FROM A PREPARATION.]
The macronucleus (*N*) is continued into the four daughter-cells, which appeared first as minute buds.

a cyst, it not unfrequently happens that each of the daughter-cells divides again, thus forming four daughter-cells within the cyst (some Mastigophora and Sporozoa). Again, as many as eight or sixteen may be formed in the same way, and from this it is not a great step to the method of reproduction by spore-formation, where a great number of young individuals are produced at one time.

Another modification of simple division is the process of budding or gemmation which is common among the Suctoria and some of the Flagellidia, less frequently observed in the Ciliata, and is possibly allied to "spontaneous division" among Sporozoa.[1] A piece of the nucleus of the mother-animal is pinched off and becomes the nucleus of a much smaller daughter-cell, which usually arises from a certain definite place on the parent (Fig. 28). When numerous pieces are thus budded off, and each piece is surrounded by a bit of protoplasm, the process is again akin to spore-formation (*Noctiluca*).

Spore-formation may thus be accomplished either by the repeated

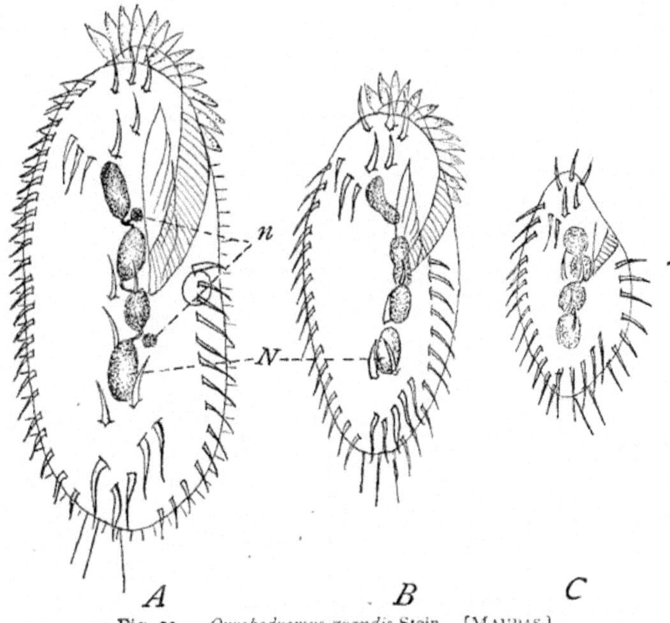

Fig. 29. — *Onychodromus grandis* Stein. [MAUPAS.]

A. Normal individual. *B.* Smaller form without micronuclei; degenerate. *C.* A still more reduced and degenerate form. *N*, macronucleus; *n*, micronucleus.

division of the entire animal, repeated division of the nucleus, the products of which are later surrounded by minute bits of cytoplasm, or by fragmentation of the nucleus without the formality of regular division. In the latter case the body of the animal breaks up simultaneously into many hundreds of small pieces, each surrounding one of the nuclear fragments and developing later into the parent form (*Paramœba*).

So-called sexual reproduction or some modification of this process is accomplished, either directly or indirectly, by the temporary or permanent union of two individuals of the same or of dissimilar size.

[1] See Chapter IV, p. 159.

From an *a priori* point of view, there is no reason why the reproduction of Protozoa by simple division should not go on indefinitely. The mechanism of metabolism, growth, and reproduction is present, and the cell appears therefore to be self-sufficient. Nevertheless, Bütschli, Maupas, and many others have shown that, in many cases at least, these divisions can be maintained only for a certain period or number of generations, after which the individuals become weaker, deformed, and finally die out, an exhausted race (Fig. 29). If two individuals conjugate while in this enfeebled condition, the result is a rejuvenescence or renewal of youth in both cases, and each of the conjugants enters upon a new cycle of cell-generations (Fig. 30). The union of two cells is accomplished in various ways. In some cases it is by the absolute and permanent fusion of two individuals; in other cases there is a union for a short time followed by a separation; in still other cases the entire cell does not conjugate, but develops a great number of swarm-spores, which again may be of equal or unequal size, and which conjugate usually by total fusion; finally, sexual reproduction, almost as highly differentiated as in the Metazoa and Metaphyta, is found in some Sporozoa and in the complex colonies of Flagellidia.[1]

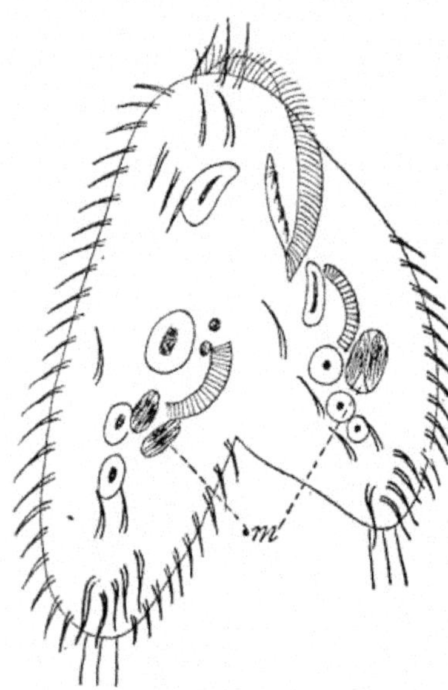

Fig. 30. — *Onychodromus grandis* St. in conjugation. [MAUPAS.] *m*, micronuclei in division.

A very interesting controversy has arisen over the question whether Protozoa ever succumb to old age. Ehrenberg was apparently the first to suggest that because of their reproduction by division, a process in which no portion of the original organism is lost, the Protozoa are potentially immortal. This conception was greatly expanded by Weismann ('84) and formed the basis of a suggestive essay in which he maintained that the Protozoa are too simply organized to die a natural death and that death is first observed in multicellular animals and plants. Dujardin ('41), opposed to so many of Ehrenberg's theories, was equally opposed to this, and although he was unable to disprove the theory, he suggested that simple division cannot continue

[1] Cf. Chapter VII, p. 229.

indefinitely, but that periods of division recur at intervals. This suggestion was confirmed by Bütschli ('76) and by Engelmann ('76), who demonstrated in connection with a number of Ciliata that after a certain number of divisions the resulting individuals become reduced in size and show other evidences of degeneration. Bütschli regarded this as evidence of old age, and he observed that the normal size and the general vitality of the reduced organisms are restored by conjugation; and he was one of the first to demonstrate that the function of conjugation is not for purposes of reproduction, but for the renewal of vitality as expressed in his term *Verjungung* (rejuvenation). Maupas ('88), finally, has confirmed the latter conception of conjugation, and in a series of brilliantly planned and carefully executed experiments has shown that Protozoa, contrary to Weismann's *a priori* assumption, may die of old age unless they be reinvigorated by conjugation. "Senescence," says Maupas, "appears to be a very general phenomenon, at least in the animal kingdom. . . . It is inherent in the organism and comes from internal causes which act independently of the surrounding conditions. . . . Its deleterious action is offset and annulled by sexual rejuvenescence or conjugation." [1]

4. *Irritability.*

Unlike the Metazoa, where the phenomena which are characterized as manifestations of consciousness are expressed through special organs of the nervous system, the Protozoa in the simplest forms have nothing, so far as we know at present, but the undifferentiated protoplasm which at the same time must be the seat of all functions.

The sensory phenomena are, however, very little known. All Protozoa are irritable, reacting in certain definite ways, although in different degrees, toward various external stimuli. All are sensitive to electrical, mechanical, thermal, and chemical irritations, and many to light, while few or none are affected by acoustic vibrations. The reactions to these stimuli are usually expressed in motion of some sort, which may be either indefinite or definite, — in the latter case, as a rule, either positive, *i.e.* toward the source of irritation, or negative, *i.e.* away from it. In many Protozoa, particularly in the lower forms, there seems to be no portion of the cell more sensitive than others; in the higher forms, however, there is a greater or less degree of sensory localization. Here, as a rule, the ectoplasm reacts energetically, and, like the cuticle of Metazoa, becomes a general sensory organ. The appendages frequently serve as special sensory organs of touch, as in the aboral cirri of the hypotrichous ciliates, while special organoids are frequently present in the form of "eye-spots," etc.

[1] ('88), p. 272.

C. SOME ECONOMIC ASPECTS OF THE PROTOZOA

The Protozoa are frequently objectionable because of the appearances, odors, and tastes which they may impart to water. In the sea great areas may be colored orange, red, etc., by incalculable numbers of *Noctiluca* or Dinoflagellidia (*Prorocentrum*, *Glenodinium*), while at night their presence is indicated by brilliant phosphorescence, the light being due to the rapid oxidization of a substance created by the organisms and thrown out by them upon irritation. In Puget Sound and in Alaska I have seen hundreds of acres of the sea surface colored orange by *Noctiluca miliaris*, although the single individuals are less than one-fiftieth part of an inch in diameter, and Haeckel ('90) graphically compares such masses to "tomato soup"! When Protozoa occur in great numbers in fresh water, and especially in drinking water, they may cause considerable annoyance; for by the color, odor, and taste which they impart they render the water unfit to drink. The colors are due in the main to the Phytoflagellida, and only those forms which are capable of making their own food are able to live in pure drinking waters. The most frequent causes of trouble in this respect are *Uroglena*, *Peridinium*, or its allies, *Euglena* and other Euglenoids, and *Synura*, all of which are flagellates. The odors and tastes, however, are more offensive than the colors, and as they are frequently misunderstood and regarded as evidence of pollution, an explanation may not be out of place. Ehrenberg noticed that certain flagellates (*Chlamydomonas pulvisculus* and *Chlorogonium*) impart a certain oily odor. Dunal ('38) and Joly ('40) described an odor like that of violets from the masses of *Hæmatococcus* which gave to a portion of the Mediterranean a distinct red color. The Massachusetts State Board of Health, dealing with this problem of the drinking waters, have obtained important results in this direction. They have shown that certain of these organisms may have definite and specific odors which, like the odors of flowers, can be recognized. An "oily odor" was traced to *Synura* and *Uroglena*, an "Iceland moss" odor to *Peridinium*, a "violet odor" to certain Euglenoids, etc.[1] The cause of these odors has been the subject of a number of investigations, and it has been found that they are "living odors" due to disintegration of the cells rather than to their decomposition, a view first advanced, I believe, by Bütschli ('84), who described a highly characteristic "fishy odor" from *Euglena sanguinea*, while the cells were found to be disintegrated, although not decomposed. The matter was considered more extensively by the writer ('92), who found that in waters infested with colonies of *Uroglena americana* the odor was not developed until the organisms had passed through the water

[1] See S. B. H. ('92).

pipes, but after such passage the odor was extremely strong and repulsive, while no colonies could be found. It was suggested at this time that the odor was one of disintegration, and due to the liberation of minute drops of oil-like substance which become disseminated through the water, giving it the characteristic *Uroglena* smell. It was also suggested that these drops of oil are analogous to the perfume oils of the fragrant plants, like them having a certain individual odor often strong enough and characteristic enough to identify the organism. Similar oil-like inclusions are found in the protoplasm of all Protozoa, but to be detected through the sense of smell, they must be present in great numbers.

Far more serious noxious effects of the Protozoa are produced through their frequently parasitic mode of life. In all classes there

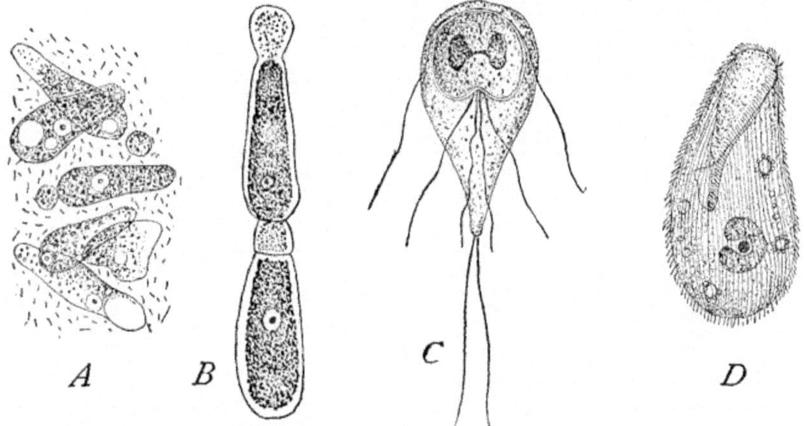

Fig. 31.—Internal parasites. [*A*, *B*, LEUCKART; *C*, GRASSI; *D*, BÜTSCHLI.]
A. *Amœba coli* Lösch, a supposed cause of dysentery. *B*. *Monocystis agilis* Leuck., a gregarine. *C*. *Megastoma entericum* Grassi, a flagellate. *D*. *Balantidium entozoön* Ehr., a ciliate.

are certain forms which live as parasites (Fig. 31), and which for convenience may be separated into two groups, the intercellular and the intracellular forms. So far as known these parasites, with few exceptions, do not produce noxious products like bacterial ptomaines, but whatever damage they may cause is due to the mechanical disturbances set up by their presence. The intercellular parasites infest the body cavities of various hosts, the cavities of blood-vessels, and ducts of various glands, or penetrate the spaces between muscle-fibres, while the intracellular parasites (which belong almost exclusively to the Sporozoa) bore into cells of epithelia (Gregarines, Coccidia), or the corpuscles of the blood (Hæmosporidiida). From the wide distribution of the intercellular parasites, it is quite possible that no animal is entirely free from Protozoa of some kind. Without entering upon a discussion of these forms, it may be stated that Rhizopoda, Flagellidia, Ciliata, and Sporozoa may be found in the various cavities and canals in man and the other vertebrates, where they usually give little or no trouble.

One form, however, *Amœba coli* (Fig. 31, *A*), has been long in dispute as the reputed cause of dysentery. If it is the specific cause of this disease, the animal occupies an interesting position amongst the intercellular parasites; for, so far as known, none other of this kind of protozoan parasites exerts a deleterious effect upon the intestinal epithelia. Nor is it proven that *Amœba coli* does this in the case of dysentery, although a belief to that effect is widespread. Briefly reviewing the history of this belief, it appears that Lambl ('60)[1] was the first to observe *Amœba* in the human intestine, although Lösch ('75), who named it, was the first to consider it in connection with dysentery. Many subsequent observers (Kartulis, Mannaberg, Cohn, etc.) found *Amœba coli* in the fæces of dysentery patients, Kartulis ('89) amongst others stating that he found them in no less than five hundred cases. The belief received a setback, however, by the observations of Cunningham ('81), Grassi ('82), and Calendruccio ('90), who found *Amœba coli* in the intestine of sound and healthy men as well as in dysenteric patients, while still other observers maintained the entire absence of such an enteric organism. Councilman ('91), in a work which is certainly as reliable as any that has been undertaken upon this subject, partially harmonized these views by showing that there are at least three forms of dysentery, of which one, at least, is characterized by certain definite symptoms and by the presence of *Amœba coli*, although it was not demonstrated that the rhizopod was the cause. The entire matter received impartial and critical treatment by Laveran ('93) in France and by Schuberg ('93) in Germany, and both came to the conclusion that the cause of dysentery was not yet known, the former basing his opinion largely upon the absence of *Amœba* in all but one of ten cases, the latter upon numerous experiments and observations upon normal and diseased individuals. Schuberg not only found that *Amœba coli* is present in normal men, but also found that there is no specific difference in the various intestinal *Amœbæ* which have been described by various observers as living freely in the intestine in the same way as the commensal ciliates and flagellates also found there and generally believed to be harmless. He pertinently says: "If the flagellates are harmless, it is certainly not impossible that *Amœba* is also. The increased number of *Amœbæ* in dysenteric patients is not necessarily evidence that they are the cause of this disease."[2] Both he and Laveran expressed the view that all experiments which had been made up to that time had not excluded the possibility of other causes, *e.g.* bacteria. That dysentery is due to some specific cause had been early demonstrated by experiment, but in none of these experiments had it been possible to isolate the Protozoa from bacteria which invariably accompany them. The reverse experiment is, however, possible, and it is singular that it has not been made more frequently. Among the first to exclude the Protozoa were Celli and Fiocca ('95), who obtained cases of dysentery by injecting cats with material from fæces in which the Protozoa had been killed by heat; the same result was also obtained by injecting material in which both *Amœba* and bacteria were absent, the cause evidently being in the poison of the sterilized matter and not in *Amœba coli*. They concluded that the poison is the product of bacteria (*Bacillus coli communis*, together with typhus-like bacteria and a streptococcus, were suggested as possible causes). This view was supported by a number of observers, amongst whom may be mentioned Gasser ('95), Cassagrandi and Barbagello ('95), and Petridis ('98). The latter especially has shown that dysentery as observed in Egypt is due to a bacillus and not to Protozoa. He found that *Strep-*

[1] Cf. Leuckart ('79). [2] Page 701.

tococcus is the most numerous of the micro-organisms and the probable cause of the disease, for he was able to isolate the bacillus and to produce dysentery in cats by injecting them with the culture obtained from it. Thus, as the matter stands, Petridis's results, the most positive that have yet appeared, together with growing evidence from the bacteriological side, make it exceedingly probable, although not definitely established as yet, that bacilli and not *Amœba coli* are the cause of this disease.

While the majority of intercellular parasites are harmless, it is quite different with the intracellular forms. These, by making their way into the interior of the cell and growing at the expense of the cell-contents, gradually cause degeneration of the tissues which may end in death of the host. These parasites belong almost exclusively to the class Sporozoa of which the Coccidiida and Hæmosporidiida are found in vertebrates, while the Gregarinida are confined to the invertebrates, where they are widely distributed.

The Coccidiida are found in nearly all of the tissues of the lower vertebrates although rarely in man, unless indeed, as many observers believe, they are the cause of various tumors and cancers. That there is some reason for this belief is shown by the fact that in the lower vertebrates, especially in fishes, the presence of Sporozoa leads to ulcers and tumors and to the ultimate death of the fish. The subject, however, as far as man is concerned, is in a very unsatisfactory state, and opinions differ widely as to the nature of certain elements found in cancerous growths. By some observers these are regarded as parasites, by others as disintegrated or pathological cells. Up to the present time no satisfactory evidence has appeared to prove the former view, and until such evidence is forthcoming the entire matter must rest in abeyance.

From the pathogenic point of view, the most important protozoön is the malaria germ (*Plasmodium malariæ*), a form belonging to the Hæmosporidiida. These organisms, in the young stages, move about by amœboid motion in the blood-vessels of men and birds. They penetrate the red blood corpuscles, which slowly hypertrophy, until in one type of the disease, at least, they attain a size three to four times that of the normal corpuscle, the parasite in the meantime growing at the expense of the hæmoglobin and finally reproducing by spore-formation. In this form alone there appears to be a poisonous substance analogous to bacterial ptomaines, which is produced by the organism and periodically discharged (at spore-formation) into the blood, thus causing the pyrexial attacks so characteristic of malaria. The recent successful results obtained by Ross, Manson, Koch, Grassi, and others, in locating the seat of the malaria germ when outside the human body, leads to the hope that some successful means of guarding against this disease may soon follow.[1]

[1] *Vide infra*, pp. 160–165.

SPECIAL BIBLIOGRAPHY II

Bütschli, O. — Protozoa. In *Bronn's Klassen und Ordnungen des Thierreichs. Leipzig*, 1883–1888.

Delage et Hérouard. — La cellule et les Infusoires. In *Traité de Zoölogie concrète. Paris*, 1896.

Ehrenberg, C. G. — Die Infusionsthierchen als vollkommene Organismen. *Leipzig*, 1838.

Entz, G. — Protistenstudien. *Budapesth*, 1888.

Lankester, E. R. — Protozoa. In *Zoölogical Articles from the Encyclopædia Britannica.* 1891.

Stein, Fr. — Der Organismus der Infusionsthiere. *Leipzig*, 1861, 1867, and 1878.

CHAPTER III

THE SARCODINA

THE term *Sarcodina*, introduced by Bütschli ('83) as the class name of the most primitive of the Protozoa, includes all forms which, like the common fresh-water type *Amœba proteus*, move by the protrusion of protoplasm in the form of broad and finger-like, or sharp and ray-like, processes called *pseudopodia*. These forms fall naturally into three groups readily distinguished by clearly marked differences in structure, — the Rhizopoda, Heliozoa, and Radiolaria.

Among the Rhizopoda are included forms of Sarcodina with blunt, finger-form or lobose pseudopodia (*Amœbida*) or with branching and anastomosing pseudopodia (*Reticulariida*). They may be naked (*Gymnamœbina*), or shelled (*Thecamœbina* or *Foraminifera*). The pseudopodia may arise from all parts of the body or they may be limited to special regions; in shelled forms they may pass through one common opening (*Reticulariida imperforina*), or through many finer openings (*Reticulariida perforina*). The body form is typically globular, but may be variable in consequence of amœboid changes, or drawn out into a monaxonic form. The material of the shell may be chitin, silica, foreign particles, or calcium carbonate.

The Heliozoa are naked or shelled forms of Sarcodina; they are usually globular with fine ray-like pseudopodia arising from all parts of the body. The rays are, as a rule, stiffened by an axial filament formed of modified protoplasm which may be readily dissolved by the organism. The shells are less compact than those of the Rhizopoda, and are usually formed of more or less loosely joined silicious spicules.

The Radiolaria are similar in form to the Heliozoa. As in the latter, the pseudopodia arise from all parts of the body and occasionally anastomose. The endoplasm is separated from the outer plasm by a firm, chitinous, perforated membrane, the *central capsule*. A test or skeleton, often of exquisite beauty, is usually present, consisting of isolated spicules of silica, or of a compact skeleton of acanthin or silica. One or more nuclei are invariably present within the central capsule.

The finer structure of the rhizopod protoplasm has already been mentioned. In many cases, especially in the monothalamous forms, the plasm is divided into a number of clearly marked zones. Schewiakoff ('88) describes three, Pénard ('90) no less than four in *Euglypha*, and Rhumbler ('98) the latter number in *Cyphoderia*. Schewiakoff ('88), apparently on very good grounds, maintained that certain specific functions characterize each of these zones, indicating, in a general way, a regional differentiation and division of physiological labor. To the outer zone, which corresponds to the ectoplasm of *Amœba*, he ascribed a locomotor function, this being the seat of pseudopodia formation ; to the second zone, which contains the nucleus, the function of assimilation, and to the third zone a reproductive function. Pénard and Rhumbler separate Schewiakoff's second zone into two on account of certain structural differences. According to these observers the

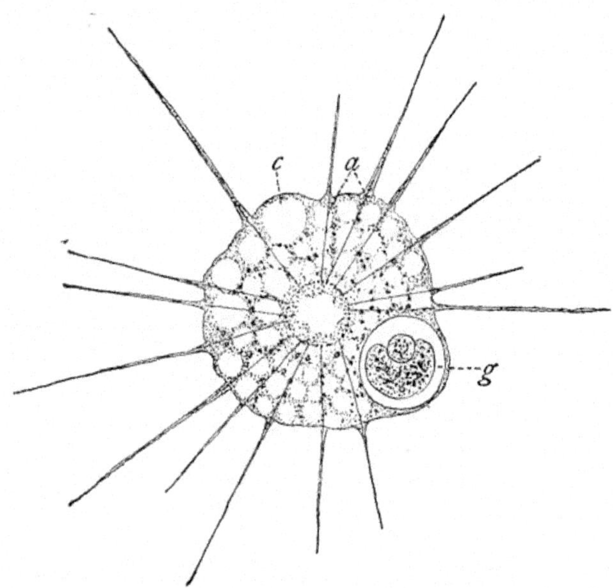

Fig. 32. — *Actinophrys sol* Ehr. [BÜTSCHLI after GRENACHER.]
The axial filaments (*a*) extend through the endoplasm to the membrane of the nucleus; *c*, a contractile vacuole in the ectoplasm; *g*, an ingested food particle in a gastric vacuole.

outermost zone is distinctly vacuolated, the second contains food-particles in the process of digestion, the third, granules which represent waste matter not determined, and the fourth, excretory granules.

The appearance of the protoplasm in Heliozoa or Radiolaria is quite different from that of the Rhizopoda. Ectoplasm and endoplasm can be distinguished, but unlike the hyaline ectoplasm of *Amœba*, the outer plasm of Heliozoa is made up of vacuoles much larger than those of the endoplasm, the walls of these vacuoles being distinctly granular (Fig. 32). The extremely vacuolated appearance, however, seems to be largely dependent upon the medium in which the animal lives. Grüber found that an *Actinophrys* when transferred from fresh into sea water soon loses its vacuoles; and, *vice versa*, when transferred back to fresh water, again acquires its vesicular appearance.

In general appearance a radiolarian resembles a heliozoön, but there is a considerable difference in the corresponding regions. A typical radiolarian can be conceived if we imagine a thick perforated chitinous membrane between the ectoplasm and endoplasm of a heliozoön. The *intra-capsular plasm* (Fig. 33, *c*) contains nuclei, fat particles, and plastids of one form or another, and is in communication with the *extra-capsular plasm* through the pores in the membrane, although, as shown by Verworn's experiments upon the isolated central capsule, it can live for a time independently. The outer or extra-capsular plasm is composed, according to Haeckel, of four parts. The outermost (*g*) is a zone of pseudopodia; the latter, however, originate in the deeper fourth zone, forming a network through the other extra-capsular parts. The second zone is of net-like (alveolar?) protoplasm, the *sarcodictyum*. A third zone, the *calymma*, is of jelly-like consistency and forms the bulk of the ectoplasm. The fourth and most important zone, the *sarcomatrix*, lies close against the central capsule, and is the go-between for the intra- and extra-capsular portions. The sarcomatrix is also the seat of digestion and assimilation, the food coming to it through the pseudopodia and the network. As the means of communication between the central protoplasm and the sarcomatrix is of vital importance to the organism, the arrangement of the apertures in the central capsule offers a good character for the classification of the Radiolaria. Hertwig ('79), who first used this character, divided the group into four legions, as follows: (1) the *Peripylea*, in

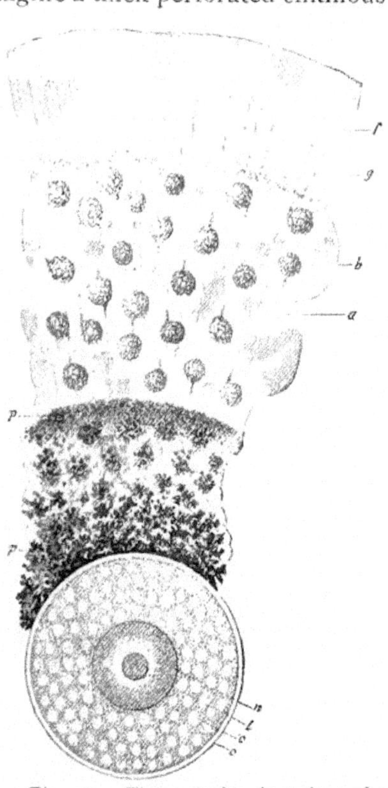

Fig. 33.—The protoplasmic regions of a radiolarian (*Thalassicolla maculata*) Haeck. [HAECKEL.]

a, large alveoli forming part of the calymma in which foreign bodies (*b*) are enclosed, and which is penetrated by meshes constituting the sacrodictyum; *c*, the central capsule and intra-capsular plasm; *f*, retracted pseudopodia. The nucleus (*n*) contains a distinct nucleolus (*l*); the sarcomatrix is darkened by pigment masses (*p*).

which the membrane of the central capsule is perforated by pores arranged regularly about the entire surface (Fig. 34, *A*); (2) the *Actipylea*, in which the pores are arranged in groups over the surface (*B*); (3) the *Monopylea*, in which there is but one such group of pores in the membrane. In these forms the perforated disk is connected with the centre of the central capsule by a conical mass of endoplasm, the *podoconus* (*D*), rich in food particles and gran-

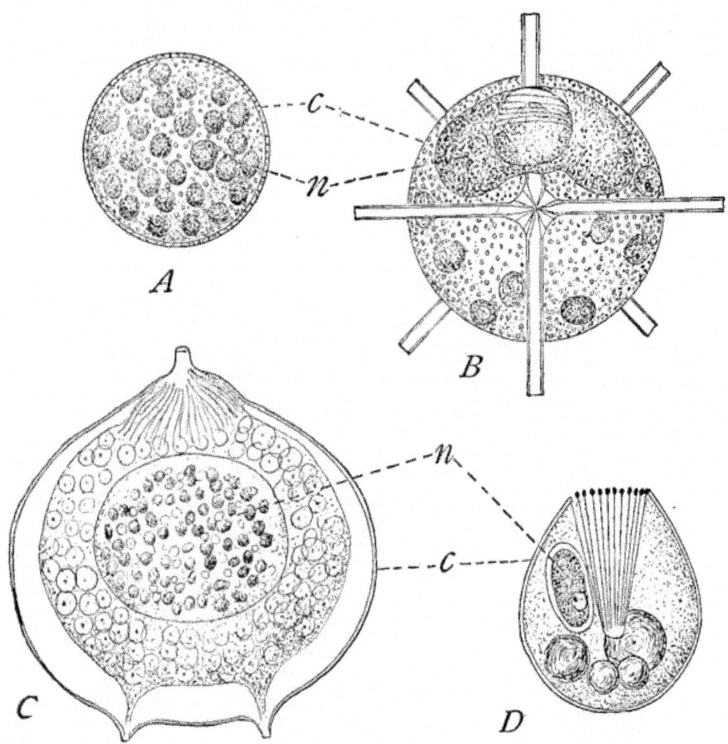

Fig. 34. — Central capsules of Radiolaria. [HAECKEL.]
A. Thalassolampe maxima Haeck., one of the Peripylea. *B. Acanthometron dolichoscion* Haeck., one of the Actipylea. *C. Aulographis candelabrum* Haeck., one of the Monopylea. *D. Tripterocalpis ogmoptera* Haeck., one of the Cannopylea. *c*, central capsules; *n*, nuclei.

ules; (4) the *Cannopylea*, in which the membrane around the pores is drawn out into funnel-like projections termed *astropyles* (*C*). The central capsule is double in these forms. Haeckel has found that certain skeletal forms accompany the structure of the membranes, and he names the above legions respectively as follows: — (1) *Spumellaria*; (2) *Acantharia*; (3) *Nasselaria*, and (4) *Phæodaria*.

In each of the orders of the Sarcodina, and especially in the Radiolaria, there are some forms with symbiotic plant-cells. The relationship between the symbionts was worked out by Cienkowsky, Brandt, Haeckel, and Entz, the latter noting that the plant-cells are invariably found just outside of the endoplasm, where they do not come in contact with endoplasm and its digestive fluids. According to the more recent observations of Le Dantec ('92), however, the digestive fluid of these animals is unable to dissolve the cellulose membranes of the plant-cells, and they remain uninjured in the endoplasm, dividing there when the conditions are favorable.

A. Shells and Tests

The ectoplasm of naked protoplasm shows a tendency to condense or stiffen when in contact with water, and a cuticle or membrane is the result. *Amœba proteus*, with its differentiation into endoplasm and ectoplasm, shows a primitive stage in the development of such membranes. Here the ectoplasm remains plastic enough to yield to the inner pressure of the organism and to form the first part of every pseudopodium; it is rapidly pushed aside, however, and the endoplasm becomes the advancing part. In *Amœba tentaculata* the outer layer has become more firm and the pressure from within expends itself upon pseudopodia which are protruded through permanent holes (Fig. 12, *A*). The membrane may become still more firm through the deposition of chitin, until, as in the radiolarian central capsule, it is an efficient means of protection. In addition to the chitin, certain Sarcodina secrete a silicious mucilaginous material, which, like the chitinous cement, is frequently the means of gluing together not only regular plates or disks which the organism also secretes, but foreign particles of various kinds. The tests thus made may be entirely of lime, as in the Reticulariida, or of silica, as in the Radiolaria and many of the Heliozoa, or of sand crystals, diatom-shells, or detritus of various kinds.

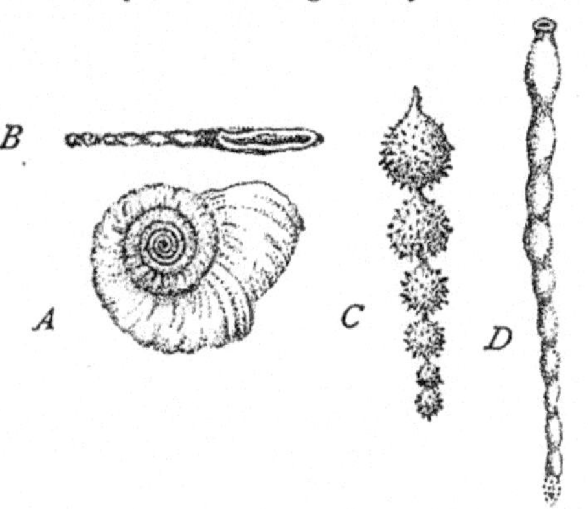

Fig. 35.—Types of marine rhizopod shells (*Reticulariida*. [Carpenter.]

A. Lateral. *B*. Ventral view of a monothalamous shell (*Cornuspira foliacea* Phillips). *C*. A simple polythalamous shell (*Nodosaria hispida* D'Orb.). *D*. *Vertebralina sp.*, a fossil form.

In the lime-shells (Reticulariida or Foraminifera) the secretion of calcium carbonate, except for the invariable presence of a mouth-opening, forms an almost complete investment like a cyst. In many cases this opening is the only means of communication with the surrounding medium (Imperforina), but in other cases the entire shell is punctured by minute openings through which pseudopodia pass to the outside (Perforina). These two types of shell are further distinguished by their appearance; the Imperforina when seen by reflected light are opaque and like porcelain, while the shells of the Perforina are almost transparent (vitreous).

Monothalamous or single-shelled Foraminifera may be either imperforate (*e.g. Squamulina, Pilulina,* or *Saccammina*) or perforate (*Lagena*). In each group a graded series of shells can be arranged, varying in complexity from the simple monothalamous to the compli-

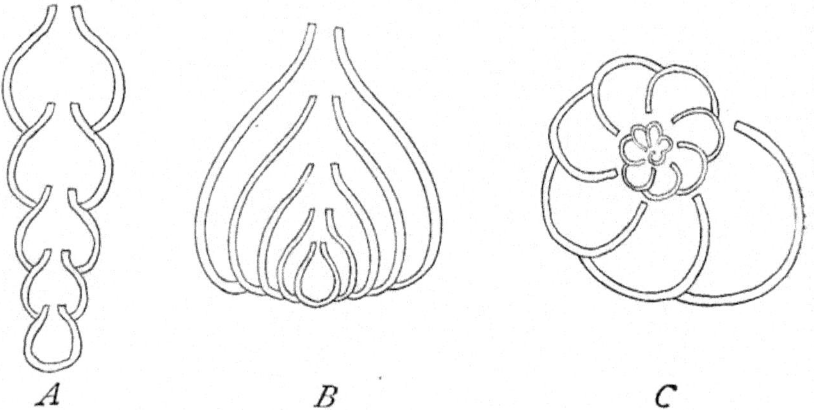

Fig. 36. — Polythalamous shell types schematized. [CARPENTER.]
A. Linear *Nodosaria* type. *B. Frondicularia* form of the *Nodosaria* type. *C.* Spiral form of the *Nodosaria* type.

cated polythalamous forms (*Polystomella, Calcarina*). One of the simplest of these shells is that of *Cornuspira*, where the plasm, as it slowly grows, constantly secretes new shell material and is capable of unlimited extension (Fig. 35, *A*). It is never divided by septa into separate chambers as in the polythalamous shells. A further step, the simplest of the polythalamous types, is found in shells where the separate chambers adhere end to end as in *Nodosaria* (*C*). Here there may be only a slight septum between adjacent chambers, but enough to indicate that growth is periodic, and not constant as in *Cornuspira*.

In these chamber-dwelling animals the plasm, as it grows, extends out of the primary shell-opening and reaches to a certain distance down the outside; new shell material is then secreted, and the process is repeated until a chain of chambers is the result (Fig. 36, *A*). If the plasm extends entirely around the shell, the new chamber almost incloses the older ones as in *Nodosarina* (*B*). In other cases the plasm may extend over one side only of the old shell, and a curvilinear axis of growth is the result (Fig. 35, *A*, *B*, and 36, *C*). The spiral thus formed may be flat or coiled around a longitudinal axis as in the mollusc *Trochus*, giving an involute shell. This type, the most highly differentiated of all of the rhizopod shells, exhibits all grades of complexity (Fig. 37). In the highest forms each new chamber has a complete wall, so that the septa between the adjacent chambers consist of two lamellæ, while between the lamellæ there is fre-

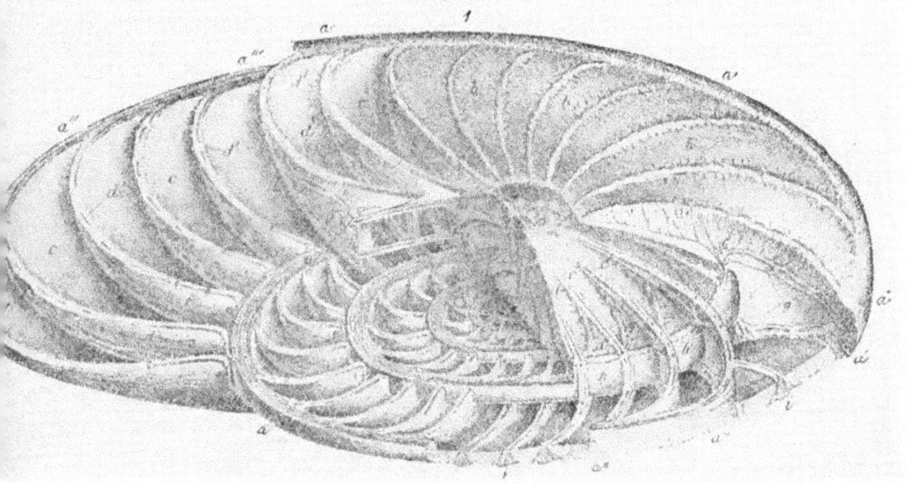

Fig. 37. — A complex polythalamous shell (*schematic*) of *Operculina*. [CARPENTER.]
The shell is represented as cut in different planes to show the distribution of the canals (*a'*, *a''*, *a'''*); *c*, *c*, *c*, the outer chambers with double walls (*d*, *d*, *d*), one of which is shown in section (*g*). The chambers communicate by apertures at the inner ends of the septa (*e*), and by minute pores (*f*). The outside (*b*) of the shell is marked by the radial septa.

quently a space filled with a calcareous deposit or what Carpenter ('62) calls the "intermediate skeleton." This inter-lamellar deposit is traversed by a complicated system of canals, and the deposit itself is frequently carried out into external processes and knobs (*Calcarina*). In the annular or discoid types a process of budding takes place around the entire circumference instead of at a localized area, and concentric circles of chambers are thus formed (*Orbitolites*).

The character of the mouth-openings between adjacent chambers depends upon the nature of the outer coating. If the lime casing is perforated by numerous pores through which pseudopodia can be thrust to collect food, then each chamber is sufficient for itself, and the so-called mouth-opening is small; but if the perforations are absent, the mouth-openings are large and allow a free communication between the youngest or external chambers and the oldest or internal. Hence there are morphological and physiological grounds for separating the Reticulariida into Perforina and Imperforina. It frequently happens that the central or original chamber varies in size in the same species, being large (*megalospheric*) in some individuals, and small (*microspheric*) in others (Fig. 38). While the relations of these two forms have been much discussed, no satisfactory conclusion has yet been reached. Lister ('95) regards the case as one of alternation of generations in which spores from individuals *A* conjugate and form individuals of the type *B*, while the latter develops spores which grow into the form *A* again. The conjugation of swarmers in these dimorphic types is a matter of inference rather than of observation, for the process has never been seen.

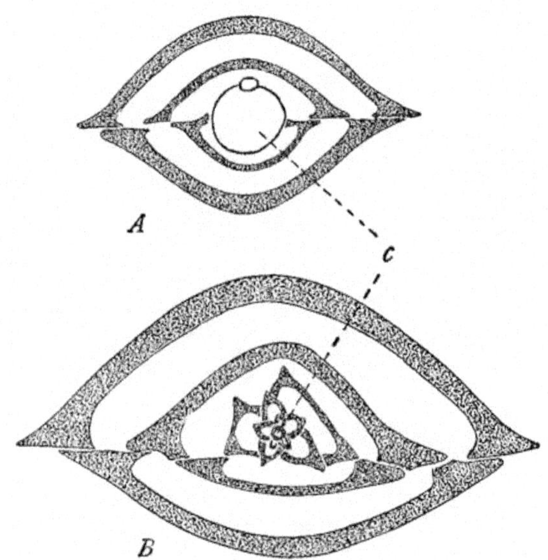

Fig. 38. — Megalospheric (*A*) and microspheric (*B*) shells of *Biloculina depressa* Lam. [SCHLUMBERGER.]
The dimorphism is shown by the central chamber *c*.

Among the Heliozoa and Radiolaria, shell formation is of a somewhat different type, consisting of the deposition of spicules and rays rather than a continuous layer of material forming a compact coating. Even naked forms of Heliozoa, such as *Actinosphærium*, secrete these spicules at certain times for the purpose of encystment, while others have them in greater or less numbers throughout life. Isolated spicules are usually retained by a gelatinous mantle, which covers the entire animal (*Nuclearia*, *Actinolophus*, etc.). These spicules are usually curved or straight rods, spindles, or blade-shaped plates, and may become firmly attached to one another, forming latticed skeletons, like those of Radiolaria (*Clathrulina*, Fig. 39). Intermediate stages are seen in such forms as *Diplocystis*, where the plates are very small and arranged without

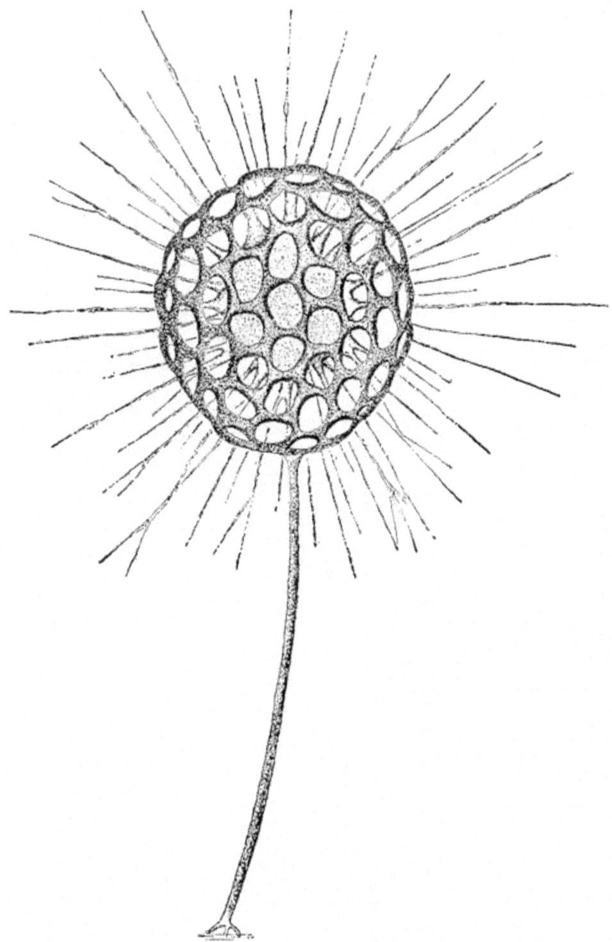

Fig. 39. — *Clathrulina elegans* Cienk. [GREEFF.]

any apparent order in the gelatinous mantle. In *Raphidiophrys* (Fig. 40) the silicious plates are much larger and more regularly arranged, while in *Pinaciophora* and *Acanthocystis* (*B*, *C*, *D*) they become so closely knit that they form an efficient shield. In *Acanthocystis*, each plate is a small rectangular prism, laid tangential to the surface with sharper spicules arranged at intervals at right angles to these, thus forming a bristling coat. *Pinaciophora* is very similar, but the spicules are not so prismatic.

Similarly the Radiolaria may have either simple isolated spicules or compact and strong skeletons. In many cases the outer plasm (calymma) is free from spicules, but in other cases isolated spicules of sharp and needle-like, or tri- or tetra-radiate form, are present. The

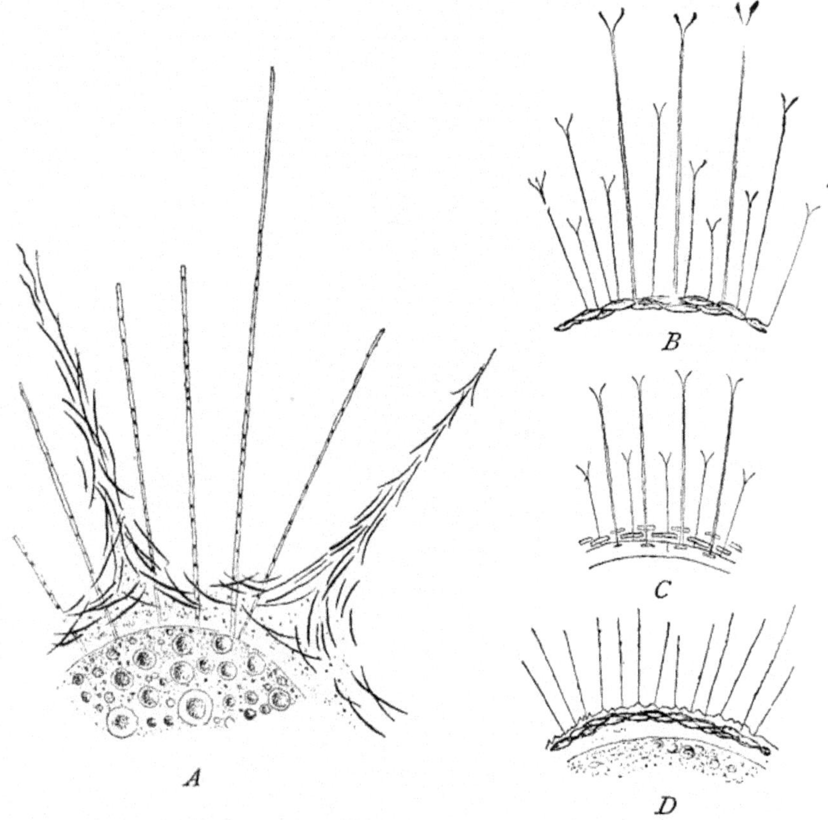

Fig. 40.—Types of spicules in Heliozoa. [PÉNARD.]
A. Raphidiophrys pallida F. E. Sch., with curved silicious rods. *B. Pinaciophora rubiconda* Hert. and Less. *C. Acanthocystis turfacea* Carter. *D. Pinaciophora fluviatilis* Greeff.

substance of the skeleton of Radiolaria is either silica or acanthin, a horn-like modification of protoplasm. According to Haeckel, the deposition of silica in many cases occurs only at certain periods, and an entire skeleton may be laid down at one time (*Dictyotic moment*, Haeckel, or *Lorication moment*, Dreyer). Again, it may be formed during the entire period of life. The material of the shell is secreted from the sarcodictyum, and as the deposition of the silica follows the outlines of the vesicles which form this zone of protoplasm, the resulting skeleton forms a reticulum. Growth may take place more rapidly, however, at certain places, and spines, spicules, or protuberances of one kind or another are the result. The usual form of the network upon which the skeleton is deposited is an hexagonal mesh, but this may become modified in numerous ways, the apertures becoming either circular, polygonal, or elliptical (Fig. 41).

When spines are formed, a secondary calymma may also be developed, carrying with it the sarcodictyum, and the latter, in turn, may give rise to a secondary skeleton outside of the first. This process may be repeated until there are as many as six or seven accessory skeletons.

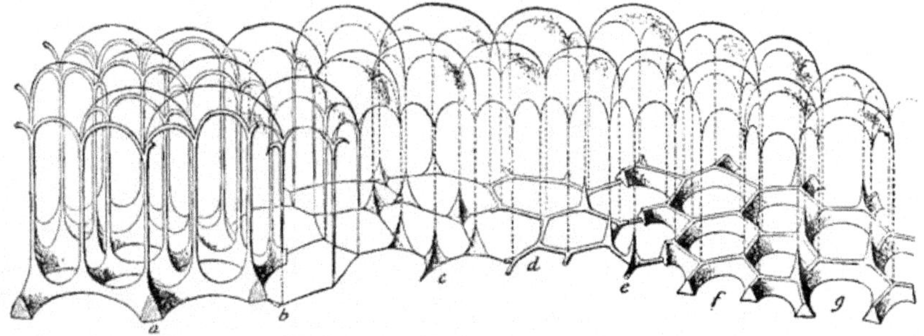

Fig. 41.—Schematic figure illustrating the modifications of skeletons according to mechanical principles of deposition. [DREYER.]

The secretion is supposed to collect in the interstices between alveoli as at (*c*), forming simple spicules, or tri- and tetra-radiate spicules (*b*). Collecting in the lines of union of six alveoli, the deposit takes the form of an hexagonal mesh (*d*), which, by the addition of more material, becomes changed as at (*a*), (*e*), (*f*), and (*g*).

A very interesting set of phenomena are connected with the acanthin skeletons where the spicules are not deposited in the calymma, but are formed at the centre of the central capsule, growing out centrifugally into the extra-capsular plasm and resulting in a skeleton of radiating spines. With a few exceptions these spines are twenty in number, and are arranged in a certain geometrical order which has been characterized as the *Müllerian law*. The points of the spines fall in five circles parallel to the equator, and there are four spines to each circle. The spines are named, according to this scheme, polar, tropical, equatorial, sub-tropical, and sub-polar (Fig. 42).

The form of the silicious skeleton is quite varied. In its least-differentiated form, as in most Heliozoa, it is a mere collection of loosely arranged spicules. In other forms a uniform covering of silica covers the meshes of the sarcodictyum. Such a generalized condition of the skeleton becomes modified in many ways, the main types being the "sagittal ring," consisting of a simple ring of silica, like a girdle around the body of the organism. A second type consists of a basal tripod, the arms of which inclose the central capsule (Fig. 43). A third modification is the simple alteration of the spherical latticed shell

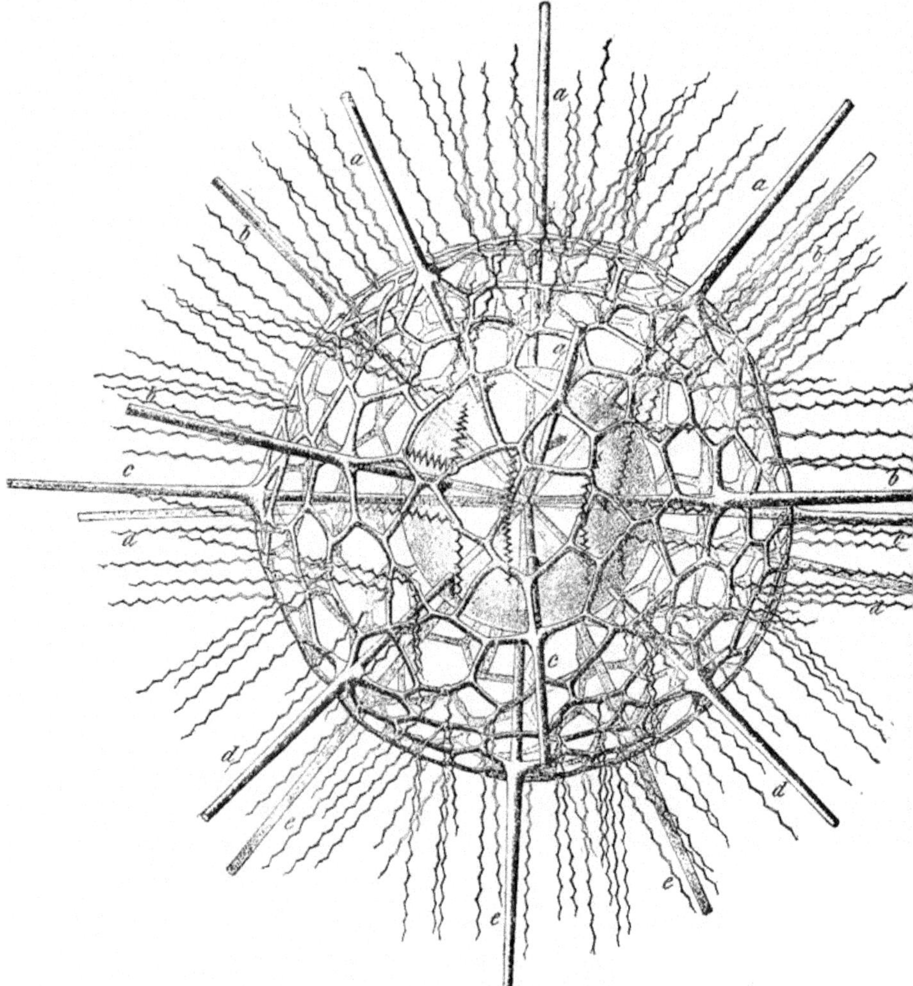

Fig. 42. — *Lichnaspis giltochii* Haeck., one of the *Acantharia* (*Actipylea*). [HAECKEL.]

The spines are arranged in accord with the Müllerian Law as follows: *a, a, a, a*, northern polar spines; *b, b, b, b*, northern tropical spines; *c, c, c*, —, equatorial spines; *d, d, d, d*, southern tropical spines; and *e, e, e*, —, southern polar spines.

into elliptical, ovate, or sub-spherical forms. Again, the skeleton may be discoid, or even bivalved, and, in still other types, there may be a combination of two or more of the above modifications.

B. Pseudopodia

A pseudopodium is a portion of the body-plasm temporarily protruded. It is most variable in form, and at any moment can be withdrawn into the body of the animal to be replaced by others. In the Rhizopoda, the pseudopodia are coarse, blunt, and finger-formed (Amœbida), or fine, and often forming a network through anastomosis (Reticulariida). In the Heliozoa and Radiolaria they are more rigid and radiate out from the body in all directions, forming a protective coating, and from their ray-like appearance suggesting the common name "sun-animalcula."

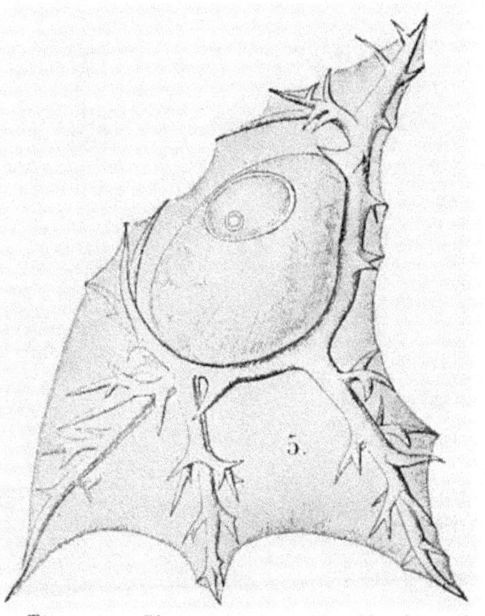

Fig. 43. — *Plagiocarpa procortina* Haeck., with tripod-like skeleton. [Haeckel.]

There is a difference in the texture as well as in the form of the lobose and reticulate pseudopodia of the Rhizopoda. In the former the hyaline ectoplasm, which goes into the pseudopodia, is apparently homogeneous and structureless, although, upon critical examination, Bütschli ('92) was able to make out a fibrous structure in some forms, and in many of them a reticular appearance was obtained upon retraction. His observations led him to the conclusion that the hyaline appearance is due to the close approximation of the walls of the alveoli, and not to their absence. The outer plasm is certainly more dense and non-granular than the endoplasm, and protoplasmic streaming is confined to the latter. The outer plasm in the reticulate type, on the other hand, is granular, while the central portion is denser and more resisting. Streaming of the granules here takes place in the ectoplasm, instead of in the endoplasm, and when two or more pseudopodia come in contact, the viscid character of this outer plasm leads to fusion. The lobose forms, on the other hand, never coalesce.

The resemblance between the central denser strand of protoplasm in the pseudopodia of the reticulate type and the axial filament of the pseudopodia of Heliozoa and Radiolaria was early recognized by M. Schultze ('63) and critically examined by Bütschli ('92) and Schaudinn ('93), and is now generally recognized.

The nature and the number of pseudopodia have frequently been used as a method of identification of certain species of Rhizopoda. *Amœba polypodia*, *A. radiosa*, and *A. proteus* have certain characteristic pseudopodial structures which are seemingly of diagnostic value, yet *A. proteus* under the influence of a constant electric current can be made to assume the forms characteristic of *A. polypodia* and then

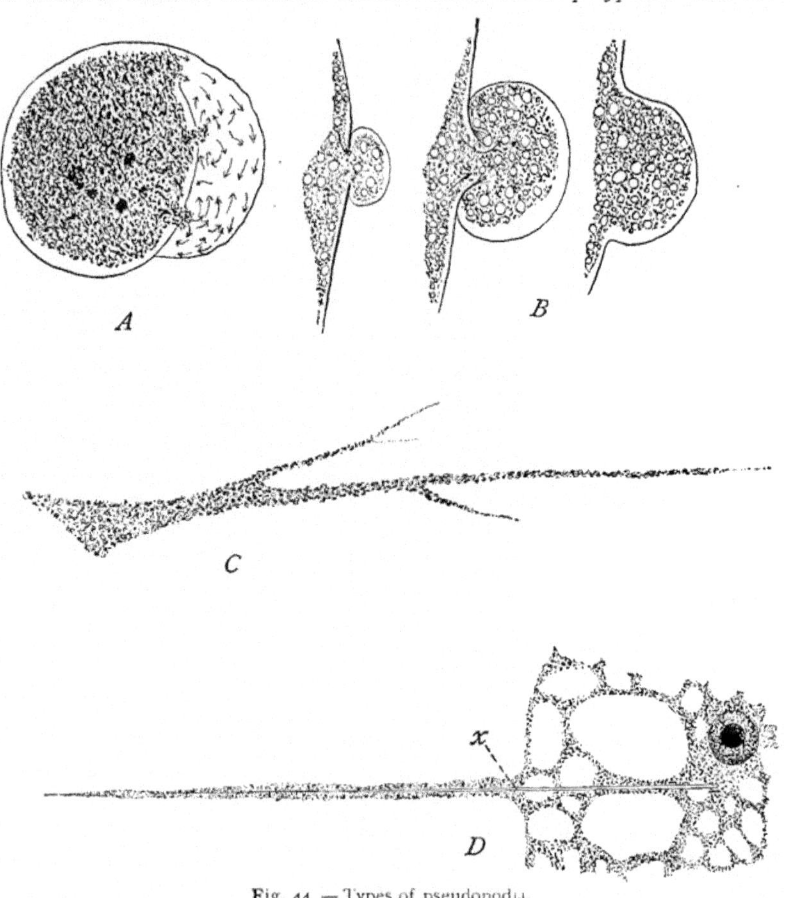

Fig. 44. — Types of pseudopodia.

A. Amœba limicola Rhmb. [RHUMBLER.] *B. Amœba blattæ* Bütsch. [BÜTSCHLI.] *C. Lieberkühnia* sp. [VERWORN.] *D. Actinosphærium Eich.* Ehr. [ORIGINAL.] *x*, axial filament.

of *A. limax*. Conversely, *A. limax*, when placed in an alkaline solution of potassium hydrate, becomes transformed into *A. proteus*, and later, into *A. radiosa* (Verworn, '94). When a change in the surrounding medium can so affect the protoplasm that the entire character of pseudopodia-formation is altered, the specific value of pseudopodia alone may well be questioned, and species based upon such a variable character can have but little value. Despite the uncertainty of the pseudopodia as a basis of classification, their structure among the Rhizopoda is frequently so characteristic that the identification of some species of *Amœba* is comparatively easy. Thus *Amœba proteus* has large and blunt pseudopodia in the adult phase, while the young form (known as *A. radiosa*)[1] has sharper, stiffer, and hyaline pseudopodia. When a pseudopodium of *A. proteus* starts from the periphery, it continues as a stream until, as a rule, a long, lobose structure results. When, however, a pseudopodium of *A. blattæ* Bütschli or of *A. limicola* Rhumbler starts from the periphery of the spherical body, it resembles a miniature eruption. A break is

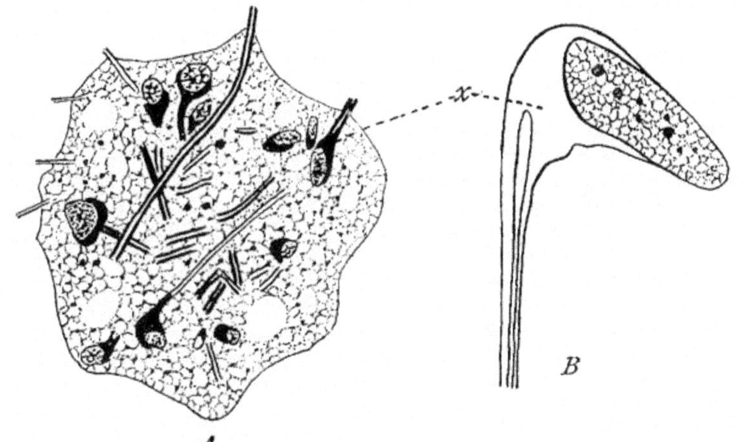

Fig. 45. — *Camptonema nutans.* [SCHAUDINN.]

The axial filaments extend throughout the endoplasm (*A*), taking their origin at the nuclear membrane (*B*). *x*, an axial filament highly magnified.

made on the periphery, and through it the granular endoplasm flows down the sides of the spherical body instead of outward into elongate pseudopodia (Fig. 44, *A*, *B*). In such cases the pseudopodia may be used to identify the organism.

The pseudopodia of the Heliozoa and the Radiolaria are far more complicated than those of the Rhizopoda. They usually have distinct axial filaments, consisting of some unknown substance, extending throughout the entire length, and even into the endoplasm, where they not infrequently abut against the membrane of the nucleus or meet at a common centre (*Actinophrys*, *Acanthocystis*). The granular protoplasm which surrounds the axial filament is in constant but slow streaming motion. The point of interest is the axial filament, which is not strictly comparable with the skeletal parts, but is probably stif-

[1] Cf. Scheel ('99).

fened protoplasm similar to the central plasm of the reticulate pseudopodia. It is easily softened by the animal, and when the latter is irritated may be withdrawn into the body. That there is some connection between the axial filament and the nucleus would seem to be indicated by their invariable propinquity, the nucleus in some cases being actually surrounded by the substance that forms the filament and which Schaudinn ('96) thinks is a soft fluid at this time (*Camptonema nutans*, Fig. 45). In other cases the filament appears to end in a peculiar crescent or spherical capsule which lies within the endoplasm (*Dimorpha*, Fig. 46). In many instances the rays pass completely across the animal's body and rest against the nucleus on the opposite side; in others they are focussed in a central or "astral"

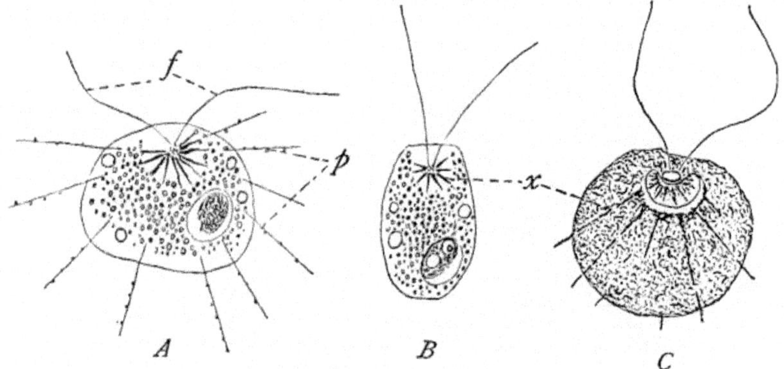

Fig. 46. — Flagella (*f*) and axial filaments of the pseudopodia of *Ciliophrys* (*Dimorpha?*) Cienk. [BLOCHMANN.]

In the Heliozoa stage (*A*) the ray-like pseudopodia (*p*) and the flagella (*f*) are present; in the flagellate stage (*B*) the pseudopodia are absent. The axial filaments (*x*) and flagella centre in the excentric nucleus (*C*).

granule (*Gymnosphæra*, *Actinophrys*, *Sphærastrum*, etc.), which in some cases has been seen to divide like a centrosome and to form an amphiaster, as in the early stages in cell-division of many cells of the Metazoa (*Acanthocystis*, *Sphærastrum*, etc.).[1]

The axial filaments have not been made out in all forms classed among Heliozoa, and it is a question whether such forms should be considered as Heliozoa or as Rhizopoda. *Vampyrella* and *Nuclearia* (Fig. 56), for example, have fine, radiating pseudopodia which change like those of the Rhizopoda and, as in many Amœbida, are formed of hyaline ectoplasm. They are placed among the Rhizopoda by some (Delage) and among Heliozoa by others (Bütschli). The pseudopodia occasionally vary in other respects from the sharp radial forms, as in *Actinolophus*, where they end in knobs; or in *Camp-*

[1] Cf. Chapter VIII.

tonema, where there is an elbow or joint which can be bent at right angles.

No entirely satisfactory explanation of pseudopodia formation and movement has yet appeared, although the subject has been attacked on many sides, and by almost all students of the Rhizopoda since the time of Dujardin. Like the early attempts to explain other phenomena in the Protozoa, the first explanations of pseudopodia-motion were based upon the analogy to higher forms. Protoplasmic contractility, the basis of locomotion in all higher animals, and probably in many Protozoa (Mastigophora and Infusoria), was early suggested as the cause of the protrusion of pseudopodia. The majority of casual observers were content with this general explanation; others, more definite, conceived the seat of contraction to be in the cortical plasm or ectoplasm (Ecker, '49; Dujardin, '41), which they compared with the dermal musculature of worms, and which they supposed forces out the pseudopodia by *backward peripheral* contraction, as water can be forced out of a rubber tube by pressure from behind. Others, again, imagined that in addition to the contractile cortex the entire mass of the amœboid body is penetrated by a contractile substance (Cienkowsky, '63), the *sarcous matter* of Brücke ('61). Still others conceived a contractile motor apparatus of even greater complexity. Amongst these, Heitzmann ('73), in working out his well-known theory of the structure of protoplasm, and adapting Brücke's view to his own interpretation, maintained that the body of *Amœba* is composed of contractile fibres and an inter-fibral "non-contractile fluid." The protrusion of a pseudopodium, he argued, is due to the local contraction or stretching of this fibrous framework. Modifications of Heitzmann's view have frequently appeared in subsequent writings. In connection with the Metazoa it still makes its appearance in the numerous theories of contractile fibres, especially in explanation of mitosis (van Beneden, Boveri, Flemming, Reinke, and many others). In connection with the Rhizopoda, it found its most ardent advocate in R. Greeff ('91), who described radial, fibrillar, contractile structures in the ectoplasm of many so-called *Earth Amœbæ*, and interpreted them as muscle-fibres whose outer ends are inserted in the ectoplasm with their inner ends attached to the protoplasmic framework of the endoplasm. Subsequent research has shown that the supposed muscle-fibres are bacteria (Bourne, '91; Israel, '94; Gould, '95).

Contractility in a somewhat different form was also brought in to explain pseudopodium formation. In connection with the Protozoa, the most noteworthy advocate was Engelmann ('79), who conceived units of contractile substance built up of molecules of protoplasm. To these hypothetical units he gave the name *inotogmata*. During rest, Engelmann assumed, each inotogma has an elongated form, becoming spherical upon contraction. If all contract at the same time, as upon a sudden shock, the animal assumes a spherical condition; if the inotogmata contract in certain groups, a pseudopodium is started, although some pseudopodia, notably the fine, thread-like forms, are due to "relaxation" of rows of contracted units. A considerable uncertainty is attached to Engelmann's theory, especially when the attempt is made to explain special cases, and Bütschli ('92) shows in a very convincing manner that it does not justify the expectations of its originator.[1]

Wallich ('63) early observed that the current in a progressive pseudopodium does not begin in the body of the *Amœba*, but at the periphery, an observation which de Bary ('64) confirmed in Mycetozoa. Bütschli ('73) drew attention to the same fact soon after, and upon the strength of his observations appeared, even at this early period, as an opponent of the contractility hypothesis.

As stated previously, Bütschli holds that protoplasm is essentially a mixture of liquids consisting of a fluid alveolar substance and an intra-alveolar fluid of different physical nature. According to this conception, which is widely accepted, a naked protoplasmic mass such as *Amœba* must be subject to the same physical laws as other fluids. The rounding-out of drops of exuded protoplasm was early interpreted by Hofmeister ('67), and by Engelmann ('69) before he adopted the theory of inotogmata, as the same phenomenon that causes the rounding-out of any liquid substance, *i.e.* to surface tension. Of late years, especially since the appearance of Bütschli's masterly work on the structure of protoplasm, there has been a general tendency to abandon the older theory of contractility and to explain the movements of amœboid bodies through the physical laws of liquids, and in particular, by the laws of surface tension. Weber ('55) compared the protoplasmic movements in plant-cells to the streaming, due to surface tension, in drops of liquid, and subsequently Berthold ('86), Bütschli ('92), Rhumbler ('98), and others, following the same line of investigation, have obtained fruitful results. An excellent account of the several interpretations along this line of reasoning may be seen in Bütschli's *Protoplasma*,[2] and it will be sufficient here to give the most recent explanation as worked out by Rhumbler ('98) upon the basis of Bütschli's earlier view. Bütschli says: "The explanation of the processes of movement in *Amœbæ* is to be found, therefore, to my mind, in correspondence with the interpretation of the phenomena of streaming movements in the drops of foam, in the fact that, by the bursting of some of the superficial alveoli, enchylema

[1] Cf. Bütschli, p. 275. [2] pp. 172–212.

is poured out upon the free surface of the protoplasmic body, where it produces a local diminution of surface tension, and in this way sets up an extension centre together with forward movement."[1]

The origin of a pseudopodium, according to this conception, is in the ectoplasm, and the rapidity of a pseudopodium-formation is increased by the peculiar "fountain currents" characteristic of most pseudopodia. As observed by Bütschli, an advancing stream of granules flows through the centre or axis of the growing pseudopodium, while near the tip back-running currents like the falling drops of water in a fountain surround the central stream (Fig. 47). "In the formation of a finger-shaped pseudopodium of *Amœba proteus*," says Bütschli, "it can be seen that the current which traverses the axis of the pseudopodium and flows away on all sides from its tip, comes to rest at a very short distance behind the tip, — a circumstance which in any case is extremely favorable to the rapid outgrowth of the pseudopodium, in contradistinction to the relations that obtain in the drops of foam, since the protoplasm that has come to rest is heaped up and the pseudopodium grows in this way."[2]

Fig. 47. — Diagram of the movements of the endoplasm granules in an advancing pseudopodium of *Amœba proteus*. [BÜTSCHLI.]

Rhumbler ('98) attempts to explain the formation of new ectoplasm and the increase in surface of an advancing pseudopodium through the hardening effect of water upon protoplasm, a fact which has long been recognized (Bütschli, Pfeffer). An advancing pseudopodium of *Amœba proteus*, if properly fixed and stained, shows an advanced mass of endoplasm broken through the walls of ectoplasm.[3]

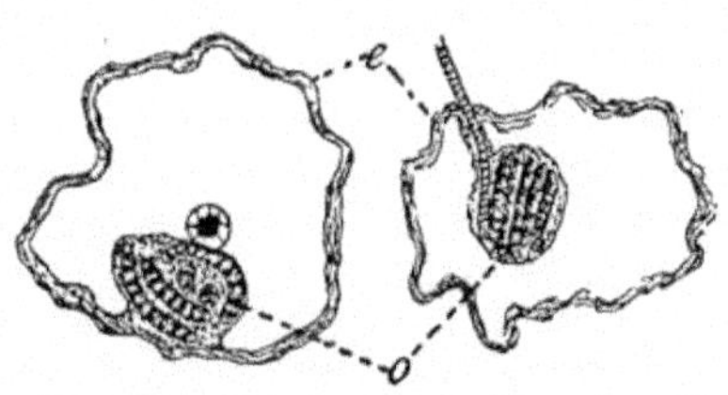

Fig. 48. — The ectoplasm (*e*) and gastric vacuoles (*o*) of *Amœba verrucosa* after treatment with *KOH*. [RHUMBLER.]

The outer ectoplasm has a firm consistency, and, as Rhumbler demonstrated by treatment with diluted caustic potash (Fig. 48), may be isolated from the endoplasm. Nevertheless, it is converted into streaming endoplasm again. The conversion of ectoplasm into endoplasm, which was early noted by Engelmann ('79) and recently by Pénard, Pfeffer, Verworn, Bütschli, and others, takes place according to Rhumbler at all times. It is particularly well shown in *Amœba limicola* Rhumbler, or *A. blattæ* Bütschli, where the eruptive pseudopodium incloses a definite portion of the old ectoplasm, which soon disappears and becomes lost in the endoplasm. Both Bütschli

[1] Bütschli, *loc. cit.*, English translation, pp. 310–311.
[2] *Loc. cit.*, p. 312.
[3] Cf. Fig. 10, *A*, p. 36.

and Rhumbler recognize that the longer the action of water is continued upon the ectoplasm, the greater the stiffening; hence, Rhumbler argues, the new ectoplasm forming at the edge of the advancing pseudopodium is less resisting than elsewhere and the forward flow continues in one direction until the surface tension is equalized. New material for the advancing pseudopodium must be supplied from endoplasm, and this in turn from the posterior ectoplasm, so the assumption is made by Rhumbler that there is a continual change of *Amœba's* protoplasm from ectoplasm into endoplasm, and from endoplasm into ectoplasm.

Explanations of this nature, based upon purely physical laws of fluid substances, seem inadequate to explain all types of pseudopodia, the reticulate and long filamentous forms in particular. Up to the present time no satisfactory and comprehensive explanation has been made, and it should be recognized that the theories advanced still remain only working hypotheses. Hofer ('89) and Verworn ('91), and many others have demonstrated that an enucleated amœboid mass soon comes to rest and assumes a spherical form. After a few days, movement recommences, and is interpreted by Hofer as an expression of the changes in surface tension. Such observations make it probable that the chemical activity, which is constantly operating between the numerous substances which make up the protoplasm, plays an important part in pseudopodia-formation, and with our present imperfect knowledge of these intra-cellular reactions, it is premature to settle upon any one cause, however suggestive and attractive it may appear, of this widely varied phenomenon.

In many cases, especially among the Heliozoa, pseudopodia-motion approximates flagella-motion. In many of the shelled Rhizopoda (*e.g. Arcella*, or some species of *Difflugia*), the hyaline pseudopodia sway backward and forward like thick, slow-moving flagella, while in some Heliozoa (*e.g. Artodiscus*) this motion is much more energetic, causing the organism to dance about like a monad. The resemblance is more noteworthy and interesting from a theoretical point of view, because both the flagellum of Mastigophora and the axial filament of Heliozoa arise in the same manner in the endoplasm, and both are apparently connected with a "division centre," a central granule, which is analogous to the centrosome of metazoan cells.[1]

C. The Nucleus

Nuclei are almost as varied in the different forms of Sarcodina as are the different types of the animals as a whole. In some cases, there is no well-defined nucleus, the chromatin being scattered in the

[1] Cf. p. 271.

form of granules throughout the entire cell, as in some of the Mastigophora; again, it is confined to a solid sphere without membrane or intra-nuclear vacuoles; or, there may be a membrane and a single compact mass of chromatin which occupies the centre of the distinct nucleus, and is separated from the membrane by hyaline matter. In other cases, there may be two or more *karyosomes* or chromatin reservoirs, or there may be a great number of granules in the nucleus without the reservoirs (*Amœba proteus*). In some of the Rhizopoda (*Euglypha*) and Heliozoa (*Actinophrys* and *Actinosphærium*), the nucleus is strikingly similar to that of metazoan cells, consisting of chromatin in the form of a reticulum and a network of linin (Figs. 14 and 54).

The uumber of nuclei is also quite variable, many forms having only one (*Amœba proteus*, *Actinophrys*, etc.); others, two (*Amœba binucleata*, *Arcella*, etc.); while some have many, the giant *Amœba*, *Pelomyxa*, having, according to Bourne ('91), about ten thousand, although, even with this large number, the proportion of nuclear substance to the total mass of the organism is about the same as in other cells (Bourne). In almost all of the shelled forms, a multiple number of nuclei is the rule, but in the many-chambered Reticulariida, every chamber does not possess a nucleus, the number of nuclei being smaller than the number of chambers, thus indicating that these forms are not colonies, but *syncytia*, or multinucleate cells.

D. The Contractile Vacuole

Contractile vacuoles are almost entirely absent in the marine forms (Reticulariida, Radiolaria), and in a few of the fresh-water forms of Sarcodina (*Protamœba*, *Pelomyxa*, *Myxodictyum*, *Protogenes*, etc.), but they are generally present in the Amœbida and Heliozoa, sometimes two or three in one organism. The number of contractile vacuoles is quite variable. In most of the naked forms there is but one; this, however, may be of large size, sometimes measuring one-quarter of the volume of the organism (*Actinophrys*, *Actinosphærium*). In the shelled forms, on the other hand, there are two or more (2–3 in *Euglypha*, 12 or more in *Arcella*).

The position of the vacuole in the naked forms is also variable, but becomes fixed in the shelled forms and in the Heliozoa. In the shelled forms they sometimes lie in the middle zone about the edge of the granular region (*Euglypha*), sometimes around the periphery of the flattened body, while in other forms they are found now in one zone, now in another. In all cases, shortly before contraction, they come to lie close to the outer edge, and in some cases they form minute wart-like excrescences.

Amœba proteus, with its comparatively clear protoplasm and freedom from pigments, is one of the most favorable objects for the study of the contractile vacuole. If a sufficiently high power is used, the formation and contraction of the vacuole and the expulsion of the contents to the exterior can be followed step by step. At first the vacuole lies near the nucleus, but as it grows, it becomes separated from the latter, and at the time of its contraction lies at the end of the body farthest from the advancing pseudopodia, at what is sometimes called the posterior end (Fig. 49, *F*). Its reappearance is always somewhere near its point of disappearance. While still small it is carried along by the streaming protoplasm back to a position near the nucleus, where it completes its development. The increasing weight of the growing vacuole causes it to lag behind the streaming granules and nucleus, until at its full growth it is widely separated from the latter organ. The vacuole may appear to move in the direction contrary to that of the protoplasmic streaming, although in reality it is quiescent; for while it remains in the field of the microscope, the main body of the animal moves well out of it, until the vacuole is surrounded only by the posterior end of the animal (*G*), which is reduced to a thin layer of granules and a hyaline layer of ectoplasm between the vacuole and the exterior. The granules later move away, passing around the vacuole, until finally there is only a thin layer of hyaline plasm between the vesicle and the exterior. Shortly after this the vacuole bursts and disappears, in most cases a distinct bulge toward the outside preceding contraction. Contraction always begins on one side of the vacuole, and is carried across it toward the outer edge (*H*).

Stokes ('93) asserts that there is no bursting of the wall, but that minute pores are formed through which the contents of the vacuole are forced to the outside. In some instances the contents of the vacuole are not completely emptied, but as much as half may be left, the vacuole then rounding out to be carried back by the streaming plasm to the nucleus, where it completes its growth. In other cases the contents of the old vacuole may be entirely discharged, with the exception of a small quantity of liquid retained in exceedingly small vesicles, each of which may grow to some extent independently, although they ultimately fuse to form the new vacuole (*J*). Thus the new vacuole does not necessarily re-form at the place of disappearance, but may be derived by the coalescence of a number of smaller ones which themselves are the remains of an old one. As many as six may unite in this manner to form the new vacuole (*A*, *B*, *C*.) These unite two by two in various parts of the plasm, and the last two may not fuse until in the neighborhood of the nucleus.

In addition to the contractile vacuoles the Sarcodina occasionally possess gas vacuoles, which were first made out in *Arcella*, but which

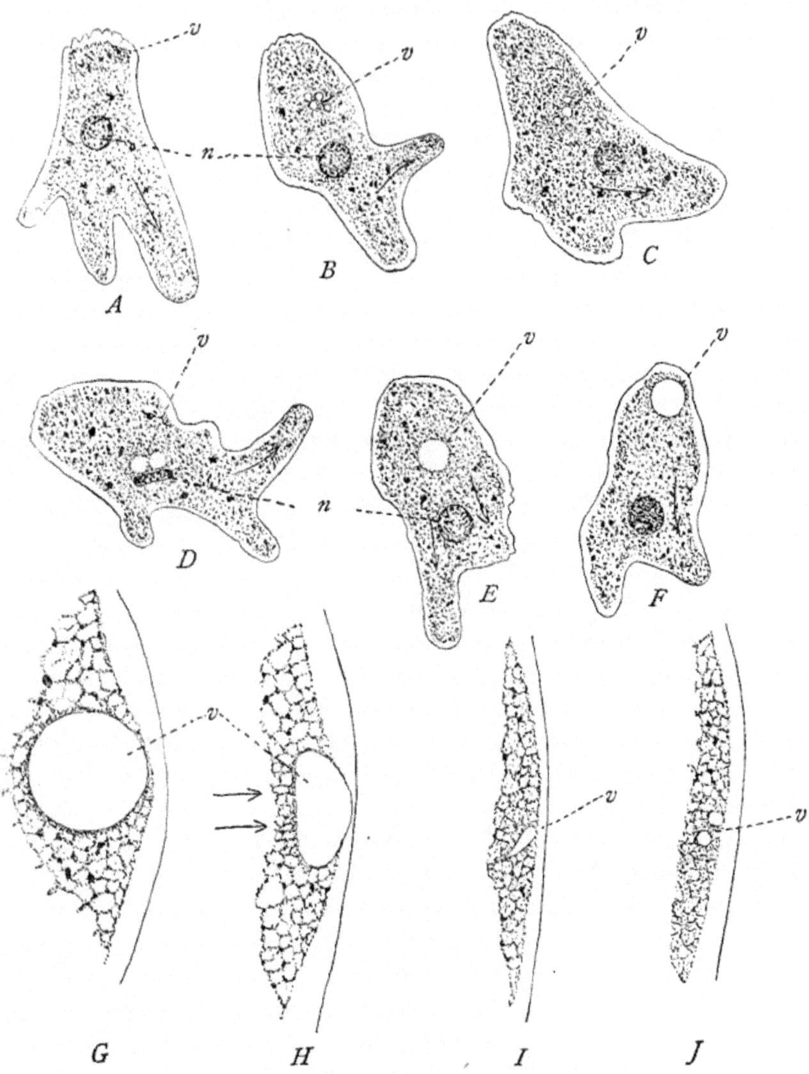

Fig. 49.—*Amœba proteus* and the contractile vacuole.

A. The vacuole (*v*) is in the form of minute vesicles in the region of contraction. *B*. Three minutes later. *C*. Two minutes later still (two vesicles have united). *D*. Two vesicles previous to union near the nucleus (*n*). *E*. The single vacuole becoming separated from the nucleus. *F*. The vacuole at the posterior end previous to contraction. *G*, *H*, *I*, and *J*. Four stages in the contraction of the vacuole.

have since been shown to be quite general in the group (Claparède and Lachmann, Engelmann, Bütschli, Entz, Rhumbler). The gas appears to be mainly carbon dioxide, and apparently serves an hydrostatic purpose, allowing the heavy forms like *Difflugia* to raise or lower themselves in the water.

E. Encystment

Encystment is undoubtedly a widespread phenomenon among the Sarcodina, although it is apparently absent altogether in the marine forms (Radiolaria and Reticulariida). With the exception of some Heliozoa, it has not been extensively studied, and the few observations are frequently contradictory. There is a general agreement, however, that its object is to protect the individual during periods of drought, cold, or during periods of reproduction. Thus a heliozoön, when its environment becomes unsuitable, draws in its pseudopodia, loses its ectoplasmic vacuoles, and secretes a double-layered coating, the inner layer being gelatinous at first, but later like a membrane. The outer layer is warty, and composed usually of silicious plates, cemented together by a silicious jelly. If multinucleate, most of the nuclei are absorbed, about 5 per cent remaining intact (Hertwig, for *Actinosphærium*). When conditions are again suitable, the animal absorbs water, swells, becomes vacuolated, bursts its membrane and outer cyst, and as a free-swimming heliozoön develops pseudopodia, and again leads an active life. Brauer ('94) and Hertwig ('98) have shown that in *Actinosphærium* encystment is accompanied by numerous phenomena of *plastogamy* and *karyogamy*.[1] In some forms of Heliozoa, as in *Vampyrella*, the outer cyst is composed of other material than silica, usually of cellulose, while in the fresh-water Rhizopoda the cyst coatings are chitinous. Here also the conditions are often changed by the presence of a shell, the cyst membrane in such cases covering over only the mouth-opening, although in one form at least (*Euglypha*) a second cyst membrane envelops the whole animal inside of the shell (see Fig. 17, p. 47).

F. Nutrition

The food of Sarcodina consists of vegetable substances, of flagellates, Infusoria, and not infrequently of larger animals, such as rotifers or small Crustacea. In the simpler forms there is no region set aside for the ingestion of food particles, but any portion of the body-surface can function as a mouth. When food particles strike the body, the stimulus causes the protrusion of specialized pseudopodia, which gradually surround the object. Zacharias ('93), upon rather uncertain data,

[1] Cf. pp. 236 and 237.

suggested that the pseudopodia in some forms are not motile, but prehensile organs, and are withdrawn after a full meal (*Actinosphæridium pedatum*). In some cases, at least, the pseudopodia apparently paralyze the prey, for flagellates or ciliates, coming in contact with the sharp pseudopodial tips, are immediately stunned and lie quiet, while either the pseudopodia lose their rigidity and bend around them, or smaller and new pseudopodia are formed from the body-substance, gradually surround the prey, and draw it into the body (see Fig. 18, p. 49). In the shelled forms the process of engulfing prey is less simple, and where there is a distinct cuticle the ingestion of the food can take place only by the softening or disappearance of some part of the membrane. In the shell-bearing Rhizopoda (Thecamœbina) food ingestion is confined to one part of the animal, the region about the mouth-opening; while in some Reticulariida the prey is not carried inside of the animal at all, but seizure, ingestion, and digestion all take place in the network of protoplasm formed by the anastomosed pseudopodia (Fig. 50). In the shell-bearing Heliozoa the outer coating must be ruptured for the entrance of the food particles (Pénard).

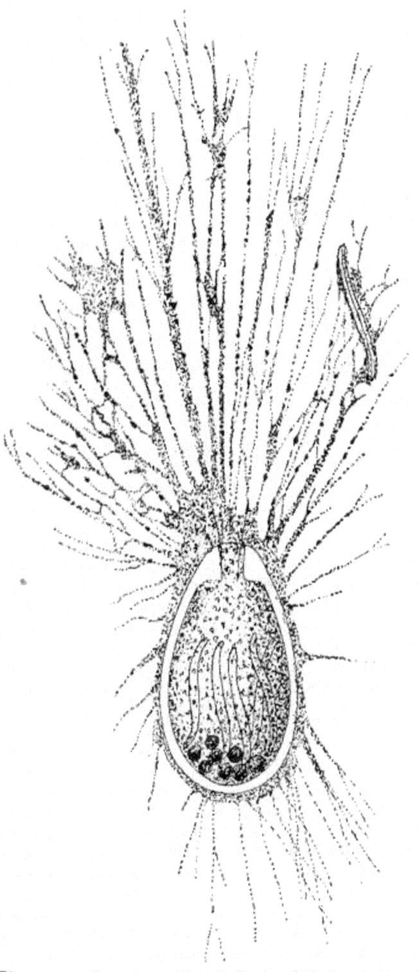

Fig. 50. — *Gromia oviformis* Duj. [M. SCHULTZE.] Some of the reticulate pseudopodia have captured a diatom.

In all cases the food substance is subsequently inclosed within a water vacuole, the liquid being taken in with the food (Dujardin, '41; Le Dantec, '90; Metschnikoff, '83). The fluid of the vacuole, at first nothing more than water similar to that in which the animal lives, gradually becomes acid, and in it the food particles are slowly disintegrated, the digestible portions being transformed into a sort of *chyle* which is distributed throughout the protoplasm. The gastric vacuole with its undigested residue is gradually left behind like a loaded contractile vacuole, until finally it is expelled to the outside (*Amœba*). The Sarcodina, apparently, digest mainly proteids, some forms of starch, and fats remaining unchanged (Meissner, '88; Greenwood, '80; Stolc, '00).

G. Reproduction

The Sarcodina reproduce mainly by simple division or spore-formation, either in the free state while active, or when quiet in the encysted state. The simplest form, consisting of a mere bipartition of the protoplasm and of the essential body-contents, occurs when the body is so large that it becomes unwieldy and it divides from sheer inertia. A well-known example is that of the division of *Amœba polypodia* (*Dactylosphæra*, F. E. Schultze). Here, as in all cell-divisions, the nucleus divides first, the body then separating into two parts. Simple division becomes more complicated when the organism is

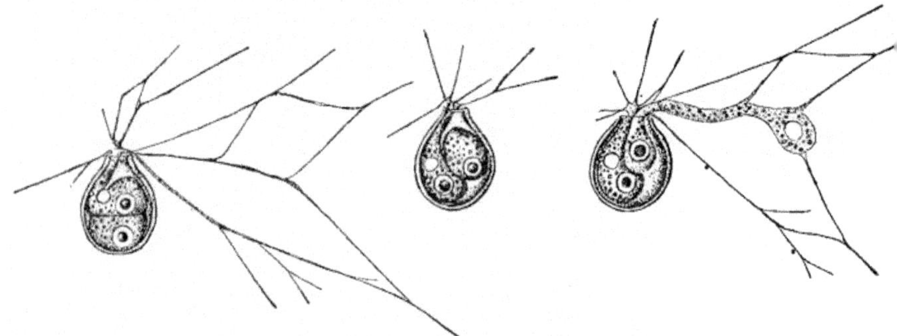

Fig. 51. — *Microgromia socialis* Hert. [Hertwig.]

Division takes place within the shell, and one of the daughter-individuals migrates, forming a new shell.

provided with an outer coating or test, although in the simplest of such cases, where the coating is flexible and plastic, as in *Vampyrella*, the process involves only the partition of the outer membrane. When the outer covering becomes hard and firm by impregnation with chitinous, silicious, calcareous, or horny materials, the operation is more complicated. The organism, while still within the shell, may divide by longitudinal division, one of the daughter-individuals then migrating from the parent shell and, after a longer or shorter time, settling down and secreting a new shell for itself, the other daughter-individual remaining in the old quarters (*Microgromia*, Fig. 51). A more complicated process is found in the majority of fresh-water shelled Rhizopoda, where division is practically a form of budding, the plasm growing out of the original shell mouth and forming a small bud on the outside. This bud grows until it has reached its definitive size (usually about that of the original cell), when the shell-coating for the new individual is deposited. The building material for the shell of the daughter-individual is formed within the protoplasm of the maternal cell. If regular plates of silica or chitin, these plates are secreted long before division and stored up in the protoplasm which surrounds the nucleus (*Euglypha*, *Quadrula*). If quartz crystals, or any other foreign bodies, these particles are picked up and stored in a similar manner, to be used later for the test of the daughter-cell. When the bud has reached a certain size, the plates or particles which are to form the shell move out through the mouth-opening of the parent shell and form around the protoplasm of the bud. In the meantime the nucleus undergoes division, and, in the case of *Euglypha* at least, the daughter-nucleus is the last element to leave the parent organism (see Fig. 23, p. 55).

Heliozoa, when preparing for division, become soft, draw in their pseudopodia, and round out into a perfect sphere, after which the nucleus divides by mitosis (*Actinophrys*), and the cell slowly separates into two parts. In *Nuclearia*, division is very rapid, the entire process taking place within one minute. In many cases, division is incomplete, the individuals remaining attached to form colonies (*Heterophrys*, *Sphærastrum*, *Raphidiophrys*).

Swarm-spore formation is widely distributed among the Sarcodina, usually taking place under the protection of a cyst. The parent organism divides into a number of daughter-cells, each containing a part of the original nucleus, and each provided with pseudopodia or flagella. A good illustration is seen in *Paramœba Eilhardi*, one of the naked Rhizopoda (Schaudinn, '96). The animal is flat and discoid, with short, lobose, finger-formed pseudopodia, and varies in size from 10 to 90 μ (Fig. 52). It usually increases by simple division, but at the end of its vegetative life it encysts, and the plasm divides into a number of pieces. Fragmentation of the nucleus is preceded by division of a peculiar cytoplasmic body, which Schaudinn terms the *Nebenkorper*. The contents of the cyst break into as many pieces as there are divisions of the *Nebenkorper*. Finally, each fragment of the protoplasm, containing a part of the original nucleus and of the cytoplasmic *Nebenkorper*, develops two flagella and breaks out of the cyst as a swarm-spore (*B*, *C*, *D*). The young organisms swim about in this condition for some time, and may increase by longitudinal division until, finally, losing their flagella, they develop pseudopodia and become young amœboid forms. Other cases of swarm-spore formation have been frequently recorded, in some cases by so many

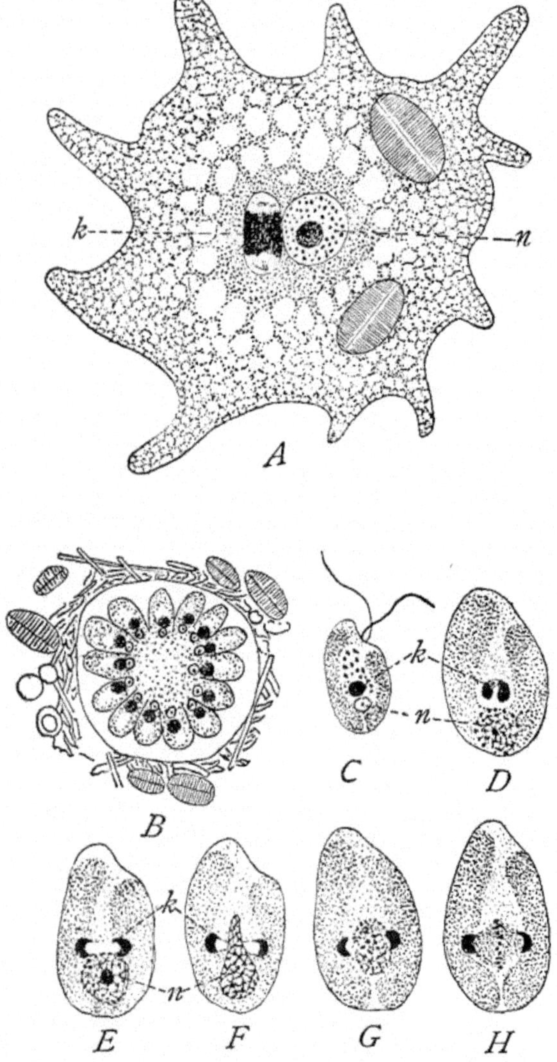

Fig. 52. — *Paramœba eilhardi* Schaud. [SCHAUDINN.]

A. Section. *B*. Sporulation. *C*. The flagellated swarm-spore. *D-H*. Stages in division of the flagellated spore. *k*, *Nebenkorper*; *n*, the nucleus.

different observers that there can be little doubt of their truth. In the Reticulariida, swarm-spore formation may be considered the typical method of increase, and it is connected with the dimorphism of the shells which Munier-Chalmas ('83) and Schlumberger ('83) have shown to be widespread throughout the group. The original nucleus divides into numerous parts, which are spread throughout all of the chambers. The protoplasm then segregates about them, and the original mass of plasm becomes divided into as many parts as there are nuclei. These leave the parent organism either by rupture of the shell or through the mouth-opening, and soon form new shells (megalospheric). After a short time they bud, and calcium carbonate is secreted around the bud, thus making a two-chambered cell. This process continues until the organism is full grown. Finally, the pseudopodia are drawn into the shell, and the protoplasm divides into numerous small swarm-spores, each with two flagella. These probably conjugate (Lister, '95; Schaudinn, '95), the copula giving rise to individuals with shells of the microspheric form. A very similar process occurs in the Radiolaria, where the endoplasm within the central capsule breaks up into swarm-spores, each with a portion of the original nucleus and each provided with flagella. These finally break out of the capsule, and, after a short free-swimming period, they lose their flagella and gradually assume the typical radiolarian form, passing through Heliozoa stages. In some cases, dimorphic spores (*anisospores*) are formed, which perhaps conjugate, as assumed by Brandt ('85) and Haeckel ('88), although the process has never been seen. Here, too, an alternation of generations is assumed by Haeckel and Brandt, an asexual or *isospore* generation alternating with a sexual anispore generation.

In addition to simple division and swarm-spore formation, some Sarcodina reproduce by bud-formation or gemmation. Buck ('77) early observed a number of small amœboid germs in the shell of *Arcella* (as many as thirty), an observation since confirmed by Cattaneo ('78), Bütschli, and recently by Hertwig ('99). Both Buck and Cattaneo traced the development of the buds up to the formation of the characteristic shell, while Hertwig has described the nuclear divisions leading to bud-formation. Bütschli found that the number of amœboid buds does not exceed nine. Le Blanc ('92) describes similar processes in *Difflugia*. Somewhat similar buds were observed inside *Pelomyxa palustris* by Weldon,[1] although neither the development nor origin was made out. Bud-formation has been repeatedly seen in the Heliozoa as well as in the Rhizopoda. The genus *Acanthocystis* in particular has been studied in this connection by Hertwig ('74), Korotneff, and more recently by Schaudinn ('96). According to the latter, the nucleus divides by amitosis, the daughter-nuclei moving toward the periphery, where they bud off with a small amount of cytoplasm; in some cases as many as twenty-four buds may be

[1] Cf. Lankester ('91).

formed by the same animal. The history of the buds is different in different individuals. In the simplest cases, the bud merely drops off the parent and remains on the bottom for some days in an amœboid state. In other cases, flagella are formed, and the buds move about like swarm-spores, although after a couple of days these, too, become amœboid. A few days later silicious spicules appear in the vicinity of the nucleus, and soon after make their way to the periphery, where the shell is formed (Fig. 53). *Clathrulina* also forms buds in a similar manner, and has been observed by Cienkowsky, Greeff, Hertwig, and Lesser; the observations thus are as well

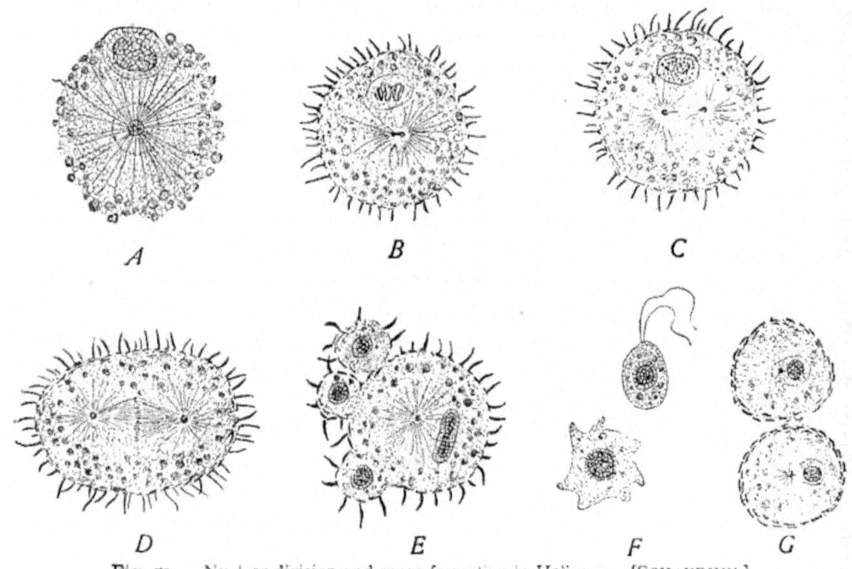

Fig. 53. — Nuclear division and spore-formation in Heliozoa. [SCHAUDINN.]
A. A vegetative cell of *Sphærastrum*, with the axial filaments focussed in a central granule (division-centre or "centrosome"). *B–D.* Division of the nucleus in *Acanthocystis*. *E, F.* Flagellated and amœboid swarm-spores formed by budding. *G.* Exit of the central granule from the nucleus.

established as in *Acanthocystis*. The number of buds is not so large as in the latter form, and the process is not unlike simple division. The body divides into three dissimilar pieces, two smaller and one larger. The latter remains within the old shell, but the former develop flagella and swim about like swarm-spores. In about half an hour they lose their flagella, settle to the bottom, throw out pseudopodia, and develop a stalk.

Conjugation has only rarely been seen among the different kinds of Sarcodina, and further observations must be made before it can be considered a widespread phenomenon. A few authentic observations, however, show that it occurs in some cases. Bütschli ('74) supposed that *Arcella*, which he found in pairs, were conjugating, and later he held the view that the phenomenon is quite widespread. Holman ('86) observed a large *Amœba* surround a small one, the two remained together for some time, and after they separated swarm-spores were formed in each. She regarded this as a possible case of conjugation. A somewhat similar process occurs in *Amœba spatula*, although here the smaller individual does not regain its identity, while the larger one appears as before, except for the presence of two nuclei (Pénard, '90). Conjugation is apparently more common among the shelled forms; Pénard ('90) says that, although he has met with conjugating animals in almost all of the species studied by him, he cannot cite a single instance where he has seen two animals, at first free, approach each other and fuse. Jickeli ('84) was more fortunate, for he saw two individuals of *Difflugia globulosa* fuse by their mouth-parts, the union being followed by lively pseudopodial movements. After twenty-four hours one shell appeared transparent, the other dense, while all movements had ceased. When the two shells separated at the end of forty-eight hours, one was empty, its contents having fused with those of the other shell. Gruber ('87) also reports a similar conjugation between two Difflugias. Conjugation has been frequently described in the Heliozoa also, although it is quite possible that many cases of so-called conjugation are only instances of plastogamy, or fusion of the cell-body, and are not followed by union of the nuclei (karyogamy), as in fertilization.

Numerous observations might be cited which seem, at first sight, to show that copulation and conjugation, in Heliozoa, are preliminary to reproduction by simple division or by spore-formation, but the evidence in most cases is incomplete, and the connection between conjugation and reproduction is still largely inferential. The ease with which large forms like *Actinosphærium* can be artificially reproduced by breaking them into pieces (Foulke, '83) is reason enough to excite caution as to generalizations on the connection between copulation and increase. The phenomenon is, nevertheless, clearly established in at least one form (*Actinophrys sol*, Schaudinn '96). In this case two free-swimming individuals come together and fuse; pseudopodia are drawn in, and the double cell sinks to the bottom, where it becomes coated by a cyst of silicious plates. Each of the two as yet ununited nuclei now prepares for division, passing through typical spireme and spindle stages as in *Euglypha*. Two of the four nuclei which are formed by these divisions round out and become normal nuclei, the others degenerate and finally disappear without playing any further rôle. The phenomenon recalls in a striking manner the formation of polar bodies among the Metazoa, and obviously represents some form of maturation (Fig. 54). The two functional nuclei now fuse, forming a cleavage nucleus, which divides by mitosis, giving rise to daughter-cysts.

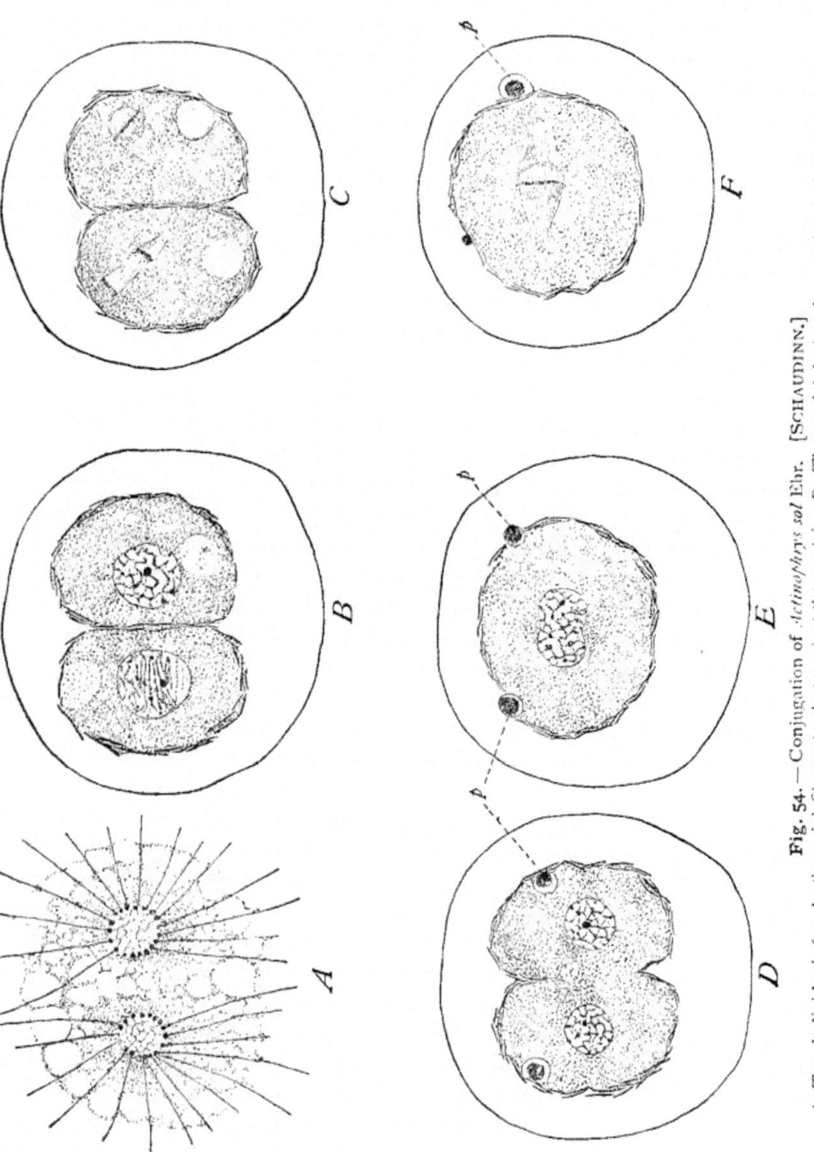

Fig. 54.—Conjugation of *Actinophrys sol* Ehr. [SCHAUDINN.]

A. Two individuals fused; the axial filaments abut against the nuclei. *B*. The nuclei during the prophase of division. *C*. For-

The conjugation of swarm-spores has been seen in a few cases. In *Vampyrella variabilis*, the animal breaks up into swarm-spores while encysted (Klein). These make their way out of the cyst to conjugate, and later form either double or multiple individuals (plasmodia).

The conjugation of swarm-spores of Reticulariida and Radiolaria has not been observed, although the diverse size of the anisospores in the latter group favors this view. In the fresh-water form, *Hyalopus*, which differs but slightly from the marine forms in regard to reproduction, the swarm-spores actually conjugate, although the development of the copula was not observed (Schaudinn, '94).

Inter-relationships of the Sarcodina

In drawing conclusions as to the most primitive group of the Protozoa, there is need of extreme caution. Bütschli long since showed that the development of the Protozoa, from spores or germs of any kind, gives but little indication of their genetic affinities, and that such affinities must be deduced from the study of the group as a whole.

In questions concerning the most primitive Protozoa, the Infusoria are immediately thrown out, for, of all groups of Protozoa, they are the most highly specialized, and along lines which have carried them to the highest point of morphological development of the single cell. The Sporozoa also have become highly specialized through their parasitic mode of life. Neither of these groups therefore can be said to have been the most primitive forms of Protozoa. Among the Sarcodina, the Heliozoa and Radiolaria show abundant evidence of descent from rhizopod-like ancestors, while a similar relation can be assumed of the Dinoflagellidia and Cystoflagellidia to the Flagellidia. It remains, therefore, to ascertain if possible which of the two groups, Flagellidia or Rhizopoda, shows the more primitive characteristics. Students of the Protozoa have differed widely on this question. Bütschli avoids the difficulty by the assumption that the beginnings of both are represented by the forms intermediate between the two, and sees in the members of the family Rhizomastigidæ (*Mast gamœba* and its allies), the common stem-forms of Flagellidia and Rhizopoda. On the other hand, the Rhizopoda, with their animal mode of nutrition, must have had other forms of life as their source of food, and from which they were possibly derived by the process of metasitism. On this account, Klebs ('92) makes the Flagellidia the original group, since here are retained forms which live to-day as the original forms probably did, with the power to manufacture their own food (Phytoflagellida). It is extremely difficult to choose between the two assumptions, although the balance apparently lies in favor of the view which Klebs advocates. From the variety of forms they assume, the Flagellidia appear to have a greater power of adaptation than the

Rhizopoda, and to this power of change may be due the fact that fewer species of Flagellidia than of Rhizopoda are known.[1] The development of a flagellate or of a rhizopod throws little light upon the question, for both flagellates with amœboid swarm-spores, and rhizopods with flagellated swarm-spores, are known. The very close relation of the two groups is also shown by the fact made out by

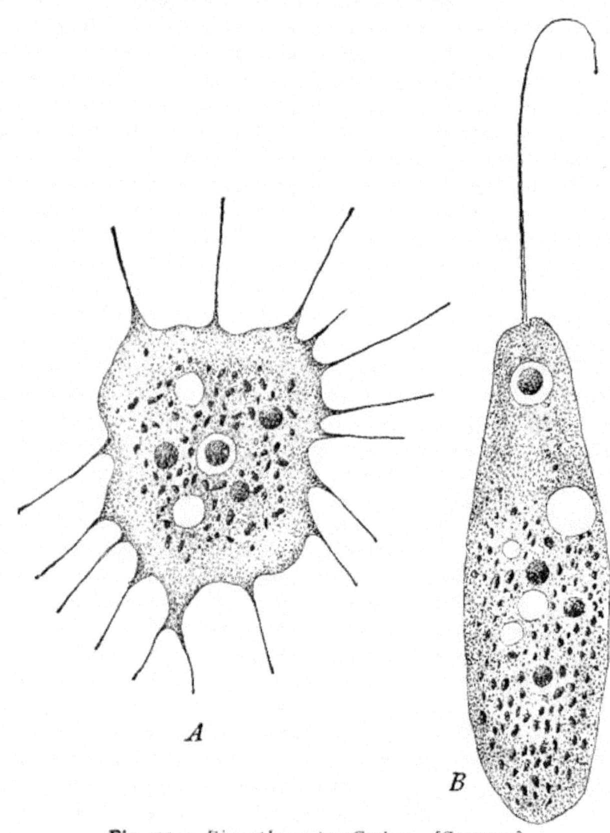

Fig. 55.—*Dimorpha mutans* Gruber. [GRUBER.]
A. Amœboid phase. *B*. Flagellated phase.

numerous observers, that in the same organism pseudopodia may change into flagella, and flagella into pseudopodia.

The relations of pseudopodia to flagella have not hitherto been sufficiently emphasized. Not only do flagella become pseudopodia, and pseudopodia flagella, in some forms, but in cases where the mutual change has never been observed, there is morphological evi-

[1] Bütschli enumerates 98 genera of Flagellidia and 174 genera of Rhizopoda; Delage and Hérouard 314 genera of Rhizopoda and 153 Flagellidia.

dence to show that pseudopodia and flagella are closely related, and this evidence is strong enough, I believe, to throw additional light upon the Flagellidia, regarded as the most primitive forms of Protozoa. It has been shown that the flagellum in most cases arises in the vicinity of the nucleus. This is also the case with the axial filaments of the Heliozoa, an extremely interesting example being shown in the form *Dimorpha*, as described by Gruber (Fig. 55, also Fig 46, p. 82). Here the axial filament is homologous with the flagellum, and there is ground for believing that the homology can be carried from *Dimorpha* to the true Heliozoa, where all of the appendages are similar to the *axopodia* of *Dimorpha*. Thus, in *Acanthocystis* and *Actinophrys*, the axial filaments radiate from a common centre in the nucleus (*Actinophrys*, see Fig. 54 *A*), or in the cytoplasm (*Acanthocystis*). In *Actinosphærium* and *Camptonema*, they arise from the nuclei and do not converge at a common point. In these particular cases, the pseudopodia have little power of vibratile motion, such as we might expect if the axial filaments are comparable with flagella. In other cases they do possess this power, however, to a certain degree, as shown by the rolling motion of *Acanthocystis*, which is able to travel a distance equal to twelve times its own diameter in one minute, or by the quick dancing motion of *Artodiscus* (Pénard). That this motion is due to the vibrations or elasticity of the axial filaments I think there can be no doubt, and these structures are comparable therefore with the flagella of the Mastigophora. Unlike flagella, however, they are covered by plastic and streaming protoplasm, which gives them their pseudopodial character. In those forms of Heliozoa which are usually regarded as more primitive, *e.g.* *Nuclearia* and *Vampyrella*, the axial filaments are not formed, and it is an important question whether these are primitive forms representing a condition before differentiation of the axial filaments in other Heliozoa, or are to be considered as degenerate forms in which the axial filaments have disappeared (Fig. 56). If this question could be answered, it might afford evidence as to whether the Flagellidia or the Sarcodina are the more primitive forms; for if degenerate, they point toward the Heliozoa or Radiolaria as the ancestors of the reticulate Rhizopoda; but if primitive, they point toward the Rhizopoda as the ancestors of Heliozoa and Radiolaria, and, through *Dimorpha*, of the Flagellidia. Evidences of the axial filaments are found in other Sarcodina than the Heliozoa and Radiolaria. In the Reticulariida the central plasm of the pseudopodia is denser and more resisting than the outer plasm, and M. Schultze, Bütschli, Schaudinn, Rhumbler, and others, assume that it has a contractile function; morphologically and physiologically, therefore, it appears to be similar to the axial filaments of Heliozoa, and to the flagella of Mastigophora.

As a matter of fact, the beginnings of the various branches of the Protozoa rest in complete obscurity, and the relationships of the sub-

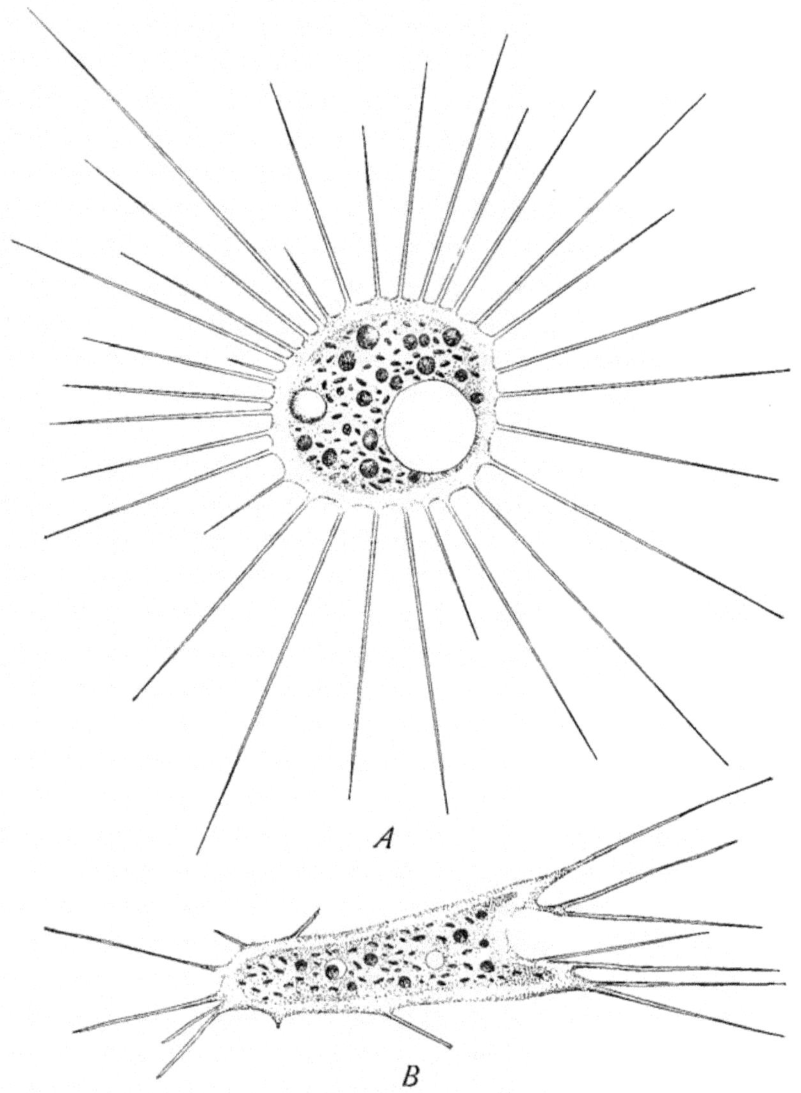

Fig. 56.—*Nuclearia delicatula* Cienk.
A. Heliozoan phase. *B.* Rhizopod phase.

groups are almost equally uncertain. There are a few interesting forms, however, which are generally given as intermediate stages between the classes, and are supposed to show genetic relationships between the groups which they simulate.

The close connection between the Mastigophora and the Sarcodina has been recognized since the discovery by F. E. Schultze ('75) and Bütschli ('78) of amœboid forms with flagella. These are unmistakably animals which take in food at any portion of the body by means of pseudopodia, and move by means of flagella or pseudopodia. In some instances they are more like an *Amœba* (*Mastigamœba*, F. E. Schultze, Fig. 57, *A*); in others they are more like a heliozoön (*Ciliophrys infusionum* of Cienkowsky, or *Actinomonas* of Kent, Fig. 57, *B*, *Mastigophrys* of Frenzel, etc.). Their undetermined position

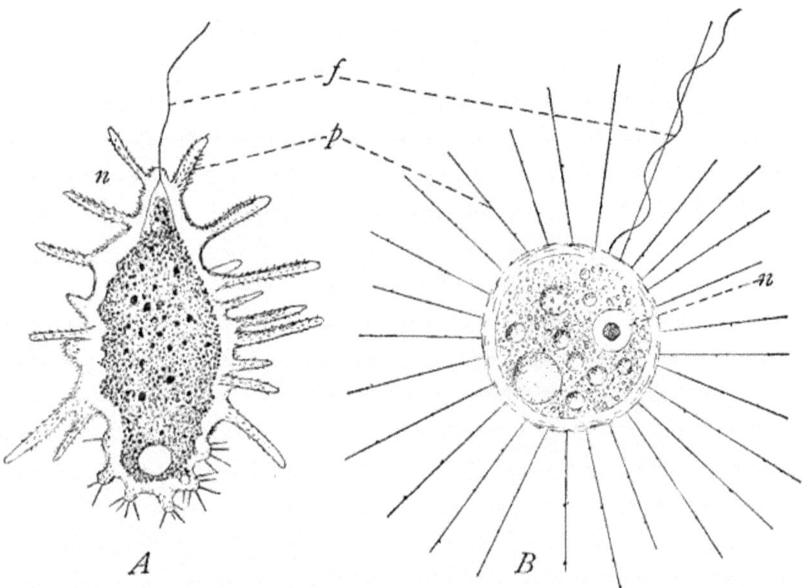

Fig. 57. — Protozoa with both pseudopodia and flagella.
A. Mastigamœba aspera F. E. Sch. [SCHULTZE.] *B. Actinomonas pusilla* S. K. [KENT.]
f, flagellum; *n*, nucleus; *p*, pseudopodia.

in classification is indicated by the fact that sometimes they are included with the Sarcodina, while at other times they are regarded as flagellates. Klebs believes that the connection between the two groups is not quite so apparent as the mere description of these intermediate forms would indicate, and places between the primitive animal flagellates, the Vampyrellidæ, and the Mycetozoa, an intermediate group, that of the Pseudosporeæ, with the genera *Pseudospora* and *Protomonas* which Bütschli includes with the Flagellidia; while the Rhizopoda are derived from the Vampyrellidæ through the Heliozoa. The forms which *Pseudospora* represents are placed between the Rhizomastigidæ as enumerated above, and the Vampyrellidæ, largely on account of their methods of reproduction and life history, which in the majority of the Rhizomastigidæ are unknown. *Vampyrella* reproduces while encysted, by dividing into a number of parts, each of which emerges as a small *Vampyrella* with pseudopodia like the parent. *Protomonas amyli* (Haeckel), *Monas amyli* (Cienkowsky), and *Pseudospora* reproduce in the same way, with the exception that the swarm-spores formed within the cyst are not amœboid, but are provided with flagella. The swarmers soon lose their flagella, however, becoming amœboid, a condition in which they fuse together to form larger or smaller plasmodia. This fusion is characteristic of the Vampyrellidæ and of Mycetozoa, but not of the Rhizomastigídæ, where it has never been observed.

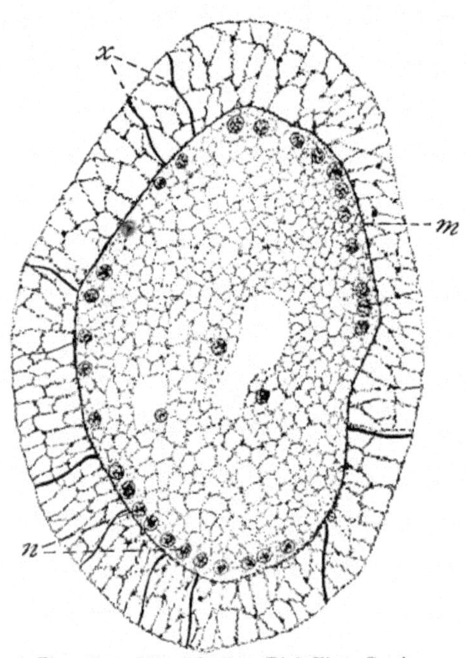

Fig. 58. — *Actinosphærium Eich.* Ehr. Section. *m*, membrane between ectoplasm and endoplasm; *n*, nuclei; *x*, axial filaments.

There are many features in this theory of Klebs to recommend it. It affords a logical and satisfactory explanation of the relations of the Mycetozoa to the Sarcodina, and from the standpoint of the botanist points out the relation of this group to the colorless plants. The close connection of the Heliozoa with the Mastigophora is shown in other ways than by the transitional forms *Dimorpha*, *Actinomonas*, etc. The finer structure of the body of the sun-animalcula, the nucleus and archoplasmic substances, show a degree of differentiation approached only by the Flagellidia and Metazoa, while the axial filaments are homologous with flagella. It may be pointed out, however, that Klebs' theory leaves unexplained the relatively simple nuclear structures and nuclear processes of division in the Rhizopoda. This objection is fatal to the view that the Rhizopoda are derived from the higher types of Heliozoa, and it must be admitted that they arose from much less specialized forms, perhaps from the Pseudosporeæ or other flagellated forms, perhaps from forms like the Vampyrellidæ. On the whole, there is no conclusive evidence to support the view that Rhizopoda are more primitive than Flagellidia, or *vice versa*. Their mutual affinities are very close, and together they stand as the most primitive forms of modern Protozoa.

The relations of the Radiolaria to the Heliozoa are extremely close, and there is abundant evidence to show that the former were derived from the latter by the acquisition of a chitinous membrane between ectoplasm and endoplasm, and the retention of a gelatinous mantle like that of *Sphærastrum* (Haeckel). As first pointed out by Brandt ('85), the young Radiolaria pass through flagellated and amœboid swarm stages, then through Heliozoa stages, until the definitive radiolarian structure is attained. Haeckel described the intermediate forms which are represented in this growth, as flagellate, amœboid, *Actinophrys*, *Sphærastrum*, and *Actissa*, the last being the simplest of the Radiolaria. Although the external appearance of a radiolarian is strikingly similar to that of a heliozoön, there is no structure in Heliozoa to compare with the chitinous central capsule of the Radiolaria. Greeff ('67, '71) described a membrane-like thickening between the endoplasm and the ectoplasm of *Actinosphærium*, and regarded it as homologous with the central capsule. Other observers, *e.g.* F. E. Schultze, Hertwig and Lesser, Bütschli, etc., have not seen it, and the latter, especially, considers Greeff's contribution of little value. However incorrect his interpretation may have been that *Actinosphærium* is a fresh-water radiolarian belonging to the Acantharia, Greeff was not mistaken in his observation, for an occasional specimen is found which shows such a membrane (Fig. 58).

CLASSIFICATION

CLASS I. **SARCODINA**. Naked or shelled Protozoa, characterized by the possession during adult life of movable or changeable processes of protoplasm, the pseudopodia, which may be finger-form, reticulate, or ray-like, and which may or may not have axial filaments. Reproduction is brought about by simple division and by spore-formation.

Subclass I. **RHIZOPODA**. Naked or shelled Sarcodina having pseudopodia of the lobose (finger-formed) or reticulate (anastomosing) type. The adult form is amœboid: the young forms are amœboid or flagellated, and are produced by spontaneous division of the cell during active phases or during encystment. The adults in some cases fuse to form plasmodia.

Order I. **AMŒBIDA**. Rhizopoda provided with lobose pseudopodia, with or without a shell, with one or more nuclei, and usually with a contractile vacuole.

Suborder I. **GYMNAMŒBINA**. Naked forms of Amœbida having lobose pseudopodia, and with or without nucleus and contractile vacuoles.

Family I. **AMŒBIDÆ**. The pseudopodia are lobose, occasionally sharp pointed and branched. Genera: *Amœba*, marine and fresh water; *Paramœba* Schaudinn ('96); *Protamœba* Haeckel ('66), marine and fresh water; *Gringa*

Frenzel ('92), lagoons; *Gloidium* Sorokin ('78), fresh water; *Chætoproteus* (*Dinamœba* Leidy) Stein ('57), fresh water; *Trichosphærium* Schneider; *Pachymyxa* Gruber; *Hyalodiscus* Hertwig and Lesser ('74), fresh water; *Plakopus* F. E. Schultze ('75), fresh water; *Dactylosphæra* Hert. & Less. ('74), fresh water; *Chromatella* Frenzel ('92), fresh water; *Stylamœba* Frenzel ('92), fresh water; *Saltonella* Frenzel ('92), fresh water; *Eikenia* Frenzel ('92), fresh water; *Pelomyxa* Greeff ('74), fresh water; *Amphizonella* Greeff ('66), fresh water; *Podostoma* Clap. & Lach. ('58), fresh water; *Arcuothrix* Hallez ('85), cultures of *Ascaris megalocephala*.

Suborder 2. **THECAMŒBINA**. Amœbida provided with a shell, with lobose pseudopodia, which may be sharp pointed and branched, and with one or more nuclei and contractile vacuoles.

Family 2. **Arcellidæ**. The shell is more or less membranous. Contractile vacuoles are numerous; the nucleus is single or multiple. Genera: *Arcella* Ehbg. ('38), fresh water, common; *Cochliopodium* Hert. & Less. ('74), fresh water, common; *Pyxidicula* Ehbg. ('38), fresh water; *Pseudochlamys* Clap. & Lach. ('58), fresh water; *Hyalosphenia* Stein ('57), fresh water; *Quadrula* F. E. Schultze ('75), fresh water; *Difflugia* Leclerc ('15), fresh water, common; *Lecquereusia* Schlumberger, fresh water.

Family 3. **Euglyphidæ**. The shell is formed of regular plates of chitin, or of silica, and is often provided with spines. The pseudopodia are sharp pointed and often branching, but do not anastomose. Genera: *Euglypha* Dujardin ('41), fresh water; *Trinema* Duj. ('36), fresh water; *Cyphoderia* Schlumberger ('45), marine and fresh water; *Campascus* Leidy ('77), fresh water; *Nadinella* Pénard ('99).

Order 2. **RETICULARIIDA**. Rhizopoda with fine branching and anastomosing, or reticulate pseudopodia, forming an irregular network around the body, which may or may not have a shell. Shells, when present, are calcareous (rarely silicious) and provided with many pores (Perforina), or without pores (Imperforina), and consisting of one chamber (Monothalamous), or of many chambers (Polythalamous).

Suborder 1. **NUDA**. Shell absent; the pseudopodia are reticulate, and the cell-body, in many cases, is apparently without a nucleus; marine. Genera: *Gymnophrys* Cienkowski ('76); *Protomyxa* Haeckel ('68); *Myxodictyum* Haeckel ('68); *Protogenes* Haeckel ('64); *Pontomyxa* Topsent ('93).

Suborder 2. **IMPERFORINA**. [With but few modifications, the following classification of the Reticulariida is taken from Brady ('84) and Lankester ('85), after Carpenter ('69)]. Shell-bearing forms; the shells are calcareous, solid, and without minute apertures, or they are made up of foreign particles cemented upon a chitinous base. They may have one, two, or many mouth openings, and are either monothalamous or polythalamous.

Family 1. **Gromidæ**. The shell is membranous and in the form of a simple sac, with a pseudopodial aperture either at one extremity or at each end. The pseudopodia are long, branching, and anastomosing; marine and fresh-water forms.

Subfamily 1. *Monostominæ*. The shell has but one aperture. Genera: *Gromia* Duj. ('35); *Lieberkühnia* Clap. & Lach. ('58), both found in fresh water, the former also marine; *Microgromia* R. Hertwig ('74), fresh water; *Platoum* F. E. Schultze ('75); *Plectophrys* Entz ('77); *Pseudodifflugia* Schlumberger ('45), brackish and fresh water.

Subfamily 2. *Amphistominæ*. With an aperture at each end of the shell. Genera: *Diplophrys* Barker; *Ditrema* Archer ('70); *Amphitrema* Archer ('70); *Shepheardella* Siddall ('80).

Family 2. **Miliolidæ.** The shell is mono- or polythalamous, usually calcareous and porcellanous, but may be covered with sand. The polythalamous forms may be linear, spiral, or a combination of the two.

Subfamily 1. *Nubecularinæ.* The shell has an irregular and asymmetrical form, with the aperture or apertures variously placed. Genera: *Calcituba* Roboz; *Squammulina* Schultze ('54); *Nubecularia* Dufrance.

Subfamily 2. *Miliolinæ.* The shell is coiled, either symmetrically or asymmetrically, on an elongated axis, with usually two chambers to each convolution. During growth the shell-mouth is alternately at each end of the shell. Genera: *Spiroloculina* d'Orb. ('26); *Biloculina* d'Orb. ('26); *Fabularia* Dufrance; *Miliolina* Williamson ('58).

Subfamily 3. *Hauerininæ.* The shells are varied, the chambers being partly milioline in their arrangement, partly spiral or linear. Genera: *Hauerina* d'Orb. ('46); *Articulina* d'Orb. ('46).

Subfamily 4. *Peneroplidinæ.* The shells are plano-spiral or cyclical, and bilaterally symmetrical. Genera: *Peneroplis* Montfort ('10); *Orbitolites* Lamarck (1801); *Orbiculina* Lamarck ('16); *Cornuspira* M. Schultze ('54).

Subfamily 5. *Alveolininæ.* The shell is spiral and elongated in the axis of the convolution; the chambers are subdivided into secondary chambers. Genera: *Alveolina* d'Orb ('26).

Subfamily 6. *Keramosphærinæ.* The shell is spherical with the chambers in concentric layers. Genera: *Keramosphæra* Brady ('84).

Family 3. **Astrorhizidæ.** The shell is invariably composite, consisting of foreign particles, such as diatom-cases, spicules, sand grains, etc. It is usually large and single chambered, frequently branched or even radiate, with usually a single pseudopodial aperture at the end of each branch.

Subfamily 1. *Astrorhizinæ.* The shells have thick walls, consisting of sand or mud, lightly cemented together. Genera: *Astrorhiza* Sandahl ('57); *Dendrophrya* Wright ('61); *Syringammina* Brady ('84); *Pelosina* Brady ('79).

Subfamily 2. *Pilulininæ.* The shell consists of one chamber, the walls being thick and composed of felted spicules and fine sand. Genera: *Pilulina* Carpenter ('70); *Bathysiphon* Sars ('71).

Subfamily 3. *Saccammininæ.* The chambers are nearly spherical, with thin walls composed of closely cemented sand grains. Genera: *Saccammina* Sars ('65); *Psammosphæra* Schultze ('75); *Sorosphæra* Brady ('79).

Subfamily 4. *Rhabdammininæ.* The shell is composed of sand grains, firmly cemented together, and often with sponge spicules intermixed. They are tubular, straight, radiate, branched or irregular, but rarely segmented. Genera: *Jaculella* Brady ('79); *Botellina* Carpenter ('70); *Haliphysema* Bowerbank ('62); *Marsipella* Norman ('78); *Rhabdammina* Sars ('65); *Aschemonella* Brady ('79); *Rhizammina* Brady ('79); *Sagenella* Brady ('79).

Family 4. **Lituolidæ.** The shell is arenaceous, and the septa which imperfectly mark the chambers are often incomplete or absent.

Subfamily 1. *Lituolinæ.* The shell is composed of coarse sand grains, is rough externally, and often labyrinthic. Genera: *Rheophax* Montfort ('08); *Haplophragmium* Reuss ('60); *Coskinolina* Stache; *Haplostiche* Reuss ('61); *Lituola* Lamarck (1801); *Bdelloidina* Carter ('77).

Subfamily 2. *Trochammininæ.* The shell is thin, and consists of a chitinous basis in which are embedded minute sand grains. The outside of the shell is smooth and often polished; the interior is smooth or occasionally reticulate, but never labyrinthic. Genera: *Thurammina* Brady ('79); *Ammodiscus* Reuss; *Trochammina* Parker and Jones ('59); *Webbina* d'Orb. ('39); *Carterina* Brady ('79); *Hippocrepina* Parker; *Hormosina* Brady ('79).

Subfamily 3. *Endothyrinæ.* The shell is more calcareous and less sandy than in the other *Lituolidæ*, and the septa between the chambers are distinct. Genera: *Nodosinella* Brady; *Endothyra* Phillips ('46); *Polyphragma* Reuss; *Bradyina* Möll. ('78); *Stacheia* Brady ('76).

Subfamily 4. *Loftusinæ.* The shell is large, lenticular, spherical, or fusiform, and deposited either in concentric layers or spirally. The chambers are occupied to a large extent by an excessive enlargement of the arenaceous cancellated wall. Genera: *Cyclammina* Brady ('76); *Loftusia* Brady ('69); *Parkeria* Carpenter ('69).

Suborder 3. **PERFORINA.** The shell wall is perforated by numerous minute openings through which the pseudopodia can pass as well as through the main openings.

Family 5. **Textularidæ.** The shells of the larger species are arenaceous, either with or without a calcareous matrix; the smaller forms are hyaline and conspicuously perforated. The chambers are arranged in alternating series, spirally or without apparent order.

Subfamily 1. *Textularinæ.* The shells are typically bi- or tri-serial, and are often dimorphous. Genera: *Textularia* Dufrance ('28); *Bigenerina* d'Orb. ('26); *Verneuilina* d'Orb.; *Cuneolina* d'Orb. ('39); *Pavonina* d'Orb. ('26); *Valvulina* d'Orb. ('26); *Chrysalidina* d'Orb. ('46); *Tritaxia* Reuss; *Clavulina* d'Orb.

Subfamily 2. *Bulimininæ.* The shells are typically spiral, the weaker forms are more or less bi-serial. The main aperture is not round, but elliptical, comma-shaped, etc. Genera: *Virgulina* d'Orb. ('26); *Bulimina* d'Orb. ('26); *Bolivina* d'Orb.; *Bifarinia* Parker and Jones.

Subfamily 3. *Cassidulineæ.* The shell consists of a series of alternating segments more or less coiled. Genera: *Cassidulina* d'Orb. ('26); *Ehrenbergina* Reuss.

Family 6. **Chilostomellidæ.** The shell is calcareous, finely perforate, and polythalamous. The segments follow each other from the same end of the long axis, or alternately from the two ends, or in cycles of three, which are more or less embracing. The aperture is a curved slit at the extremity of the final segment. Genera: *Ellipsoidina* Seguenza; *Chilostomella* Reuss; *Allomorphina* Reuss.

Family 7. **Lagenidæ.** The shell is calcareous and very finely perforated; it is monothalamous or polythalamous. In the latter the chambers may be joined together in a straight, curved, spiral, or branching series. The aperture is terminal, and may be simple or radiate. The shell is not complicated by interseptal skeletons or by canal systems.

Subfamily 1. *Lageninæ.* Shell monothalamous. Genera: *Lagena* Walker and Boys (1784); *Nodosaria* Lamarck ('16); *Lingulina* d'Orb. ('26); *Vaginulina* d'Orb. ('26); *Rimulina* d'Orb. ('26); *Frondicularia* Defrance; *Marginulina* d'Orb. ('26), etc.

Subfamily 2. *Polymorphininæ.* The segments composing the shell are arranged spirally or irregularly around the long axis; they are rarely biserial and alternate. Genera: *Polymorphina* d'Orb. ('26); *Uvigerina* d'Orb. ('26); *Sagrina* Parker and Jones.

Subfamily 3. *Ramulininæ.* The branching shell is composed of long tubulariform tubes. Genera: *Ramulina* Rupert Jones.

Family 8. **Globigerinidæ.** The shell is free, calcareous, and perforated. The conspicuous shell-aperture may be single or multiple. There is no supplementary skeleton or canal system. The animals are normally pelagic in habit. Genera: *Globigerina* d'Orb. ('26); *Orbiculina* Lam.; *Hastigerina* Thompson ('76); *Candeina* d'Orb. ('26); *Pullenia* Park. & Jones ('62); *Sphæroidina* d'Orb. ('26).

Family 9. **Rotalidæ.** The shell is calcareous, perforated, free, or adherent; it is typically spiral in form, but irregular forms may be outspread or flaring, acervuline or irregular. Some of the higher types have double walls, with supplemental skeleton and a canal system.

Subfamily 1. *Spirillinæ.* The shell is a flat spiral, without septa; it may be free or attached. Genera: *Spirillina* Ehbg ('41).

Subfamily 2. *Rotalinæ.* The shell is spiral, rotaliform, and rarely evolute or irregular. Genera: *Discorbina* Lamarck ('04); *Planorbulina* d'Orb. ('26); *Truncatulina* d'Orb. ('26); *Anomalina* d'Orb. ('26); *Rotalia* Lamarck (1801); *Calcarina* d'Orb. ('26); *Patellina* Williamson ('58); *Carpenteria* Gray ('58); etc.

Suborder 4. **TINOPORINÆ.** The shell consists of irregularly heaped chambers, usually with a more or less spiral primordial portion; a main pseudopodial aperture is usually absent. Genera: *Tinoporus* Carpenter ('57); *Polytrema* Risso ('26); *Gypsina* Carter; *Thalamopora* Roemer; *Aphrosina* Carter.

Family 10. **Nummulinidæ.** The shell is calcareous and finely tubulated; it is typically polythalamous, free, and symmetrically spiral. The higher forms possess a supplementary skeleton and a well-developed canal system.

Subfamily 1. *Fusulininæ.* The shell is bilaterally symmetrical, with chambers extending from pole to pole, so that each convolution completely incloses the preceding whorl. The septa between the chambers are single as a rule. Genera: *Fusulina* Fischer ('29); *Schwagerina* Möller ('77).

Subfamily 2. *Polystomellinæ.* The shell is bilaterally symmetrical and nautiloid. The simpler forms are without supplemental skeleton; the more complex forms have a skeleton, and canals leading to the outside at regular intervals along the external septal depressions. Genera: *Polystomella* Lamarck ('22); *Nonionina* d'Orb. ('26).

Subfamily 3. *Nummulitinæ.* The shell is lens-shaped or flattened. Genera: *Archeodiscus* Brady; *Amphistegina* d'Orb. ('26); *Operculina* d'Orb. ('26); *Nummulites* Lamarck (1801); *Heterostegina* d'Orb. ('26).

Subfamily 4. *Cycloclypeinæ.* The shell is flat, with a thickened centre, or lens-shaped, and consists of a disc of chambers arranged in concentric annuli with peripheral thickenings. The septa are double, and furnished with a system of interseptal canals. Genera: *Cycloclypeus* Carpenter ('56); *Orbitoides* d'Orb.

Subclass II. **HELIOZOA.** These are naked or shelled forms of Sarcodina of typically spherical form, with but little tendency to change form by amœboid motion. The pseudopodia, radiating from all parts of the body, are fine and ray-like, rarely changeable, and usually provided with an axial filament.

Order 1. **APHROTHORACIDA.** Heliozoa, without a skeleton, but provided with a more or less developed power of amœboid motion, and with plastic (myxopodia) or stiff (axopodia) pseudopodia, the latter possessing axial filaments. Genera: *Vampyrella* Cienk. ('65); *Nuclearia* Cienk. ('65); *Monobia* A. Schneider ('78); *Myxastrum* Haeck. ('70); *Actinophrys* Ehr. ('30); *Actinosphærium* Stein ('57); *Actinolophus* F. E. Schultze ('74).

Order 2. **CHLAMYDOPHORIDA.** Heliozoa, with a soft gelatinous or felted fibrous covering. Genera: *Heterophrys* Archer ('69); *Sphærastrum* Greeff ('73); *Astrodisculus* Greeff ('69).

Order 3. **CHALARATHORACIDA.** Heliozoa, with a silicious coating composed of separate and loosely-jointed spicules. Genera: *Pompholyxophrys* Archer ('69); *Raphidiophrys* Archer ('70); *Pinacocystis* Hert. & Less. ('74); *Pinaciophora* Greeff ('73); *Acanthocystis* Carter ('63); *Diplocystis* Pénard ('90); *Cienkowskya* Schaudinn; *Wagnerella* Mereschkowsky ('81).

Order 4. **DESMOTHORACIDA.** Heliozoa, with a shell of one piece perforated by numerous openings. Stalked or unstalked forms. Genera: *Orbulinella* Entz ('77); *Clathrulina* Cienk. ('67).

Subclass III. **RADIOLARIA**. Marine forms of Sarcodina, similar to Heliozoa in having ray-like pseudopodia (axopodia and myxopodia), but provided with a chitinous capsule which incloses the nuclei. They may or may not have a skeleton; when present the skeleton is formed of acanthin or of silica. The group is subdivided into 4 legions, 20 orders, 85 families, 739 genera, and 4318 species. (Haeckel, 1885.)

Legion 1. **SPUMELLARIA** (or **PERIPYLEA**). The central capsule is perforated by numerous fine pores. A skeleton may or may not be present.

Order 1. **COLLOIDIDA**. Without skeleton. Families: **Thalassicollidæ** (solitary forms); **Collozoidæ** (colonial).

Order 2. **BELOIDIDA**. The skeleton consists of loose silicious needles. Families: **Thalassosphæridæ** (single); and **Sphærozoidæ** (colonial).

Order 3. **SPHÆROIDIDA**. The skeleton consists of from one to many concentric globular shells. Families: **Liosphæridæ** (single); **Collosphæridæ** (colonial); **Stylosphæridæ** (single); **Staurosphæridæ** (single); **Cubosphæridæ** (single); **Astrosphæridæ** (single).

Order 4. **PRUNOIDIDA**. With ellipsoidal to cylindrical latticed shells and similar central capsule. Families: **Ellipsidæ**; **Druppulidæ**; **Sponguridæ**; **Artiscidæ**; **Cyphinidæ**; **Panartidæ**; **Zygartidæ**.

Order 5. **DISCOIDIDA**. Shell and central capsule are discoidal or lenticular. Families: **Cenodiscidæ**; **Phacodiscidæ**; **Coccodiscidæ**; **Porodiscidæ**; **Pylodiscidæ**; **Spongodiscidæ**.

Order 6. **LARCOIDIDA**. The skeleton is irregularly lenticular or discoid. Families: **Larcaridæ**; **Larnacidæ**; **Pylonidæ**; **Tholonidæ**; **Zonaridæ**; **Lithelidæ**; **Streblonidæ**; **Phorticidæ**; **Soreumidæ**.

Legion 2. **ACANTHARIA** (or **ACTIPYLEA**). The skeleton is formed of acanthin arranged in radiating spines, usually twenty in number.

Order 7. **ACTINELIDA**. The spines are more than twenty in number. Families: **Astrolophidæ**; **Litholophidæ**; **Chiastolidæ**.

Order 8. **ACANTHONIDA**. With twenty spines arranged according to Müller's law (four equatorial, eight tropical, and eight polar). Families: **Astrolonchidæ**; **Quadrilonchidæ**; **Amphilonchidæ**.

Order 9. **SPHÆROPHRACTIDA**. With twenty equal quadrangular spines and a complete, fenestrated shell. Families: **Sphærocapsidæ**; **Dorataspidæ**; **Phractopeltidæ**.

Order 10. **PRUNOPHRACTIDA**. With ellipsoidal, flat, or double-coned shell, through which twenty spines radiate according to Müller's law. Families: **Belonaspidæ**; **Hexalaspidæ**; **Diploconidæ**.

Legion 3. **NASSELLARIA** (or **MONOPYLEA**). The skeleton is silicious and rarely absent. The central capsule has a single, limited, perforated area at one pole; the extracapsular plasm has no pigment.

Order 11. **NASSOIDIDA**. Monopylaria without a skeleton. Families: **Nasselidæ**.

Order 12. **PLECTOIDIDA**. A complete latticed shell is never formed, but the skeleton consists of three or more spines radiating from one point below the central capsule, or from a central rod. Families: **Plagonidæ**; **Plectanidæ**.

Order 13 **STEPHOIDIDA**. The skeleton consists of one or two fused rings which may be connected by a loose network. Families: **Stephanidæ**; **Semantidæ**; **Coronidæ**; **Tympanidæ**.

Order 14. **SPYROIDIDA**. The skeleton consists of a single sagittal ring and a latticed shell which is furrowed in the sagittal plane. Families: **Zygospyridæ**; **Tholospyridæ**; **Phormospyridæ**; **Androspyridæ**.

Order 15. **BOTRYOIDIDA**. The skeletons are similar to the preceding, but ornamented by one or more wing-like processes. Families: **Cannobotryidæ**; **Lithobotryidæ**; **Pylobotryidæ**.

Order 16. **CYRTOIDIDA.** Similar to the preceding, but without the sagittal furrow. The skeleton is helmet-shaped. Families: **Tripocalpidæ**; **Phænocalpidæ**; **Cyrtocalpidæ**; **Tripocyrtidæ**; **Anthocyrtidæ**; **Sethocyrtidæ**; **Podocyrtidæ**; **Phormocyrtidæ**; **Theocyrtidæ**; **Podocampidæ**; **Phormocampidæ**; **Lithocampidæ**.

Legion 4. **PHÆODARIA** (or **CANNOPYLEA**). The central capsule has a double membrane, with a spout-like main opening at one pole, and frequently with accessory openings on each side of the main axis at the opposite pole. The central capsule may be multiple in number. There is always a pigmented mass on the outside of the central capsule (the *phæodium*) and covering the main opening. The skeleton, which is rarely absent, is silicious and always outside of the central capsule.

Order 17. **PHÆOCYSTINIDA.** Skeletal structures may or may not be present; the central capsule is the centre of the spherical body. Families: **Phæodinidæ**; **Cannoraphidæ**; **Aulacanthidæ**.

Order 18. **PHÆOSPHERIDA.** The skeleton is a simple or a double latticed covering; the central capsule is in the centre of the shell. Families: **Orosphæridæ**; **Sagosphæridæ**; **Aulosphæridæ**; **Cannosphæridæ**.

Order 19. **PHÆOGROMIDA.** Radiolaria provided with a simple latticed shell, having a mouth opening at one (the main) pole. The central capsule is in the aboral half of the shell. Families: **Challengeridæ**; **Medusettidæ**; **Castanellidæ**; **Cercoporidæ**; **Tuscaroridæ**.

Order 20. **PHÆOCONCHIDA.** The shell consists of two latticed valves, one dorsal, the other ventral (right and left according to Bütschli). Families: **Concharidæ**; **Cælodendridæ**; **Cælographidæ**.

SPECIAL BIBLIOGRAPHY III

Brady, H. B. — Report on the Foraminifera dredged by H. M. S. Challenger: *Challenger Reports, Zool.*, IX., 1882-4.

Brandt, K. — Koloniebildenden Radiolarien (Sphaerozœën) des Golfes von Neapel: *Fauna and Flora*, 1885.

Bütschli, O. — Protozoa; Sarcodina. In Bronn's Klassen und Ordnungen des Thierreichs. *Leipzig*. 1883.

Carpenter, W. B. — Introduction to the Study of the Foraminifera: *Ray Society, London*, 1862.

Haeckel, E. — The Radiolaria: *Challenger Reports. Zool.*, XVII. and XVIII., 1888.

Hertwig, R. — Der Organismus der Radiolarien. *Denkschrift. Jenaisch. Akad.*, II., 1879.

Hertwig und Lesser. — Ueber Rhizopoden und denselben nahe stehende Organismen: *Arch. f. mik. Anat. Bd.* X., Suppl., 1874.

Leidy, Jos. — Fresh-water Rhizopods of North America. *Washington*, 1879.

Pénard, Eug. — Études sur les rhizopodes d'eau douce: *Mém. phys. hist. nat. Genève*, XXXI., 1890.

CHAPTER IV

THE MASTIGOPHORA

THE Mastigophora are provided with a motile apparatus in the form of flagella, which may vary in number from one to many. In the majority of cases, the body is of well-defined and constant form, and covered with a cuticle, membrane, or shell. They abound in infusions, in stagnant pools, in clear water, and in the sea, while many of them are found as parasites in higher animals, where they live in the cavities and cells of the body.

In this class are found many diverse types of unicellular organisms, including, at one extreme, primitive forms whose allies are undoubtedly among the bacteria and the lowest plants (monads), at the other extreme, colonial forms, which in the complexity of their structure and functions are little lower than some of the Metazoa and Metaphyta. It includes forms whose bodies are naked; others that are clothed with complex membranes, or incased in chitinous, silicious, or cellulose shells. It includes organisms with very different methods of food-taking: in some forms the food, like that of the green plants, consists of products made from simple compounds by the organism itself; in others, the food, like that of the fungi, consists of dissolved organic matters; and in still others, the food, as in the higher animals, consists of solid particles of proteid and other matters.

Notwithstanding these many structural and functional differences, there are some well-defined structural characteristics according to which the Mastigophora may be subdivided into a number of more or less homogeneous groups. These groups are the Flagellidia, Dinoflagellidia, and Cystoflagellidia. The first comprises the least homogeneous forms; they consist usually of minute cells with a simple naked body, which may become more or less amœboid, and with one, two, or several flagella. In some cases, there is a complicated cell-membrane, in others a shell, while colony-formation is frequently seen. The Dinoflagellidia are distinguished by the presence of one or two furrows, in which the flagella find their origin, one to pass around the organism transversely, the other to vibrate freely in the surrounding water. The majority are covered by a cellulose shell, consisting frequently of several plates. The Cystoflagellidia, a group consisting of only two genera, *Noctiluca* and *Leptodiscus*, are distinguished by the peculiar parenchymatous structure and by the presence of a tentacle and a collar.

A. Protoplasmic Structure

The alveoli, forming the structural basis of the protoplasm, vary in size from minute and scarcely visible spaces to large vacuoles. In the majority of forms, they are arranged in a typical outer layer (*Rindenschicht*) of small-sized alveoli, surrounding an inner mass of larger ones (*e.g.* *Chilomonas*). The protoplasm is not equally dense in all

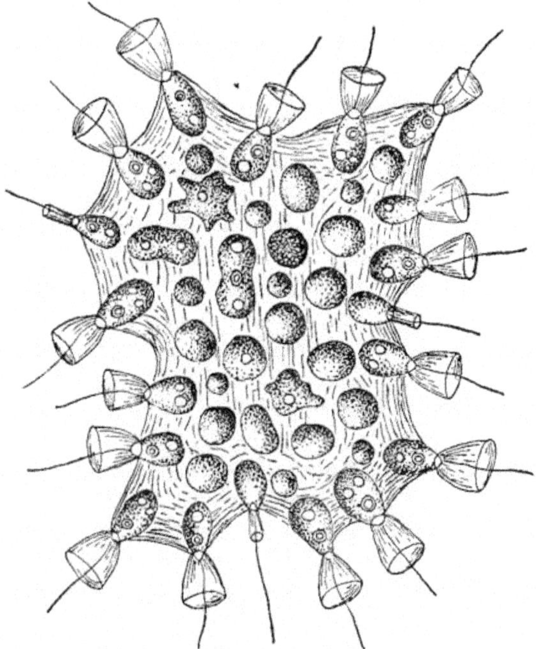

Fig. 59. — *Proterospongia Hæckeli* S. K. [S. Kent.]

cases, but, as in the Rhizopoda, may be of variable consistency. It may be so soft and flexible that, as in *Amœba*, the periphery will give way, and pseudopodia may be formed at any point in response to local changes in the surface tension (Euglenoids and forms of *Astasia*). There is but little tendency to the differentiation into zones, so frequently seen in Rhizopoda, and only rarely is there a differentiation into ectoplasm and endoplasm (*Mastigamœba*).

Klebs ('92) distinguishes two types of peripheral structures, the *periplasts* and outer coats, stalks being included with the latter. The periplasts include all cuticular differentiations which are a living part of the organism, and even diverse modifications of protoplasm, such as the fine peripheral layer of alveoli (*Pellicula* of Bütschli), and the complex membranes of *Euglena* and *Astasia* (cf. Fig. 10, *B*). The outer coatings as in all Protozoa, serving probably for the purpose of protection, include houses and tests of all kinds which are not a living part of the animal. In many cases they are simply jelly-like coverings, which in many colony-forms also serve to keep the individuals together (*Uroglena*, many Choanoflagellida, Fig. 59; see also Fig. 25, p. 56). In other cases, the gelatinous mantle becomes a tube, into which the organism can completely withdraw (some Choanoflagellida). In still other cases, the jelly is apparently hardened into a well-defined goblet or beaker-shaped cup with the consistency of chitin (*Codonœca, Epipyxis, Dinobryon, Salpingœca*, etc.). The relations of the firm case to the gelatinous mantle are shown in forms like *Codonœca*, where the chitin-like urn-shaped cup may become gelatinous (Fig. 60). The organisms are attached to the bottoms of such goblet-shaped cups by a protoplasmic process, and in no case does the cup fit the organism as tightly as a membrane. Colony-forms also are frequent in these types, arising, in the simplest cases, by a young individual attaching itself to the edge of the parent test and there secreting its own covering (*Dinobryon*, Fig. 61). The majority of these colonies are attached, but *Dinobryon* is a free-swimming form, usually found in the clearest waters.

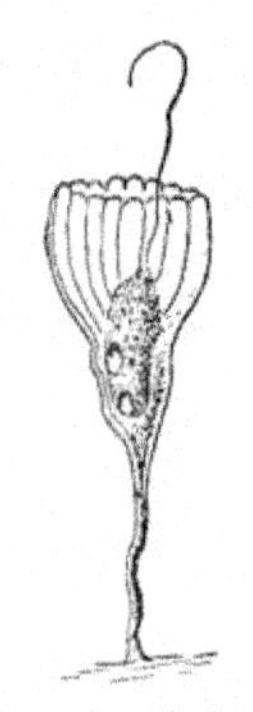

Fig. 60. — *Codonœca costata* J. Clark. [JAMES CLARK.]

Shells are distinguished from tests or houses by the fact that they completely inclose the animal, the so-called mouth-opening where the flagellum is inserted being the only aperture. Both tests and shells are usually transparent and colorless, although they may be colored by the presence of iron, as in *Trachelomonas, Rhipidodendron*, etc., where the shells, when present in any quantity, give a distinctly red color to the water. The simplest shells are the cellulose coverings of many Phytoflagellida, which, although lifeless, have the same general appearance as membranes. The shell, which is frequently protected by sharp spines (*Trachelomonas*), may be separated from the plasm by a considerable space. It is bivalved in *Phacotus*, the two parts being easily separated (Fig. 62). In one form only, *Distephanus speculum* Stöhr, there is a silicious skeleton which recalls the latticed skeletons of Radiolaria (Fig. 63).

The most highly differentiated of these outer coatings are found in the Dinoflagellidia, where the cellulose shells are often composed of separate plates fitted together with the greatest nicety and often

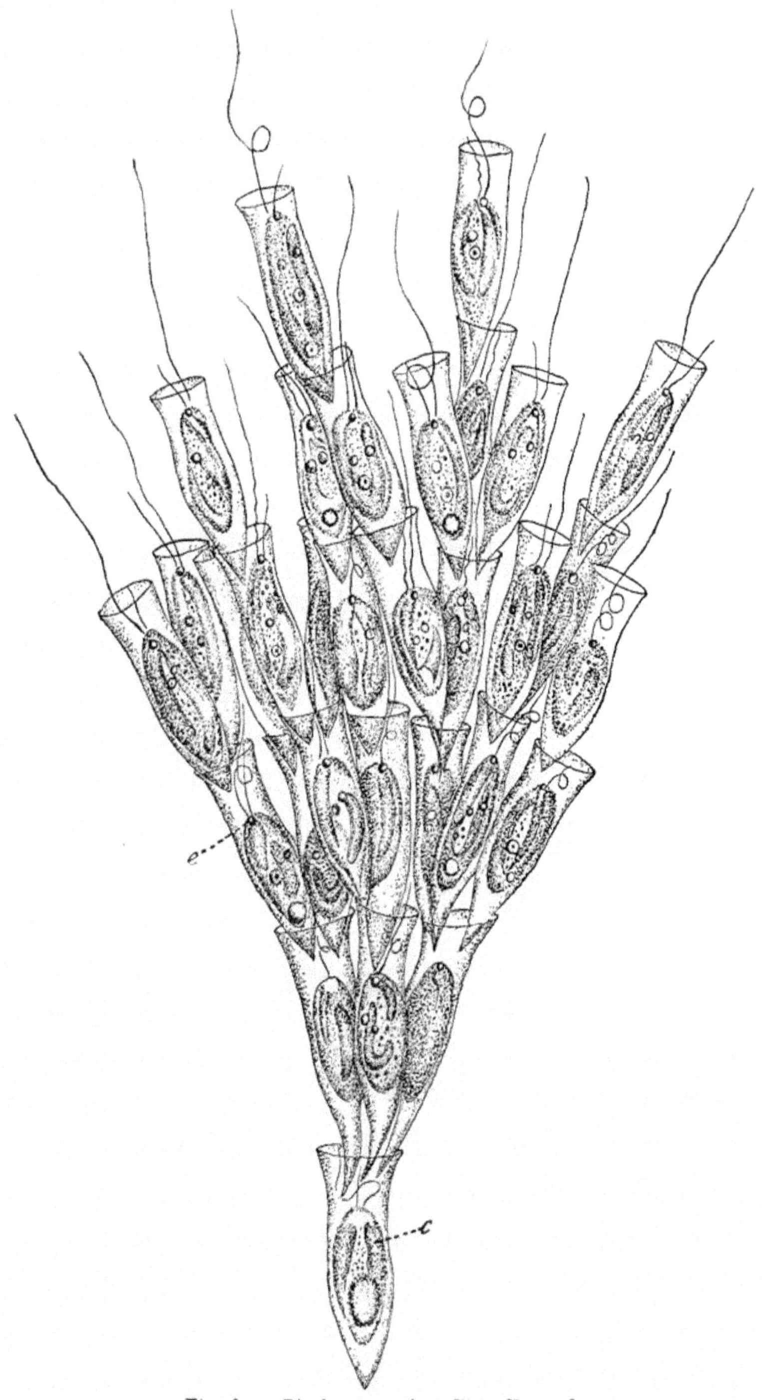

Fig. 61. — *Dinobryon sertularia* Ehr. [STEIN.]
c, chromatophore; *e*, eye-spot or stigma.

complicated by the presence of spines, wing-like processes, and other appendages, or they may be pitted by minute depressions or pores. After the death of the animal the plates can, as a rule, be separated by gentle pressure. The substance of the shell is not true plant cellulose, but a modification, the exact nature of which has not been definitely determined.

The furrows in the shells of the Dinoflagellidia, in which the two flagella lie, are perhaps the most characteristic feature of these forms.

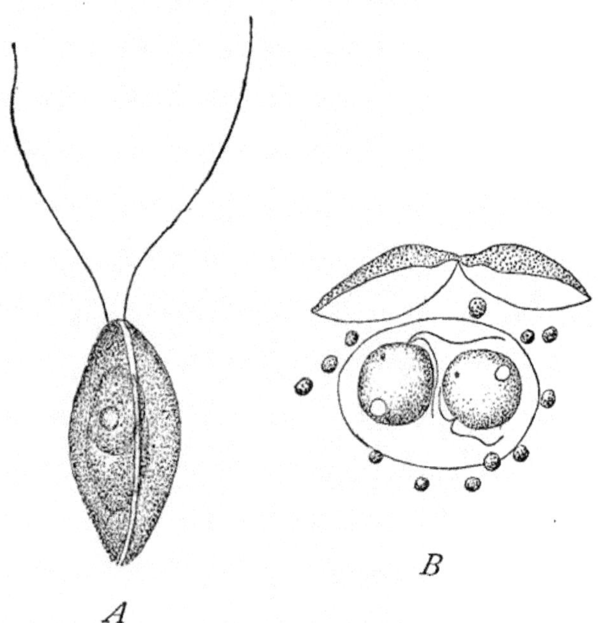

Fig. 62. — *Phacotus lenticularis* Ehr. [BÜTSCHLI.]
A. Individual within its bivalved shell. *B*. Spore-forming individual.

One runs across the organism, while the other, which may often, however, be obliterated, is at right angles to this, usually in the direction of the longitudinal axis. There may be one, two, or many transverse furrows, the number determining the family to which the organism belongs. In the genus *Hemidinium*, the single transverse furrow begins on the ventral side and runs as far as the middle of the dorsal side, where it disappears. In the genera *Gymnodinium*, *Glenodinium*, and *Peridinium*, it runs completely around the organism; while in *Ceratium* it may be broken in its course (Fig. 64). The longitudinal furrow, on the other hand, is invariably confined to the ventral side, usually to the lower half, but in some cases (*Glenodinium*, *Peridinium*) it traverses the cross furrow and stretches some distance along the upper half of the shell. The two flagella which lie in these grooves pass from the body-plasm to the outside through a distinct aperture in the shell, which Stein ('78) called the "mouth-opening"; but as it serves no purpose in food-taking, Bütschli has substituted the better term of *flagellum fissure*.

The protoplasm of the Mastigophora usually contains chromatophores in which one or more deeply staining bodies — the *pyrenoids* — may be found, and these are frequently covered by a shell of amylum or starch. *Paramylum*, a food product allied to starch, and various particles of oil-like substance are widely distributed. The latter are frequently so numerous that the cell is fairly filled with them. Upon diffluence, these oil-like bodies run together, forming globules of large size; or they become finely divided, giving to the surrounding liquid the appearance of an emulsion. Not infrequently the oils have a characteristic odor and taste, comparable to the scent of oils of plants (*Uroglena americana*, *Synura uvella*, Fig. 65).[1]

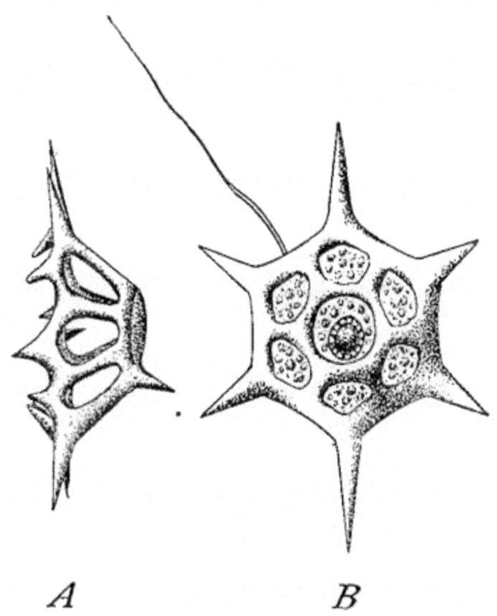

Fig. 63. — *Distephanus speculum* Stöhr. [BORGERT.] *A*. Lateral view of skeleton. *B*. Surface view.

Chromatophores are widely distributed among the various Flagellidia. They consist of clearly defined, thickened bodies, usually of definite size and shape and of different shades of green, yellow, and brown. Clear green chromatophores, colored by chlorophyl as in the plants, occur in Euglenidæ, Peranemidæ, Chlamydomonadidæ, and Volvocina. Yellow chromatophores (colored by *diatomin*, as in Diatomaceæ) occur in Chrysomonadidæ, Cryptomonadidæ, among the Flagellidia, and possibly in some Dinoflagellidia; but the yellow color, when present in the latter group, frequently shades off into brown. Bergh ('81), Klebs ('84), and others regarded the coloring matter of the Dinoflagellidia as pure or slightly mixed diatomin, which supported the popular view that the Diatomaceæ and the Dinoflagellidia are closely related. Schütt ('90), however, who has made the most complete study of the coloring matter in these

[1] Cf. p. 62.

forms, disproved this view by showing that the coloring matter is quite distinct from diatomin and is peculiar to the Dinoflagellidia. He succeeded in extracting three substances: (1) *phycopyrrin*, similar to the brownish red coloring matter of the Florideæ and Phæophycaceæ among the plants, and like this, soluble in clear water; (2) *peridinin*, like chlorophyl soluble in alcohol, but of quite different spectrum; and (3) *chlorophyllin*, a substance more like chlorophyl, but difficult to isolate.

The shape and size of the chromatophores vary considerably in different species, but are fairly constant for the same species. They increase by simple division. The pyrenoids, which seem to be the centre of starch formation, are sometimes quite naked (*Euglena*),

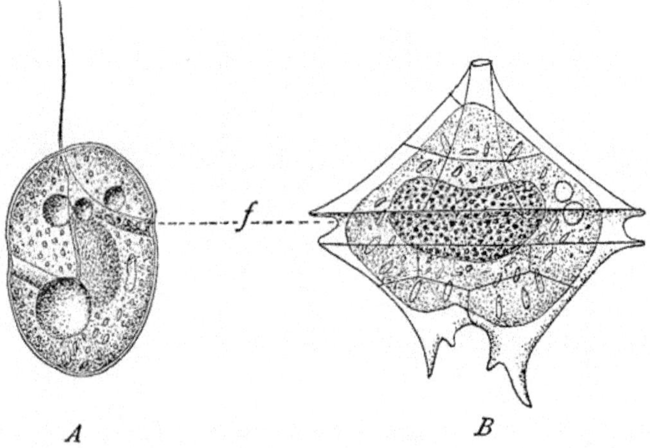

Fig. 64. — *A. Gymnodinium ovum* Schütt. *B. Peridinium divergens* Ehr. *f*, transverse furrow with (*A*) flagellum. [SCHÜTT.]

sometimes covered by a shell of paramylum, which apparently differs from starch only in its reaction to iodine. The paramylum granules are round, rod-like, or ring-form bodies. Pure starch is also recorded as a product of non-colored, saprophytic forms (*Chilomonas*, *Polytoma*, etc.). In many of the Mastigophora, especially in those holding chromatophores, there may be an intense red coloring matter, in the form of fine drops, scattered throughout the protoplasm. These consist of oil particles impregnated with a deep red pigment, — *hæmatochrome*, — and the same substance is found in the so-called "eye-spots," or stigmata, which are supposed to be more sensitive to light than other parts of the protoplasm, although Engelmann's ('82) results show that, in *Euglena* at least, the clear plasm just in front of the stigma is more definitely involved. In many cases the structures accompanying the stigmata are so strikingly analogous to the visual organs of higher forms that there is apparently good reason for supposing them to play a similar physiological rôle. In the green flagellates there are often distinct concretions, regarded by some observers as lenses; and if Pouchet ('86) is correct, a still more striking differentiation is found among the Dinoflagellidia. The so-called eye of *Gymnodinium* consists of a transparent, highly refracting lens, rounded at its free extremity, and always directed forward (Pouchet). The inner surface is embedded in a hemispherical, cap-like mass of red or black pigment, which Pouchet considered a choroid. The lenses develop by the union of from six to eight refringent corpuscles, while the organism is still encysted or while undergoing fission. The choroid likewise results from the union of several of the pigment granules. Considerable doubt has, however, been thrown upon these observations by subsequent observers.[1]

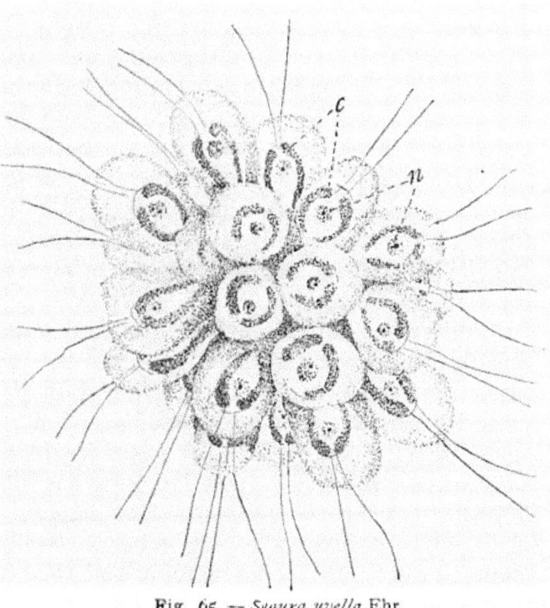

Fig. 65. — *Synura uvella* Ehr.
Each individual of the colony is surrounded by a gelatinous membrane, and possesses two chromatophores (*c*) and a nucleus (*n*).

Other inclusions of interest are the thread-like structures which are common among holotrichous ciliates, and which occur sporadically in other Protozoa. Among the Mastigophora they are found in only two cases (*Gonyostomum* Blochmann and *Polykrikos* Bütschli). In the former they are trichocysts similar to those of the ciliate *Paramæcium* and allied forms, but in the latter they are true nematocysts, comparable to those of the Cœlenterata.

B. The Flagella

The most characteristic part of a flagellate is its motile organ, the flagellum. This consists of a vibratile filament usually tapering to a fine point, although in some cases (*e.g.* in all Choanoflagellida) it is

[1] Cf. Pénard ('88).

of one thickness throughout. In either case the base is inserted in the vicinity of a pharyngeal depression and usually at the end of the body. There is good reason to believe with Klebs, Frenzel, Blochmann, and others, that like the axial filaments of the pseudopodia in Heliozoa and Radiolaria, the flagellum originates at or near the nuclear membrane, and does not consist exclusively of the outer or peripheral plasm. Dallinger ('78) asserts that the newly forming flagella are smooth and uniform, arising in or near the nucleus. Fischer ('94), however, claims that many of them are provided with branches like the cilia of a typhoid germ. He finds others which are not vibratile throughout their entire length, but are rigid and uniform for a certain distance and then taper to the extremity. Such flagella resemble a whip stock and its lash, the relative proportion of stock and lash varying in different flagella, the stock sometimes running nearly to the end, and again only a short distance from the body. Other forms, especially in the Dinoflagellidia, have spirally rolled flagella of various kinds, while some have flattened or band forms (*Peridinium tabulatum* and *P. divergens*).

A difference of opinion exists as to the ability of the organism to absorb or retract its flagellum into the body-protoplasm. Most observers agree with the early observation of Dujardin that there is a close relation between pseudopodia and flagella, numerous observations having been recorded of cases where, under certain conditions, the pseudopodia change into swinging flagella, and flagella into pseudopodia. There is no doubt that flagella can be absorbed after changing to pseudopodia. Whether the fully formed flagella can be changed over into plastic material and then withdrawn, is still a subject of dispute; Fischer ('94) holds that they are invariably discarded upon irritation, and Schütt ('95) shows that the longitudinal flagellum in the Dinoflagellidia is thrown off upon irritation, while the horizontal flagellum is flattened into a band form. A general rule, therefore, cannot be formulated in regard to the disposal of flagella. In some cases they are absorbed; in others, thrown off.

The action of the flagella varies with the type of structure. In the simple, straight, or tapering forms the tip moves in a circle while waves pass from the base to the extremity. In the whip-like flagella the basal portion moves back and forth or in a circle, while the distal region vibrates or undulates like the snapper of a whip. The band-formed flagella move by simple undulations.

The position of the flagella is extremely variable. When there is but one, it is found at the anterior end of the cell, — that is, the end which is directed forward when in motion. When there are two flagella, they may both be directed forward (*Chilomonas*, *Cryptomonas*, etc.), and may be of equal (*Cryptomonas*, etc.) or unequal length

(*Uroglena*, etc.). Again, they may be of equal length but turned in opposite directions (*Bodo*, etc.). When there are numerous flagella, they may be distributed about the body, regularly or irregularly or aggregated at certain points (*Multicilia*, *Tetramitus*, etc.).

In the Dinoflagellidia, the longitudinal flagellum, a long, fine thread, is invariably directed backward or forward, while the other, the transverse, lies around the body in an equatorial groove. This flagellum has a simple undulating motion resembling a row of moving cilia, for which it was at first mistaken.[1] So strictly does the transverse flagellum adhere to the usual direction of motion, that even when the groove is absent, as in *Prorocentrum*, where the flagellum no longer surrounds the body, the motion is retained, the flagellum being directed outward from the end of the body for a short distance and then turned at right angles to form a circle, with the customary undulatory motion, as though still encircling the body (Fig. 66).

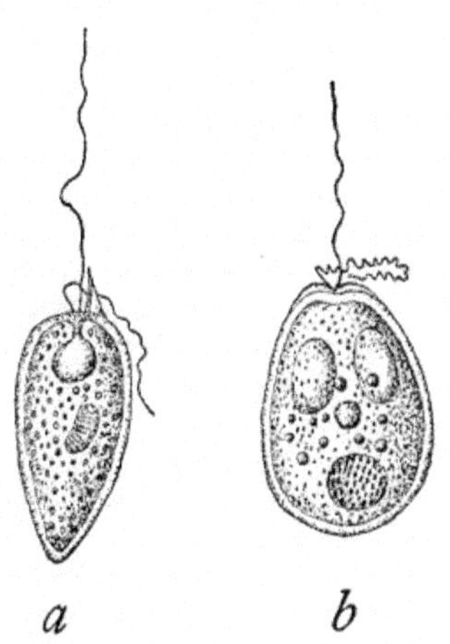

Fig. 66. — Primitive forms of Dinoflagellidia. [BÜTSCHLI.] *a*, *Prorocentrum micans* Ehr. *b*, *Exuviælla lima* Ehr.

With the exception of the Choanoflagellida, which swim like a spermatozoön with the flagellum behind (James-Clark, '66), the Mastigophora swim with the flagellum in advance. The forward movement of most flagellated organisms is, therefore, exceedingly difficult to interpret. It is very curious to see the comparatively large body of *Peranema*, for example, drawn steadily forward by the minute tip of its rather long flagellum. No satisfactory mathematical demonstration of the application of the force necessary to produce this motion has been given. Lankester ('91) compared it to the force produced by a man's arm and hand when swimming upon his side; Bütschli ('83) offered a simple and apparently reasonable explanation, showing that the resistance, which is directed at right angles to the advancing undulation, can be reduced, through the parallelogram of forces, to a force of rotation and one of translation, but Delage ('96) holds that while this explanation is perfectly consistent with the mechanism of certain mechanical contrivances, it is incompatible with the structure of the flagellate body, and that the explanation is much more complicated. Delage's interpretation involves the principles of conic sections, the resisting force being

[1] Hence the name of the group, — *Cilioflagellata*, — which was in use until a comparatively recent date.

applied in a very indirect manner, and he calls the resulting movement "conical translation."[1]

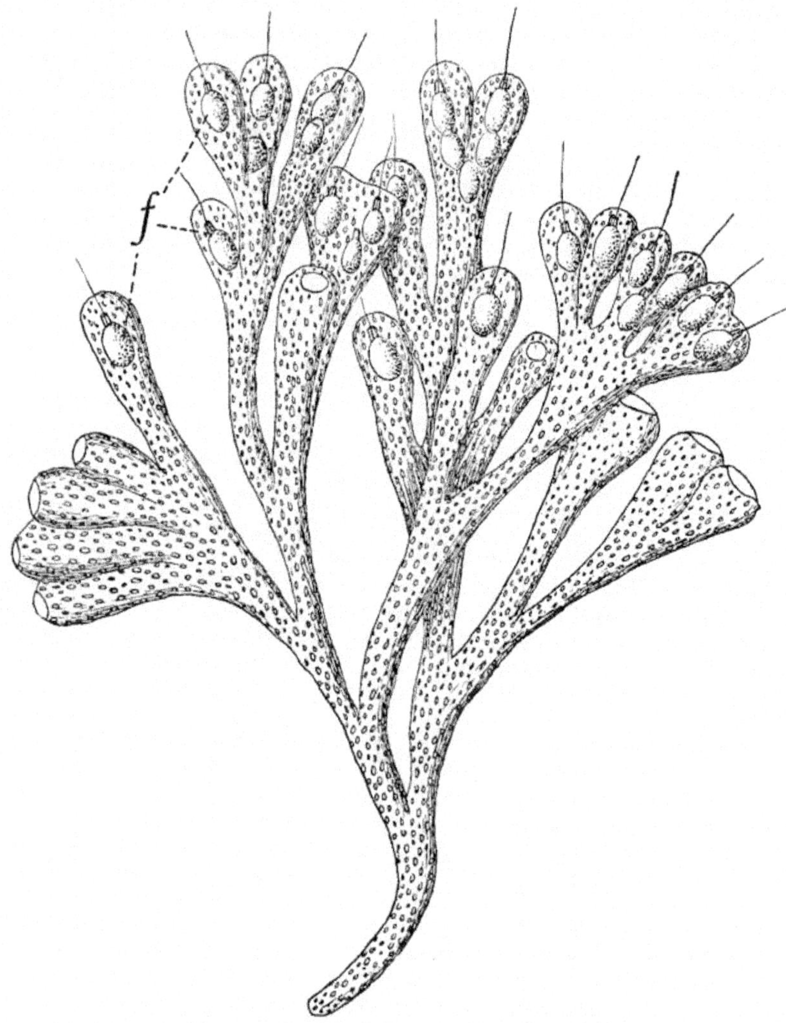

Fig. 67. — *Phalansterium digitatum* St. [S. KENT.]
f, Collared cells.

A peculiar pseudopodial process, the *collar*, is found in one order of the Flagellidia, the Choanoflagellida. This collar, which forms a cup around the base of the flagellum, is extremely thin, delicate, and transparent, and like a pseudopodium can be altered in shape and

[1] Cf. Delage ('96), p. 305, for a very full discussion.

either wholly or partly withdrawn into the protoplasm of the cell. It is occasionally so small and inconsiderable that, as in *Phalansterium*, it can have little or no use (Fig. 67). Again, it may be fully or even twice as long as the body. In shape it is either like a bowl or else like a truncated cone (Fig. 68, *A*, *C*). There may be two of the collars, one within the other (*Diplosiga* Frenzel, *B*). Bütschli described a vacuole which appeared to him to move rapidly around the base of the collar and to disappear for a short time when a food particle

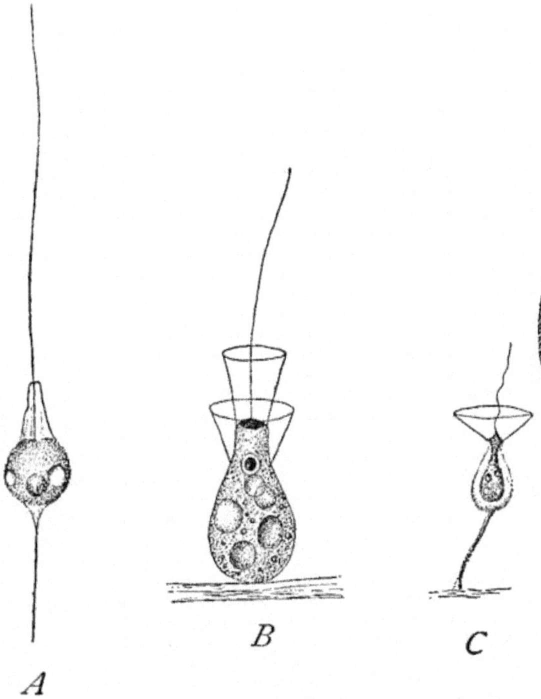

Fig. 68. — Types of collars.
A. Codosiga pulcherrimus Jas. Cl. [J. CLARK.] *B. Diplosiga socialis* Frenz. [FRENZEL.] *C. Salpingœca marinus* Jas. Cl. [J. CLARK.]

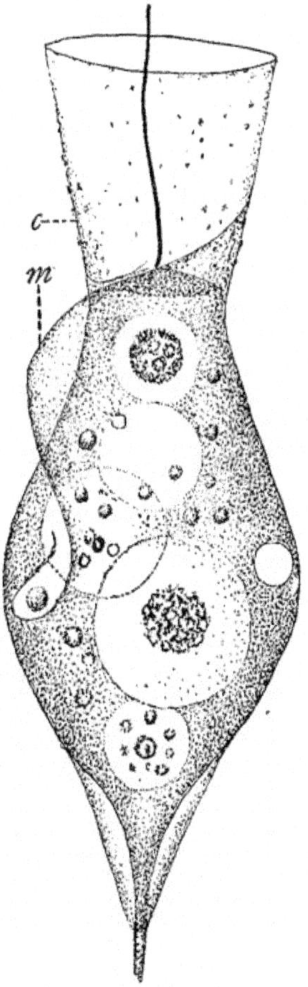

Fig. 69. — A Choanoflagellate type. [FRANCÉ.]
c, collar; *m*, swinging membrane.

is taken in. Entz ('83) and Francé ('94) claim, however, that this "mouth-opening" is not a vacuole at all, but the edge of a swinging membrane. According to their view the collar is not a continuous structure with an unbroken wall, but is like a conical roll of paper with a free edge capable of motion (Fig. 69).

C. The Nucleus

Inclosed in the protoplasm in all Mastigophora is a more or less clearly defined nucleus. It is variable in position, and although a multiple number may occur, is generally single. The structure also

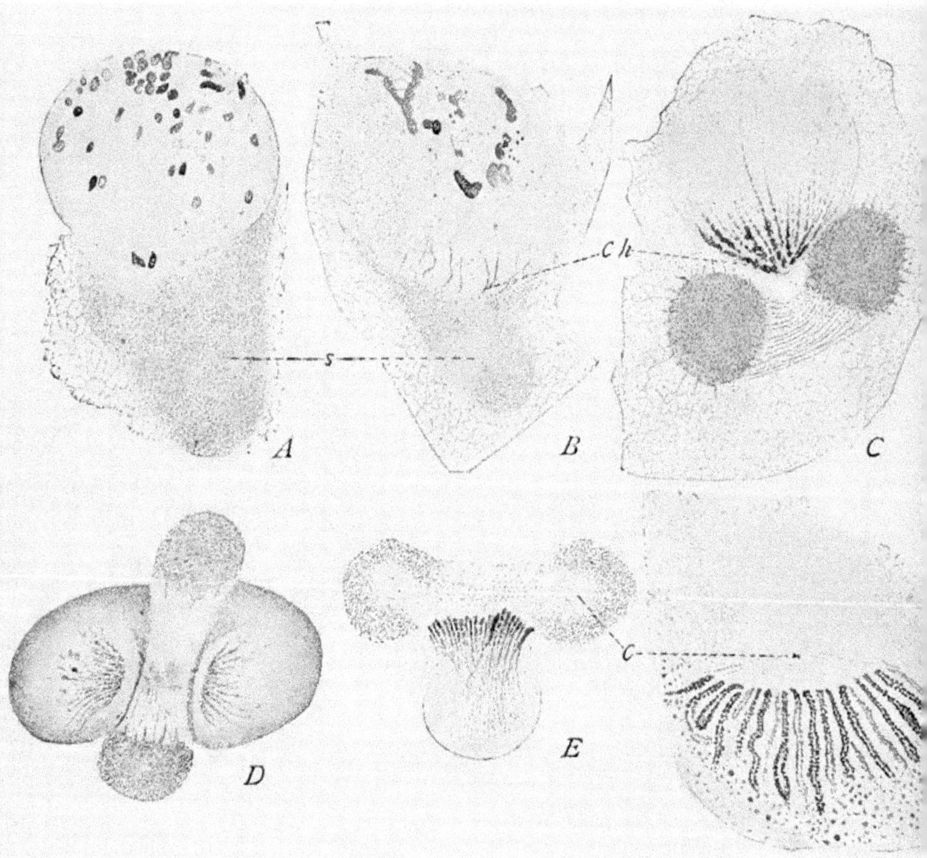

Fig. 70. — Nuclear division in *Noctiluca miliaris*.

A. The first changes of the chromatin from the large karyosome condition; concentration of the substance sion centre (*s*). *B.* Further disintegration of the chromatin and arrangement of granules to form the chromos *C.* Amphiaster and completed chromosomes. *D.* The central spindle in the hollow of the nucleus; the nucle chromosomes is thus wrapped around the division-centre. *E.* A section through the centre of the long axis of th centre before division of the chromosomes. *F.* A section through the dividing chromosomes. *c*, the centre radiating mantle-fibres.

is extremely variable. The chromatin may be either in the form of granules distributed throughout the cell and not confined by a distinct nuclear membrane (*Tetramitus*), or the granules may be aggregated without a membrane (*Chilomonas*), or again, the chromatin may be inclosed by a definite membrane (euglenoids and the majority of Mastigophora; cf. Figs. 14, *C*, *D*, *E*, and 10, *B*). Again, it may be in the form of a homogeneous mass in which no granular structure can be seen (many Phytoflagellida), or, as in many Rhizopoda, it may be massed in several such aggregates (*Noctiluca*). Still another arrangement is seen in the Dinoflagellidia, where the chromatin is arranged in the form of a twisted thread or threads. Finally, in some forms the resting nucleus closely resembles that of the Metazoa in having a linin network in which the chromatin granules are suspended.

An integral part of the nucleus is the so-called "nucleolus," which, however, is not analogous to the nucleolus of the Metazoa, but functions as a sphere or the division centre during mitosis.[1]

Nuclear division in all forms of Mastigophora may be regarded as more or less simplified mitosis, or indirect division. In the simplest types the chromatin masses merely draw out and divide into equal parts, but in the more complicated types, the process closely resembles that in the Metazoa, the complete mitotic figure consisting, as in the higher forms, of chromosomes, mantle-fibres, centrosomes, and spheres (Fig. 70).

D. Food-taking

Closely dependent upon the mode of living is the manner of taking food. Some forms, which live in foul water, are saprophytic like the colorless plants, and absorb, through the body walls, the substances which are dissolved out of decaying vegetable matter. Those which live in pure and clear water generally have chromatophores, colored by chlorophyl, diatomin, or some allied substances, and have the power of manufacturing their food from carbon dioxide, water, and salts; like the green plants, their nutrition is holophytic. Parasitic forms live upon the juices of other living organisms, which are absorbed through any part of the body wall (Fig. 71). Finally, some take in solid food, which is acted upon and digested by the fluids of their inner plasm, the indigestible portions being excreted as in the higher animals.

In the holophytic forms there is frequently an unbroken shell about the animal which makes it impossible for solid food to enter (*Hæmatococcus;* many Dinoflagellidia). Many of the holozoic forms have a distinct mouth and œsophagus. In its simplest form the mouth is merely a softened area about the base of the flagellum, against which the solid food particles strike (*Oikomonas*, see Fig. 18, *B*). Others have a distinct mouth-opening leading into a gullet, which in turn opens into the fluid endoplasm (*Peranema; Petalomonas*, see Fig. 1, *B*).

[1] See *infra*, p. 258.

Where the flagella are in separate groups, there is a mouth at the base of each group. The collar of the Choanoflagellida is especially adapted for the collection and direction of the food particles into the interior.

Perty ('52), Kent ('81), Stein ('67), Entz ('88), and others have described a number of types which they claim have both kinds of nutrition, and are intermediate between the holozoic and holophytic forms; but Bütschli, although he admits that food-taking may be either holozoic or holophytic in one form at least (*Chromulina flavicans*), which lives equally well when one or the other mode of nutrition is prevented, is inclined to doubt the wide distribution of this double function. Meyer ('97) maintains that in one form (*Ochromonas granulosa*) the organism may be either holozoic, saprophytic, or holophytic in nutrition. In many of the holophytic forms, there is a distinct gullet, which Perty and Kent regarded as a food-taking organ and, therefore, evidence of holozoic nutrition. Bütschli, however, maintained that it is a part of the excretory system and connected with the contractile vacuole (*Euglena*, *Cryptomonas*, etc.).

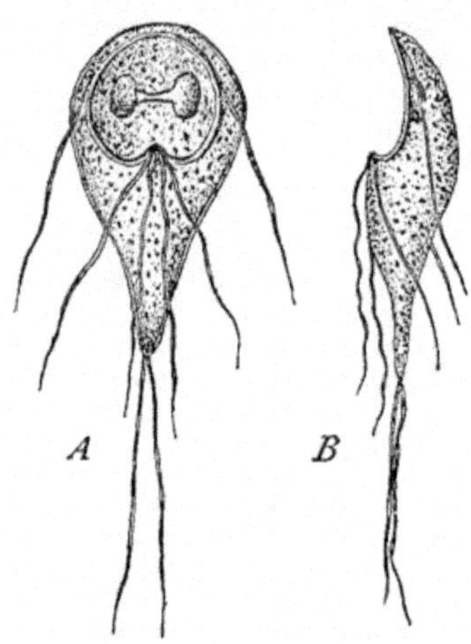

Fig. 71. *Megastoma entericum* Grassi. Ventral and side views. [GRASSI.]

A close connection between holozoic and holophytic forms is found, not only in Flagellidia, but in Dinoflagellidia as well. Possessing chromatophores and a coating of modified cellulose, these organisms were for a long time regarded as plants, but some forms among them are known to move about like animals and to ingulf solid food. Such forms may be either naked, as in *Gymnodinium* (Fig. 64, *A*) and *Polykrikos*, where food-taking has been actually seen by a number of observers (Schmarda, Stein, Bergh, Schilling, Dangeard), or shell-bearing, as in *Glenodinium edax* (Schilling). It is probable that they are much more closely related to the animal Flagellidia than to the Diatomaceæ (as Warming maintains), or other plants, although no hard and fast line can be drawn about any of these groups.

The food of the holozoic flagellates consists of bacteria and minute bits of disintegrated proteid matter. These in the Rhizomastigidæ, as in the Rhizopoda, are surrounded by pseudopodia, and are subsequently drawn into the body. In other Mastigophora, the flagellum is the chief factor in alimentation, and by its vibrations a current is created toward the base, where the mouth or its equivalent is situated. The particles of food brought with the current find their way into the body-plasm, where an indefinite cyclosis carries them hither and thither until the digestible portions are separated from the indigestible, and the latter are finally thrown out. James-Clark ('66), Kent ('81), and most observers have maintained that, in the Choanoflagellida, the food particles strike against the collar, subsequently working down on the inside to the mouth, but Entz ('83) and Francé ('93, '97) claim that the mouth is not within the collar, but that the so-called vacuole described by Bütschli ('84) is a soft ingesting area at the base of the overlapping edge of the collar (see Fig. 69). In *Noctiluca*, the flagellum brings a current of food toward the collar, while the tentacle, which constantly beats down into the bottom of the collar area, drives it into the mouth situated at the bottom of the pharyngeal groove. The particles are then received into a gastric vacuole, which, in the vicinity of the relatively large nucleus, performs its function of digestion.

E. Vacuoles

Some of the vacuoles which make up the protoplasm of the Mastigophora are gastric, while others are contractile. The former are formed about the food particles, which are probably digested in the same way as in the Sarcodina, although in this group no experiments have been made to test the digestive fluids.

The contractile vacuoles, acting possibly as respiratory and excretory organs, pulsate rhythmically and at definite rates, varying from one or two pulsations per hour to five or six a minute, according to the temperature and nature of the surrounding medium. They are typically small, single or double in number (multiple in *Chlorogonium*), and are situated at either end of the body, or near the centre, while in some cases they move with the granules in cyclosis. In some of the more complicated types of Euglenidæ, the vacuole is connected with the so-called gullet by a minute canal. This canal in some cases receives its supply of waste matter from a reservoir, which is the receptacle for the contents of numerous small vacuoles surrounding it, and which pulsate at regular intervals.

F. Reproduction

Binary fission, the typical method of reproduction among the Mastigophora, and the simplest of all modes of increase, is invariably preceded by division of the nucleus. When chromatophores, eyespots, and pyrenoids are present, they also may be halved and equally represented in the daughter-cells, or they may remain whole, going to that daughter-cell to which they are nearest, the other cell forming a new set. The flagellum, in some cases, is also divided throughout the entire length, although in other cases it is thrown off before division takes place, new ones being formed by the daughter-cells. In many cases new flagella, as well as all of the important structures, are pre-formed before division. Such divisions may take place while the organisms are moving freely about in the water, or

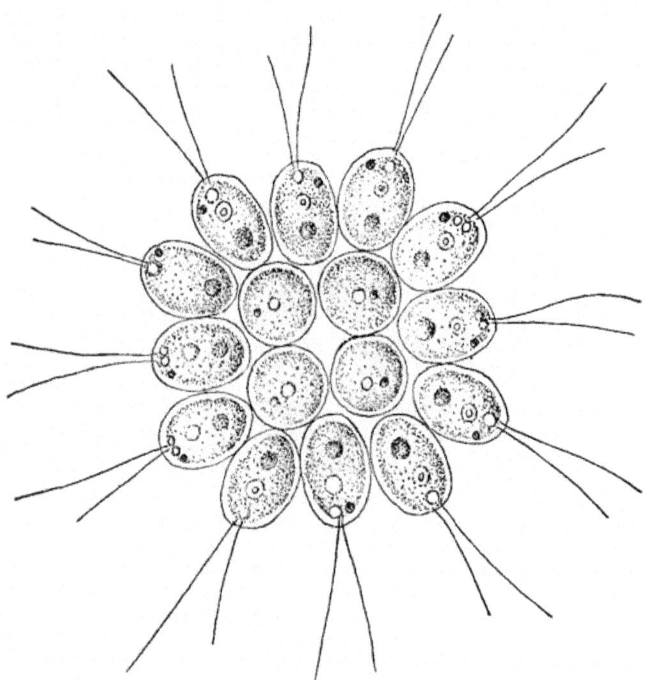

Fig. 72. — *Gonium pectorale* O. F. M. [STEIN.]

while they are quiescent and inclosed in a firm cyst. As a rule, division is longitudinal, but cases are well known where it is transverse (in most of the Dinoflagellidia, in *Epipyxis*, *Stylochrysalis*, *Oxyrrhis*, etc.).

Some forms (*e.g.* *Trachelomonas*) reproduce by simple division while still within the shell, one half making its way out through the neck or flagella-opening, leaving the other in possession of the original home.

Colony-formation is closely connected with the process of simple division, and nowhere among the Protozoa does it reach such high grades of differentiation as in this class. The colonies of this group are composed, for the most part, of descendants of one ancestor and are formed by incomplete division or by subsequent attachment. In these colonies, which are wonderfully varied, the individual monads in some cases are embedded in a transparent cellulose jelly secreted by the cells, and in which they lie freely, or attached to one another

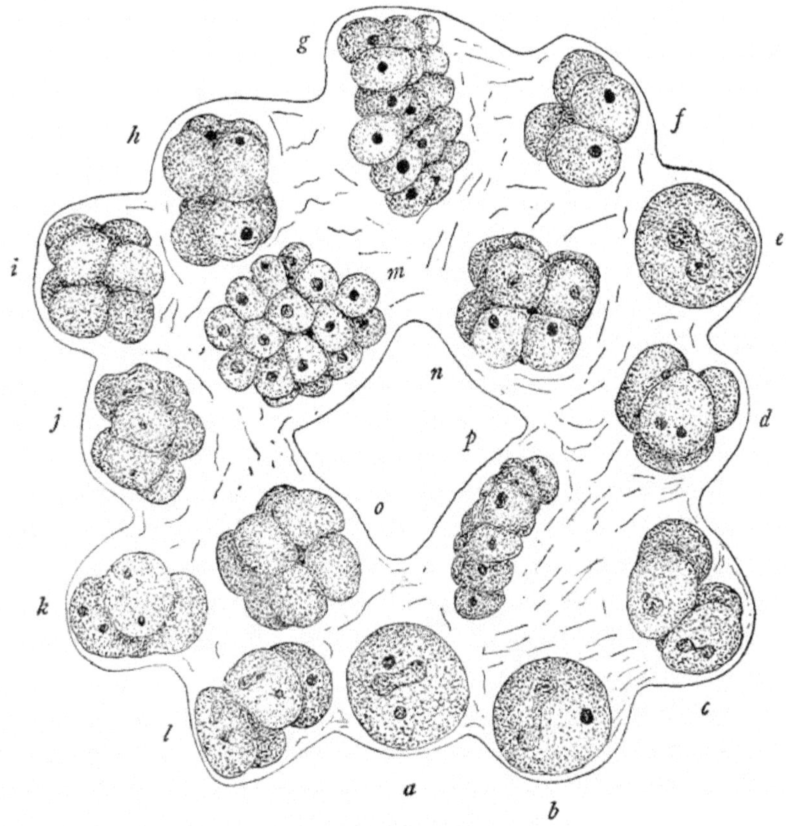

Fig. 73. — Division of *Gonium pectorale* O. F. M.
a, *b*, and *e*, undivided cells; *c*, *d*, *f*, *k*, and *l*, 4-celled stages; *h*, *i*, *j*, *n*, *o*, 8-celled stages; *g*, *m*, and *p*, 12–16-celled stages.

by stalks, while in other cases there is no surrounding matrix, the individuals remaining connected through incomplete division or by attachment subsequent to division. *Gonium* (Fig. 72), *Pandorina*, *Uroglena*, *Proterospongia*, *Volvox*, etc., have the cellulose jelly, while *Dinobryon*, *Anthophysa*, etc., are formed by attachment subsequent to division. In some of the more complicated colony-forms, especially in the Phytoflagellida, the adult condition is attained through cleavage stages as regular as in any metazoan egg. The formation of such a colony never varies, and the number of individuals is constant. (In *Gonium sociale* there are 4 individuals; in *G. pectorale* 16, while in *Eudorina* there are from 16 to 32, and in *Pandorina* 32.) A *Gonium* colony lies in one plane, but this arrangement is brought about by a secondary shifting of the cells (Fig. 73), while a *Eudorina* colony retains the spherical form. *Pandorina*, a similar compact and definite colony, is derived as in *Eudorina*, by the regular cleavage of a single cell.

In some colonies the individuals are connected in the centre by protoplasmic strands, as in *Synura*, while in one genus (*Uroglena*) connecting strands may or may not be present. Ehrenberg ('38) described *U. volvox* as a colony form whose peripheral individuals are connected in the centre by tail-like processes which, except for a much greater length, are similar to those of *Synura*. He was confirmed in this by Zacharias ('95) and Kent ('81). Bütschli, however, regarded this central attachment as extremely doubtful. In one form, *U. volvox*, this connection does actually exist, but in another, *U. americana*, the posterior ends of the cells are rounded and have no trace of a central filament. The genus *Uroglena* may afford, therefore, a clue to the phylogenetic relations of the relatively huge gelatinous colonies which, save for the surrounding matrix, have no means of connection.[1] *Proterospongia*, in its general form and structure, agrees with *U. americana*. In both cases, as far as known, there is an indefinite number of individuals and no typical method of increase, as in *Pandorina*, *Eudorina*, and *Gonium*.

The most highly differentiated colonial forms are the genera *Volvox* and *Magosphæra*, which should perhaps be considered simple multicellular forms rather than Protozoa.

In *Volvox* the monads (often as many as 12,000 in a single colony) are arranged as in *Uroglena*, around the periphery of a gelatinous mass, and no organized connections with the centre of the cell can be traced, although they are connected with one another by definite protoplasmic strands.

In *Magosphæra* the individuals are connected not only by the jelly matrix, but also, as in *Synura*, by protoplasmic stalks, and they are in close contact at the periphery. In both *Magosphæra* and *Volvox* the appearance of the peripheral cells is strikingly similar to a pavement epithelium, and the comparison which is so often made between such colonies and the blastula stage in the development of Metazoa is certainly justifiable.

Stalked colonies have an entirely different mode of origin, being formed by repeated longitudinal division, the daughter-cells remain-

[1] Cf. Calkins ('91).

ing attached to the stalk by their basal ends. In *Dinobryon*, a free-swimming colony of variable size, each monad occupies a small cup of cellulose (see Fig. 61). They increase by simple longitudinal division, one daughter-cell remaining in the original house, while the other moves out to the edge of the parent cup, where it attaches itself by the posterior end. A cellulose cup is then secreted about the daughter-cell, remaining firmly attached, however, to the parent cup. The mother-cell may divide again and again, the daughter-cells attaching themselves to the edge of the cup already formed until there are three or more individuals around the edge of the original one. At the same time the daughter-cells may be dividing in a similar manner, and a much-branched bush-like colony is the result. Other forms have stalks which in some cases are much longer than the individuals themselves (*Codonocladium*, *Dendromonas*, *Codonosiga*, etc.).

Still another type of colony-formation is found among the Dinoflagellidia, where from two to eight individuals are connected end to end by their shell processes (*Ceratium*). The significance of this chain-formation (catenation) is not clearly established, many regarding it as the result of incomplete division (Pouchet, '85), others as preparatory to conjugation (Bütschli, '83).

The colonies as well as the individuals may increase by division, a purely mechanical process, however, and probably due to the unwieldy size of the overgrown aggregate. Zacharias ('95) and others have seen large colonies of *Uroglena* break into two portions through the asynchronous action of flagella in different regions. If the flagella of one half of the colony vibrate in one direction while those of the other half vibrate in an opposite direction, the result is a twisting of the entire mass which must ultimately give way. Such division cannot be regarded as reproduction in a strict sense.

Closely allied to simple division is the formation of swarm-spores or microgonidia. This may occur either in the free motile condition as in *Polytoma* or *Chlorogonium*, or in the encysted and protected state, as in many Monadida. The simplest form is seen in such cases as *Polytoma*, where, instead of dividing into two portions, the organism divides into four, eight, or, according to Dallinger and Drysdale, into sixteen smaller forms. These develop new flagella, make their way through the parent membrane, and grow to full size. The formation of similar gametes has been observed in most of the Mastigophora, either in their resting or in their encysted stages. The flagella are drawn in, a mantle or cyst is secreted, within which the protoplasm divides into a number of spores. In some cases swarm-spores, like those of the Radiolaria, are of different sizes (macro- and micro-gametes), and these may conjugate.

So far as known, the formation of gametes is not accompanied by complex nuclear changes. As in the Reticulariida and some Sporozoa, the chromatin is reduced to minute granules which are spread throughout the cell, but in the Flagellidia they are so small that their further history is not known. Not all forms, however, are of this primitive type; some, as for example *Noctiluca miliaris*, and some of the Dinoflagellidia, undergo a complicated mitotic process which in *Noctiluca* is repeated until five or six hundred spores are formed (Fig. 74).

The formation of spores or gametes may or may not be preceded by the conjugation of individuals. In those species of Mastigophora in which spore-formation is preceded by conjugation, a very interesting series of forms may be selected, showing the gradual development of sex from types in which there is a union of individuals of similar form, size, and, apparently, of condition, to the union of specially developed male and female reproductive elements. Cienkowsky ('56) was the first to observe the fusion of similar monads, but the most complete observations are those of Dallinger and Drysdale ('73), who watched the fusion of several individuals of *Bodo* (*Cercomonas*) *crassicauda*, the encystment of the fused mass, and the subsequent divisions of the plasm up to the formation of an immense number of minute spores (Fig. 75). In another form (*Oikomonas Dallingeri*) similar spores are formed, but without the preliminary fusion of two or more small individuals. The gametes move about until they come in contact with the adult individuals with which they fuse. The fused mass then encysts and finally breaks up into minute spores.

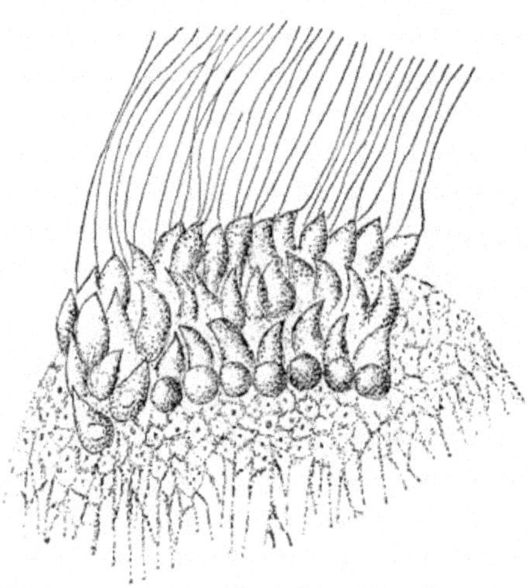

Fig. 74. — *Noctiluca miliaris* Sur. Spore-formation. [ROBIN.]

An advance toward sexual differentiation is seen in *Pandorina* (Pringsheim, '69), where, after a long period of asexual reproduction resulting in numerous colonies, the cells separate and begin to form swarm-spores which may be of the same or of different size. These spores then swim about until two of them meet and fuse by the colorless ends into a common body (Fig. 76). Fusion may take place between two small gametes, or between a large and a small one.

Pringsheim regards the larger ones as females, while the smaller ones may be either male or female.

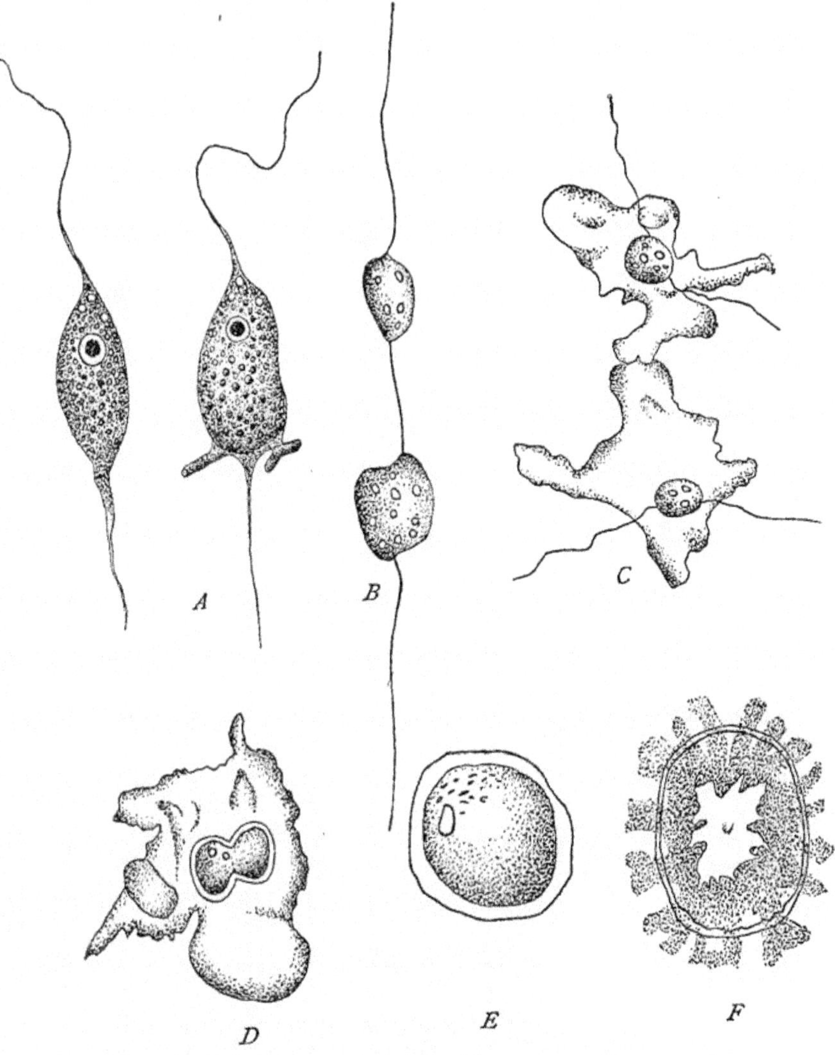

Fig. 75. — *Cercomonas crassicauda* Duj. [DALLINGER and DRYSDALE.]
A. Ordinary forms. *B.* Division stage. *C.* Conjugation of two individuals in amœboid condition. *D–E.* The copula. *F.* Sporulation.

The fused mass (zygospore) encysts and dries, the color changing from green to red. When remoistened, the contents again turn green and break open the cyst, usually as a single swarm-spore, although two or more may be formed. These gametes soon divide and form the typical sixteen-cell *Pandorina* colony. Thus in *Pandorina*, each of the cells forms both sexual elements, but an advance in differentiation is seen in *Eudorina elegans*, where, according to the rather incomplete observations of Carter ('58), the thirty-two cells forming the colony have a different fate when the conjugation period comes around. Four of the thirty-two cells situated at the end of the colony form gametes by repeated divisions in one plane, while the other twenty-eight cells merely develop more amylum granules and turn darker. The gametes which are formed from the upper four cells are elongate and spindle-shaped, with two flagella, a red eye-spot, and a long tail. The fate of the different cells was not made out, but there seems reason to believe that if the observations were correct, the gametes represent male elements, the other cells female.

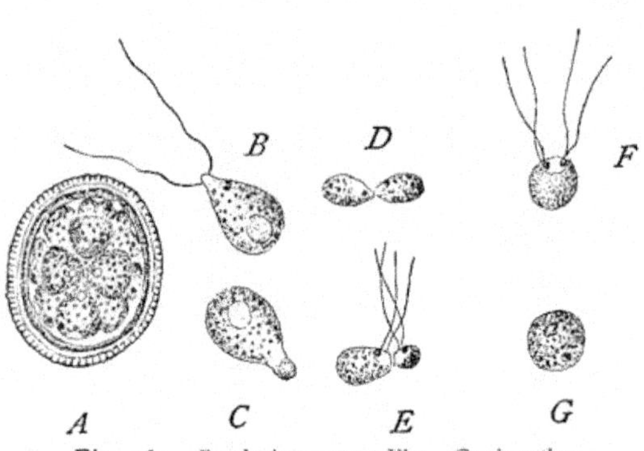

Fig. 76. — *Pandorina morum* Ehr. Conjugation. [PRINGSHEIM.]

A. 16-celled colony. *B.* Macrogamete. *C* and *E.* Fusion of macrogamete with microgamete. *D* and *F.* Fusion of microgametes. *G.* Copula.

A still more decided advance is shown by the colonies of *Volvox*. *Volvox* can scarcely be regarded as a unicellular organism, for differentiation has gone so far that the cells if separated, with the exception of the reproductive elements, cannot live. The individuals forming the peripheral layer (in *Volvox globator* about 12,000, Cohn, '75) form the sterile vegetative or somatic cells of the aggregate. A few of these cells, which reproduce asexually, are found upon the inside of the peripheral layer, which protects them like a mantle. Stein finds eight of these asexual cells or *parthenogonidia* in *V. globator*, and Cohn, one to nine in *V. minor*.

The parthenogonidia, by repeated division, form daughter-colonies from one-quarter to two-fifths the size of the parent colony, which finally make their escape from the latter by rupture of its walls. After a considerable period of such asexual reproduction, sexual elements are formed. These are at first similar to the parthenogonidial cells, but are more numerous. Later they can be distinguished as male (*androgonidia*, Cohn) and female (*gynogonidia*, Cohn). In *Volvox globator* there may be from twenty to forty *gynogonidia* and from two to five *androgonidia*, but there may be one hundred androgonidia in *V. minor* while there are only eight *gynogonidia*. Each female cell becomes at first flask-shaped, then withdraws inside of the colony and becomes a mature egg. The male cell, at the beginning about three times the size of the sterile cells, soon begins to divide in one plane until a bundle of from sixty-four to one hundred and twenty-eight flagellated, spindle-formed elements results. These, the spermatozoids, gather around the egg-cells, which are fertilized exactly as in higher animals or plants. Fertilization takes place while the egg is still in the parent colony; the copula forms two membranes about itself, while the color changes from green to orange. After a considerable resting period, the egg undergoes regular cleavage, forming the adult colony, in which, even before the embryo leaves the egg, the cells are differentiated into somatic cells and parthenogonidia. After these cleavage stages the outer cyst wall is ruptured and the young colony swims out.

G. Inter-relations of the Mastigophora

Transitional forms between the Mastigophora and the Sarcodina show how closely the two classes are related. The development, both in Sarcodina and in Mastigophora, throws little or no light upon the question as to the more primitive nature of one or the other. While many Sarcodina have flagellated swarm-spores, many Flagellidia have amœboid spores, and even in the same species both amœboid and flagellated swarm-spores are formed at the same time (*Acanthocystis* Schaudinn, see Fig. 53, *F*).

The most primitive forms of Flagellidia suggest the long-disputed question over the boundary-line between animals and plants.[1] Unquestionably, the most primitive flagellates are those forms which, while actively motile, possess chromatophores and chlorophyl, and are able to make their own food, or which, like the bacteria, can support themselves upon simpler substances than proteid. The living flagellates which come closest to these primitive types are the monads, while the Choanoflagellida are probably not far removed. Klebs ('93) takes the ground that the collar-like process of the Choanoflagellida is not a sufficient taxonomic differential save for ordinal distinctions, and recent zoölogists are inclined to accept this view.

Stein, Bütschli, and Bergh have been the most active in formulating views as to the origin of the Dinoflagellidia, although none of these is wholly satisfactory. Some authors derive them from the Phytoflagellida directly (Haeckel), others place them as a group of the

[1] Cf. p. 22.

Diatomaceæ (Warming). The majority of observers are agreed, however, that a connecting link with the Flagellidia is seen in certain species of the family Prorocentridæ and represented among living forms by *Exuviælla* and *Prorocentrum* (Fig. 66). Bütschli ('83) and Klebs ('93) agree that these might be called true Flagellidia; for, as Delage happily expresses it, they are little more than a chromomonad in the shell of *Phacotus*. The shell is bivalve, and perforated by minute apertures characteristic of the Dinoflagellidia, and there is an entire absence of longitudinal and transverse furrows, while the flagella are directed outward from the anterior end. Bütschli, Bergh, and Klebs derive them from forms like *Cryptomonas*, where the two flagella are pointed in the same direction and the chromatophores are yellow. The transition from these primitive forms of Dinoflagellidia to the more complex types with a shell composed of nicely articulated plates, is much more difficult than the connection between the main groups. Stein maintained that the simplest form is the unshelled *Gymnodinium*, but Bütschli showed that this view is not in harmony with the other forms of Dinoflagellidia, it being much more obvious to consider the shell of the Peridinidæ, for example, as arising by the splitting up of the bivalve shell of the primitive type, than by the loss of this shell and the subsequent formation of the articulated forms. Evidence in Bütschli's favor is seen in the forms where there are but few plates (*e.g. Ceratocorys*), although we are inclined to agree with Klebs that a polyphyletic origin of the group is possible and that *Gymnodinium* might have been derived from the Rhizomastigidæ.

The origin of the Cystoflagellidia, composed of *Noctiluca* and *Leptodiscus*, is to-day generally conceded to be from the Dinoflagellidia, and is supported by direct evidence in the development of *Noctiluca*, where the swarm-spores are strikingly similar to *Peridinium*. The relationship to the Dinoflagellidia, as first pointed out by Allman, was based upon superficial resemblances only, and the first conclusive observations must be credited to Pouchet ('83) and to Stein ('83), while Bütschli ('85) first applied the theory on the basis of the swarm-spore as described by Cienkowsky ('73) and Robin ('78). The interesting form which Pouchet later described ('92) as *Peridinium pseudonoctiluca* is now considered a young stage of *Noctiluca*.

CLASSIFICATION

CLASS II. **MASTIGOPHORA**. Protozoa of definite or indefinite form; naked, or provided with a well-defined membrane. The nutrition is holozoic, parasitic, holophytic, or saprophytic. The motile organs are flagella, which may vary in number from one to many. Mouth, contractile vacuole, and nucleus are usually present. They are usually small forms with a widespread tendency to colony-formation.

Subclass I. **FLAGELLIDIA**. These are small organisms possessing usually a sharply defined, mononucleate body with a definite anterior end in which are inserted one or more flagella. They are actively motile during the greater period of life, but all have the power of encystment. Reproduction occurs by longitudinal division, usually during the flagellated stage, although it may take place during resting phases. Nutrition is holophytic, holozoic, parasitic, or saprophytic.

Order 1. **MONADIDA**. Small forms of Flagellidia having a simple structure. The body is frequently amœboid, with one or two flagella at the anterior end. There is no distinct mouth-opening, but a localized area about the base of the flagella serves for the ingestion of food particles.

Family 1. **Rhizomastigidæ**. Simple, mouthless forms with one or two flagella and an amœboid body capable of putting out lobose pseudopodia like a rhizopod, or stiff radial pseudopodia like a heliozoön. The contractile vacuole is frequently at the posterior end. Food particles may be ingested at any part of the body by the aid of the pseudopodia. Genera: *Mastigamœba* F. E. Schultze ('75); *Ciliophrys* Cienk. ('76); *Dimorpha* Gruber ('81); *Actinomonas* Kent ('80); *Trypanosoma* Gruby ('43); *Mastigophrys* Frenzel ('91).

Family 2. **Cercomonadidæ**. Oval or elongated forms which are frequently amœboid or changeable, but unable to form pseudopodia. There is one large flagellum with a mouth area at its base. The family includes small forms, saprophytic, or holozoic, or sometimes parasitic in nutrition. Genera: *Cercomonas* Dujardin ('41); *Herpetomonas* Kent ('80), parasitic. *Oikomonas* Kent ('80); *Ancyromonas* Kent ('80); *Phyllomonas* Klebs ('93).

Family 3. **Codonœcidæ**. Small colorless monads which secrete and remain in a gelatinous or membranous cup. Genera: *Codonœca* James-Clark ('66); *Platytheca* Stein ('78).

Family 4. **Bikœcidæ**. Small monads of peculiar form. They are provided with a cup, to which they are attached by a slender thread. The basal portion is broader than the upper part, which bears a curious tentacle-like process. Nutrition is holozoic; the individuals are single or colony-forming. Genera: *Bicosœca* James-Clark ('67); *Poteriodendron* Stein ('78).

Family 5. **Heteromonadidæ**. Small colorless monads which have, in addition to the chief flagellum, one or two accessory flagella. They frequently form colonies upon a common stalk. Increase of the individuals is by longitudinal division. Genera: *Monas* Stein ('78); *Dendromonas* Stein ('78); *Cephalothamnium* Stein ('78); *Anthophysa* Bory d. St. Vincent ('24); *Epipyxis* Ehr. ('38); *Amphimonas* Kent ('81); *Spongomonas* Stein ('78); *Cladomonas* Stein ('78); *Rhipidodendron* Stein ('78); *Diplomita* Kent ('80).

Order 2. **CHOANOFLAGELLIDA**. Flagellidia with one or more collar-like processes about the base of the single flagellum.

Family 1. **Phalansteridæ**. Colony-forming Choanoflagellida. Each individual is situated in a granular gelatinous tube. The gelatinous tubes form either a discoid colony in which the single tubes are arranged radially, or a dichotomously branched aggregate. Genera: *Phalansterium* Cienk. ('70).

Family 2. **Craspedomonadidæ**. Solitary or colonial forms. The individuals are naked, or lie in an incomplete cup, or in a gelatinous mass. Genera: *Monosiga* Kent ('80); *Codosiga*, James-Clark ('67); *Codonocladium* Stein ('78); *Hirmidium* Perty ('52); *Proterospongia* Kent ('81); *Sphærœca* Lauterb. ('99); *Salpingœca* James-Clark ('67); *Polyœca* Kent ('81); *Diplosiga* Frenzel ('91).

Order 3. **HETEROMASTIGIDA**. A small group with various kinds of flagellated organisms, which are sometimes naked and amœboid, sometimes provided with a complex membrane. The essential character is the possession of two or more flagella, one being directed forward and used in locomotion, the others directed backward and trailed after the organism. Nutrition is holozoic, and all of the forms included are colorless.

Family 1. **Bodonidæ.** Small naked forms in which there is only a slight difference, if any, between the flagella. Genera: *Bodo* Stein ('78); *Phyllomitus* Stein ('78); *Colponema* Stein ('78); *Oxyrrhis* Dujardin ('41).

Family 2. **Trimastigidæ.** With two accessory flagella. Genera: *Dallingeria* Kent ('81); *Trimastix* Kent ('81).

Order 4 **POLYMASTIGIDA.** The body is invariably without a shell, and is provided with a delicate membrane, which allows more or less amœboid movement. The number of flagella varies from three to many, and the number of mouth openings, or food-taking areas, likewise varies. Nutrition is holozoic. They increase by longitudinal division.

Tribe 1. **Astomea.** Polymastigida with many flagella and without a mouth opening. Genera: *Multicilia* Cienk. ('81); *Grassia* Fisch ('85).

Tribe 2. **Monostomea.** The anterior part is provided with a large mouth opening at the base of the four or six flagella. Genera: *Collodictyon* Carter ('65); *Tetramitus* Perty ('52); *Monocercomonas* Grassi ('82); *Trichomonas* Donné ('37); *Megastoma* Grassi ('81).

Tribe 3. **Distomea.** The flagella are separated into two symmetrical groups, with a mouth area at the base of each group. Genera: *Trigonomonas* Klebs ('93); *Hexamitus* Dujardin ('38); *Trepomonas* Dujardin ('39); *Spironema* Klebs ('93); *Urophagus* Klebs ('93).

Tribe 4. **Trichonymphinea.** Polymastigida, of unknown affinities, provided with numerous flagella. They are parasites in the rectum of various hosts (Termites). Genera: *Lophomonas* Stein ('78); *Leidyonella* Frenzel ('91); *Trichonympha* Leidy ('77); *Jœnia* Grassi ('85); *Pyrsonympha* Leidy ('77).

Order 5. **EUGLENIDA.** Large forms having one or two flagella, a contractile or firm body-wall, a mouth and pharynx at the base of the flagellum, and with a contractile vacuole opening into the pharynx. They frequently form colonies and are usually provided with chromatophores. Nutrition is holozoic, holophytic, or saprophytic.

Family 1. **Euglenidæ.** Elongate forms, with a more or less pointed end and usually with one flagellum. The cuticle is marked with spiral stripings: the contractile vacuole, or vacuoles, open into a reservoir, which in turn opens into the pharynx. A red eye-spot, or stigma, and green chromatophores, are usually present. Within the body there are discoid, or, occasionally, band-formed chromatophores. Paramylum granules are always present. Genera: *Euglena* Ehr. ('30); *Colacium* Ehr. ('33); *Eutreptia* Perty ('52); *Ascoglena* Stein ('78); *Trachelomonas* Ehr. ('33); *Lepocinclis* Perty ('49); *Phacus* Nitsch ('16); *Cryptoglena* Ehr. ('31).

Family 2. **Astasiidæ.** The body is elongate and usually has a striped membrane. The anterior end is similar to that of *Euglena*, but there is no eye-spot. The body is invariably colorless. Nutrition is saprophytic. Genera: *Astasia* Ehr. ('38); *Distigma* Ehr. ('31); *Rhabdomonas* Fres. ('58); *Menoidium* Perty ('52); *Atractonema* St. ('78); *Sphenomonas* Stein ('78).

Family 3. **Peranemidæ.** The body is either stiff or plastic, and usually symmetrical. The anterior end bears either one or two dissimilar flagella, which are more or less deeply sunk in the body. A distinct mouth is found at the base of the flagella. Nutrition is holozoic. Genera: A. With plastic body and one flagellum: *Euglenopsis* Klebs ('93); *Peranema* Dujardin ('41); *Urceolus* Mereschkowsky ('77). B. With a plastic body and two flagella: *Heteronema* Dujardin ('41); *Dinema* Perty ('76); *Zygoselmis* Duj. ('41). C. With a constant body form and one flagellum: *Scytomonas* Stein ('78); *Petalomonas*

Stein ('59). D. With a constant body form and two dissimilar flagella: *Tropidoscyphus* Stein ('78); *Anisonema* Duj. ('41); *Entosiphon* Stein ('78); *Thaumatomastix* Lauterb. ('99).

Order 6. **PHYTOFLAGELLIDA**. Flagellated unicellular organisms with chlorophyl and holophytic nutrition, or without chlorophyl, and saprophytic in nutrition. They are sometimes classified as plants, sometimes as animals.

Suborder 1. **CHLOROMONADINA**. The body is somewhat plastic and without a distinct membrane; with numerous discoid chromatophores but without stigmata. Genera: *Vacuolaria* Cienk. ('70); *Cœlomonas* Stein ('78).

Suborder 2. **CHROMOMONADINA**. Small forms with strong tendency to colony-formation. They are often inclosed in a gelatinous mass, or occupy cups. They may or may not have chromatophores, which, if present, are yellow or yellowish brown in color. Nutrition is usually holophytic, but holozoic and saprophytic forms are occasionally present. There may be one or two flagella, which are invariably directed forward.

Family 1. **Chrysomonadidæ**. The body is rarely naked, but usually covered by a gelatinous mass or by a hyaline cup. With one or two flagella at the anterior end and with or without stigmata. One or two yellowish chromatophores are invariably present. Nutrition is holophytic or holozoic, sometimes both. Genera: A. With naked body which may be inclosed during resting stages in a gelatinous mass. Nutrition either holozoic or holophytic. *Chrysamœba* Klebs ('90); *Chromulina* Cienk. ('70); *Ochromonas* Wysotzki ('87); *Stylochrysalis* Stein ('78). B. With a shell or lorica in which the individuals are attached. Nutrition is holophytic. *Chrysococcus* Klebs ('92); *Dinobryon* Ehr. ('38); *Chrysopyxis* Stein ('78); *Nephroselmis* St. ('78); *Hyalobryon* Lauterb. ('99). C. Individuals protected by a close-fitting membrane. *Hymenomonas* Stein ('78); *Microglena* Ehr. ('31); *Mallomonas* Perty ('76); *Synura* Ehr. ('33); *Syncrypta* Ehr. ('33); *Uroglena* Ehr. ('33); *Chrysosphærella* Lauterb. ('99).

Family 2. **Cryptomonadidæ**. The body has a firm cuticle and is never amœboid. There are two similar flagella, a peculiar œsophagus-like canal, and a contractile vacuole in the anterior end. Two chromatophores of variable color may or may not be present. Nutrition is holophytic or saprophytic. Genera: *Cryptomonas* Ehr. ('31); *Chilomonas* Ehr. ('31); *Cyathomonas* Fromentel ('74).

Suborder 3. **CHLAMYDOMONADINA**. Body-form more or less changeable. Color usually green, and due to the presence of a large, single chromatophore containing chlorophyl. A firm shell is usually present. The body has two or four flagella, one or two contractile vacuoles, and a stigma at the anterior end. Reproduction takes place by continued division within the shell either during active or resting phases. Macro- and micro-gametes may be formed.

Family 1. **Chlamydomonadidæ**. With a stiff coating perforated only by minute apertures for the flagella. Genera: *Chlamydomonas* Ehr. ('33); *Chlorogonium* Ehr. ('35); *Polytoma* Ehr. ('38); *Hæmatococcus* Agardh ('28); *Carteria* Diesing ('66); *Spondylomorum* Ehr. ('48); *Chlorangium* Stein ('78).

Family 2. **Phacotidæ**. The body of the flagellate corresponds to that of *Hæmatococcus*, and is surrounded by a thick shell membrane which the body does not fill. The shell is frequently bivalved. Genera: *Coccomonas* Stein ('78); *Mesostigma* Lauterb ('94); *Phacotus* Perty ('52); *Tetratoma* Bütschli ('85); *Pyramimonas* Schmarda ('50); *Chloraster* Ehr. ('48).

Suborder 4. **VOLVOCINA**. Colony forms. The individuals possess two flagella and chlorophyl-bearing chromatophores. The number of individuals composing the colony may be constant or variable; when constant, the colony is formed by regular cleavage, as in the eggs of Metazoa. Reproduction asexual by division or sexual. Genera: *Gonium* O. F. Müller (1773); *Stephanosphæra* Cohn ('53); *Pandorina* Bory de St. Vincent ('24); *Eudorina* Ehr. ('31); *Volvox* Leeuw. Ehr. ('38); *Plæodorina* Shaw ('94); *Platydorina* Kofoid ('99).

Order 7. **SILICOFLAGELLIDA.** A single genus, *Distephanus* Stöhr ('81), characterized by the presence of a silicious latticed skeleton like that of the Radiolaria. There is no mouth nor modifications of the plasm whatsoever, but the animal is colored yellow by (probably) diatomin. Parasitic on Radiolaria.

Subclass II. **DINOFLAGELLIDIA.** Naked or shelled Mastigophora. There are usually two flagella, of which one is directed away from the body, the other around the body; the shell usually has two furrows, one running transversely around the body, the other vertically. Marine and fresh-water forms. The nutrition is holophytic or holozoic.

Order 1. **ADINIDA.** The transverse furrow is absent, and the two flagella arise from the anterior end of the body. The shell may be bivalved.

Family 1. **Prorocentridæ.** With the characters of the order. Genera: *Exuviælla* Cienk ('82); *Prorocentrum* Ehr. ('33).

Order 2. **DINIFERIDA.** Dinoflagellidia with two transverse furrows.

Family 1. **Peridinidæ.** The cross furrow is near the middle of the body, which may be with or without a shell. The form is extremely variable. Genera: *Podolampas* Stein ('83); *Blepharocysta* Ehr. ('73); *Diplopsalis* Bergh ('82); *Peridinium* Ehr. ('32); *Goniodoma* Stein ('83); *Gonyaulax* Diesing ('66); *Ceratium* Schrank (1793); *Amphidoma* Stein ('83); *Oxytoxum* Stein ('83); *Pyrophacus* Stein ('83); *Ptychodiscus* Stein ('83); *Protoceratium* Bergh ('82); *Glenodinium* Ehr. ('35); *Gymnodinium* Stein ('78); *Hemidinium* Stein ('78); *Steiniella* Schütt ('95); *Monaster* Schütt ('95); *Amphitholus* Schütt ('95).

Family 2. **Dinophysidæ.** The cross furrow is above the middle of the body, and its edges are raised into characteristic ledges. Marine. Genera: *Phalacroma* Stein ('83); *Dinophysis* Ehr. ('39); *Amphisolenia* Stein ('83); *Citharistes* Stein ('83); *Histioneis* Stein ('83); *Ornithocercus* Stein ('83); *Amphidinium* Clap. and Lach. ('59); *Ceratocorys* Stein ('83).

Order 3. **POLYDINIDA.** The order consists of the single genus *Polykrikos* Bütschli ('73), which is characterized by a naked body, by several transverse furrows and flagella, by macro- and micro-nuclei, and by nematocysts. Nutrition is holozoic.

Subclass III. **CYSTOFLAGELLIDIA.** Mastigophora of considerable size, with a single nucleus, parenchymatous protoplasm, and a firm membrane. Nutrition is holozoic. Marine. Genera: *Noctiluca* Suriray ('36); *Leptodiscus* R. Hertwig ('77).

SPECIAL BIBLIOGRAPHY IV

Bütschli, O.— Protozoa, Mastigophora. In Bronn's Klassen und Ordnungen des Thierreichs. *Leipzig*, 1883–1887.

Francé, R. H.— Der Organismus der Craspedomonaden. *Budapesth*, 1897.

Kent, W. Saville.— A Manual of the Infusoria. *London*, 1881.

Klebs, G.— Flagellatenstudien I. *Zeit. f. wiss. Zool.* Bd. LV., pp. 265–445.

Schütt, F.— Die Peridineen der Plankton-Expedition. Th. I. *Kiel and Leipzig*, 1895.

Senn, G.— Flagellata. In A. Engler's Die naturlichen Pflanzenfamilien. 202 and 203 Lieferung. *Leipzig*, 1900.

Stein, Fr.— Der Organismus der Infusionsthiere. *Leipzig*, 1878.

CHAPTER V

THE SPOROZOA

THE Sporozoa are unicellular animal parasites living in the cells, tissues, and cavities of various hosts and, as the name indicates, characterized by reproduction through spore-formation. If we except the bacteria, they are the most widely distributed of all parasites, and are found in every class of animals, frequently in Vermes, Arthropoda, Mollusca, and Vertebrata, rarely in Protozoa, Cœlenterata, and Echinodermata. They may infest the alimentary tract, and all

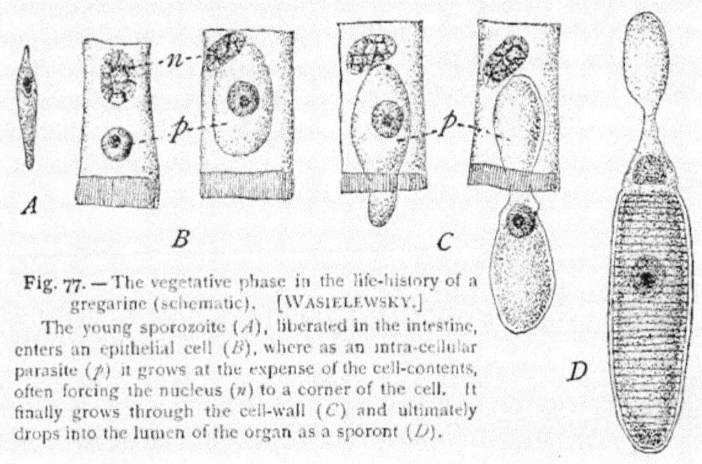

Fig. 77.—The vegetative phase in the life-history of a gregarine (schematic). [WASIELEWSKY.]

The young sporozoite (*A*), liberated in the intestine, enters an epithelial cell (*B*), where as an intra-cellular parasite (*p*) it grows at the expense of the cell-contents, often forcing the nucleus (*n*) to a corner of the cell. It finally grows through the cell-wall (*C*) and ultimately drops into the lumen of the organ as a sporont (*D*).

of the connecting organs and ducts; the kidneys and their ducts; the blood-vessels and the blood; the muscles and connective tissues; while even the skin is not exempted. In most instances they are harmless, but they may produce morbid and even fatal results, either indirectly, by increasing to such numbers that the lymph-spaces and cavities are filled with them, thus preventing nutrition of the cells and tissues, or directly, by causing atrophy and death of the cells in which they live. They are usually taken into the system in the spore-stage with the food of their host, although infection may take place through the gills or lungs, or even by inoculation from insects. The spore-membranes are soon dissolved by the fluids of the host, and one or more germs are thus liberated. These germs, the *sporozoites*, then bore into the epithelial cells, where they grow (Fig. 77). All forms, apparently, begin life as intra-cellular parasites, where, at first, they do little harm, but as they grow by the absorption of fluids contained within their cell-hosts, the latter are improperly nourished and, unless the parasites leave them, they degenerate and die (Fig. 78). The duration of intra-cellular life varies in different kinds of Sporozoa: some are permanently intra-cellular (*monophagous* forms, so-called *Cytosporidia*, etc.); others are intra-cellular only in the young or immature phases (Gregarinida); while still others pass different phases of their life-history in different cells (*polyphagous* forms). The mature parasites finally may leave the cell-host and sporulate in the digestive cavity or cœlom, and the spores are then carried to the outside with the fæces or other excreta.

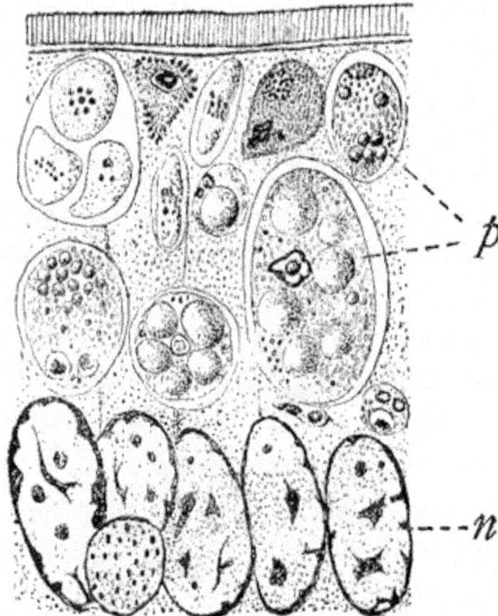

Fig. 78. — Coccidia in the epithelial cells of *Triton cristatus*. *n*, nuclei of the tissue cells; *p*, the intra-cellular parasite *Pfeifferia tritonis*. [LABBÉ.]

A. PROTOPLASMIC STRUCTURE

A typical sporozoön consists of protoplasm and one nucleus. It has no mouth, anus, excretory pore, or other openings. It has neither gastric nor contractile vacuoles, and has at most a sluggish movement in the adult stage, although the young forms may be amœboid or flagellate. Owing to the number of cytoplasmic granules which make up the bulk of the animal, the protoplasmic structure of adult forms can be made out only with the greatest difficulty. Apart from these granules, however, which are regarded as reserve nutriment, it is probable that, as in all Protozoa, the protoplasm is alveolar. This is certainly the case in Coccidiida (intra-cellular Sporozoa), especially in the young forms, where Labbé ('96) describes the cytoplasm as alveolar; in some forms of Gregarinida (Fig. 87), and in Myxosporidiida as described by Thélohan ('95) and Doflein ('98) (Fig. 79). The granules, which are so characteristic of the group, completely fill the alveolar network, and give to the protoplasm its peculiarly dense appearance. They differ somewhat in size and shape, and apparently in chemical composition, and are generally regarded as food substances reserved for use during the spore-producing period.

Wasielewsky ('96) enumerates the following kinds: (1) *Paraglycogen.* These form the bulk of the granules in the Gregarinida; they are distinct refringent granules of variable size and are usually oval or spherical in form, consisting of a peculiar amyloid substance which Bütschli ('84) regarded as similar to amidon or glycogen. They give characteristic reactions, staining brown to violet with dilute sulphuric acid, and dissolving in potassium carbonate and strong mineral acids. (2) *Carminophilous granules.* These granules, which were first made out by Schneider ('75), are less numerous than the paraglycogen granules, but like them variable in size and strongly refractive.

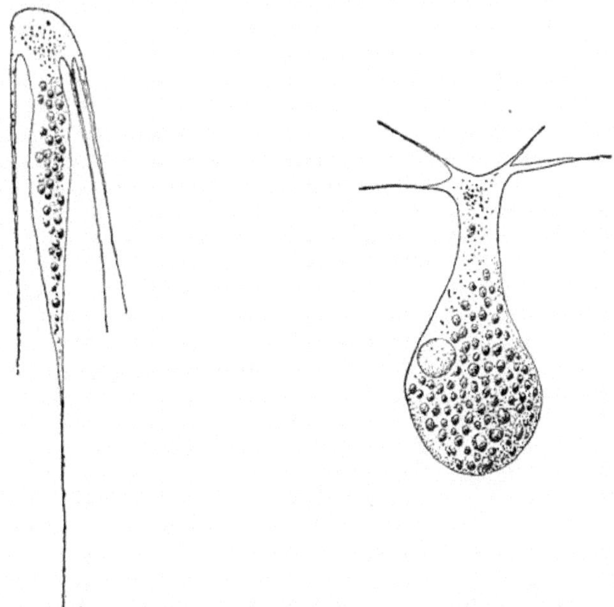

Fig. 79. — *Leptotheca agilis* Dof., one of the Myxosporidiida. [DOFLEIN.]

They are easily soluble in ammonia, but are not destroyed by alcohol, embedding in paraffine, etc., and are easily stained by carmine and many aniline colors, but not at all by hæmatoxylin. They consist, apparently, of albumen. (3) *Fat.* These granules are widely distributed throughout the entire group, and have about the same appearance in all types, although they are colored differently in different species. They are soluble in alcohol, ether, and chloroform, and are stained black by osmic acid. In addition to the above granules, which are found in most Sporozoa, there are others which have been found hitherto only in certain subdivisions. In the Gregarinida, *pyxinine* granules and *protein crystals* have been observed in certain species, the former by Frenzel ('85) in *Pyxinia*, where they apparently take the place of the paraglycogen granules, although of similar chemical nature, but slightly different in their reactions. The latter are also rather questionable inclusions in certain *Didymophyes*. In the Coccidiida all adult forms are characterized by the presence of so-called *plastic granules*. These are globular, strongly refractive granules of slightly variable size which react differently from the glycogen granules, remaining unchanged in sulphuric acid and staining yellow with iodine. Here also are found the so-called *chromatoid granules*, which are distinguished by their affinity for hæmatoxylin, and are probably albuminoid in nature. In the Hæmosporidiida or blood-infesting Sporozoa, the effect of the intra-corpuscular life is shown by the presence of pigmented granules (*melanin*) of black, yellow ochre, or red color resulting from the disintegration of hæmoglobin.

In the majority of the Sporozoa the cell-body consists of a more or less sharply differentiated ectoplasm and endoplasm, while even a third layer, *mesoplasm*, is said to have been observed in some forms (Cohn, '96). It is possible that these different zones are functionally specialized, a supposition first made by Labbé ('96) in connection with the Coccidiida, and repeated by Doflein ('98) in connection with the Myxosporidiida. As in the Sarcodina and Mastigophora, the ectoplasm may be plastic, yielding to the pressure from within and thus giving rise to pseudopodia (*Monocystis ascidiæ*, Siedlecki, '99, Myxosporidiida), or it may be modified into a hard and tough cuticle, which offers a good protection for the cell-body within. Again, it may be modified into a complex membrane, plastic and capable of various kinds of motion and similar to that of the higher types of Flagellidia. The most highly differentiated ectoplasm is found in the Gregarinida, where it forms a dense cortical layer about the body, while its outermost part is transformed into a complex membrane. In some cases the inner cortical layer of the ectoplasm is carried across the cell, forming a partition dividing the organism into two portions which are known as the *protomerite* and the *deutomerite*, the nucleus being in the latter. The non-nucleated portion is often further differentiated into an apparatus called the *epimerite*, which usually develops hooks or anchors used for attaching the animal to its cell-host. The organism thus appears to be multi-chambered, and the presence or absence of such chambers was formerly regarded as a good basis for classification; but it has been shown, especially by Léger ('92), that the partitions vary considerably in the same species, and even in the same individual, at different times, and in the recent systems of classification this feature has been discarded in determining the limits of the larger divisions. The epimerite, which is so important in holding the lumen-dwelling parasites in place, may be simple or branched; plain, like a knob or a rod; branched with filiform, or flat with digitiform, appendages. It may be closely attached to the protomerite, or carried on a long neck, while variations in all types are numerous (Fig. 12, *E*, *F*, *G*, p. 39). Under certain conditions, prior to reproduction, the animal throws off the epimerite which may be left in the cell-host, and drops into the lumen of the organ in which it lives. Here it encysts, the protomerite and deutomerite forming one spore-producing individual. As the attached and the detached stages in the life-history of the Gregarine are each important, they have received special names, the former being known as a *cephalont*, the latter as a *sporont*.

Between the cortical ectoplasm and the inner endoplasm there is a layer of *myonemes*, or muscular fibrils similar in all respects to those of the Ciliata (Fig. 80). These are occasionally found in the Hæmosporidiida, but are much more characteristic of the Gregarinida, where, except in the epimerite, they form a network about the entire animal. On the outside of this network, according to Schewiakoff ('94), there is, at times, a layer of gelatinous matter apparently secreted by the ectoplasm, and this, in turn, is covered by the membrane proper. The membrane is longitudinally striated by riblike projections, while the canals or furrows between them are filled with jelly from the gelatinous layer below (see Fig. 82, *B*). Schewiakoff believes that the active secretion of jelly in these furrows accounts for the peculiar gliding motion of certain kinds of Gregarinida. In the region of the epimerite, the membrane is plain, the ribs and furrows stopping with the protomerite. The hooks or spines are formed from the cortical plasm.

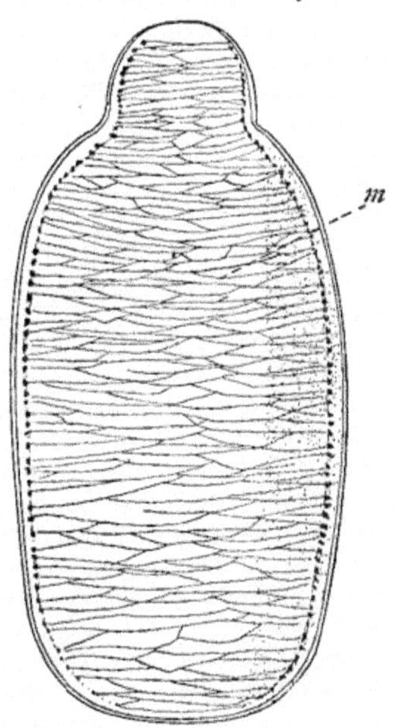

Fig. 80.— Schematic figure of the myonemes of *Clepsidrina munieri*; *m*, the myonemes. [SCHNEIDER.]

In one group of Sporozoa, the Sarcosporidiida, the protoplasmic body is inclosed in a peculiar pouch which appears to be a secretion from the protoplasm rather than a true cellular membrane. The mass slowly enlarges by regular growth until it reaches a considerable length, in some cases several millimetres. It then undergoes spore-formation. These organisms, known as *Rainey's Tubes*, are parasites of sheep, swine, deer, horses, rats, etc., where they infest the muscle-tissues, causing morbid symptoms, similar to those in trichinosis.

B. The Nucleus

With the exception of the multinucleate Myxosporidiida, the Sporozoa are mononucleate. Schneider ('81) abandoned the attempt to compare the nuclei in Sporozoa with those of ordinary animal and plant cells, because of their peculiar structure. In most cases they consist of a firm and resisting membrane containing a single large chromatin reservoir or *karyosome*, and are apparently without a linin reticulum, such as is found in the nuclei of Metazoa. In some forms the nucleus is similar to that of the Sarcodina and Mastigophora, consisting of membrane, reticulum, and one or more chromatin reservoirs. Recent observers have found that the different appearances of the nucleus are characteristic of different stages of nuclear activity, and that the reticulum, and even the nuclear membrane, are derived from the karyosome, which in the sporozoite appears as a solid homogeneous sphere of chromatin.[1] In the active phases the nuclei of the Sporozoa differ widely from those in other Protozoa, the most striking point of difference being the disappearance of the nuclear membrane during division. The chromatin reservoirs may divide directly, thus simulating the entire nucleus, or they may break down into small chromatin granules, resembling the first stages of chromosome-formation in the flagellate *Noctiluca*. In the former there is no distinct spindle, in the latter the completed spindle-figure has two sets of fibres, although, according to Wolters's ('91) description, the fibres seem to have a different function from those in the mitotic figures of the Metazoa, since there is no connection between them and the chromatin.[2]

C. Food-taking

Like all endoparasites, the Sporozoa absorb fluid food through the body-wall, even when, as in the Myxosporidiida, pseudopodia are present. There is probably no specialized area devoted to food-taking, but all parts are equally receptive. It is believed that in some cases, notably in the Gregarinida and Myxosporidiida, minute pores perforate the membrane between the outer markings. Although the taking of food has never been observed, the indirect effects are seen in the rapid growth of the parasite when in a suitable medium. Thus a young gregarine, when it penetrates an epithelial cell (Fig. 77, *A*), is a minute ball of protoplasm; but it rapidly grows until it occupies the greater part of its host, often forcing the nucleus to one side. As it continues to grow, the front wall of the cell is pushed outward until it finally breaks, and the lower portion of the parasite

[1] Cf. *infra*, p. 253. [2] See *infra*, p. 259.

is left exposed in the lumen of the digestive organ (Fig. 77, *C*). If it is a polycystic or multi-chambered form, the exposed portion becomes differentiated into protomerite and deutomerite, while the intra-cellular portion remains as the epimerite. After growth, the surplus food is stored in the endoplasm in the form of granules as described

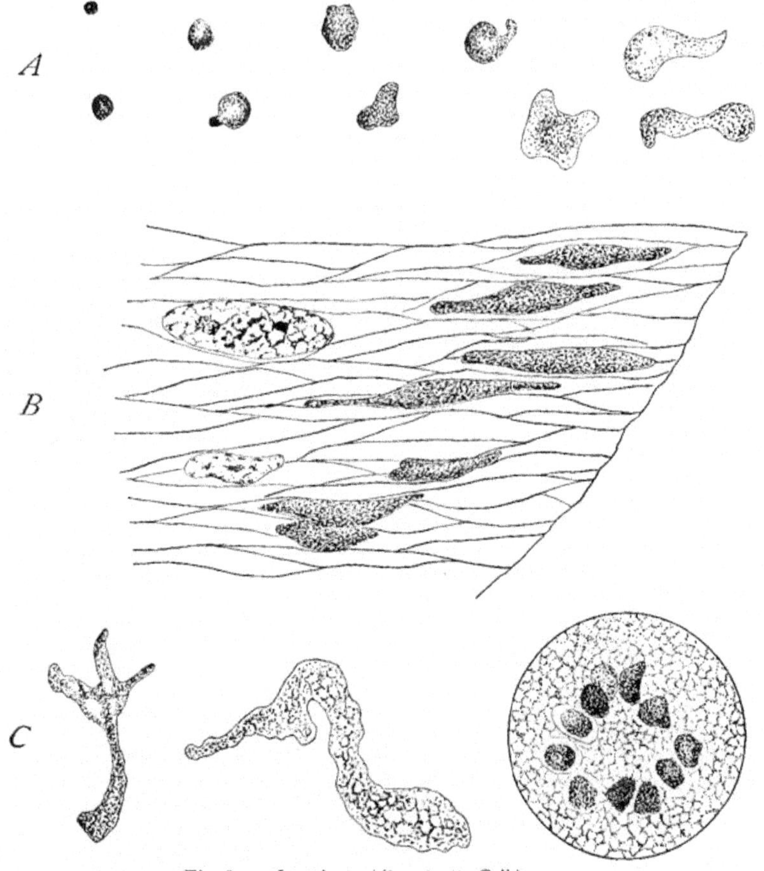

Fig. 81. — *Lymphosporidium truttæ* Calkins.
A. The young sporozoite and its development. *B.* Older forms in the muscle-bundles surrounding the intestine. *C.* Still older amœboid form prior to, and during, spore-formation.

above, to be used during the process of spore-formation and encystment.

In Sarcosporidiida and other muscle-infesting Sporozoa, growth takes place at the expense of the muscle-cells, although the organisms are not intra-cellular parasites. Thus, *Lymphosporidium truttæ* begins to grow in the lymph surrounding the intestine. The sporozoite develops into a small amœboid form which penetrates the muscle-bundles, and there grows to adult size by absorbing food destined for the muscles. When mature, it leaves the muscle-bundles and returns to the lymph-spaces, where it sporulates (Fig. 81).

D. Motion

In addition to the amœboid motion which has already been mentioned, there are various movements, due to the contraction of the myonemes or of the entire ectoplasm. Among these may be men-

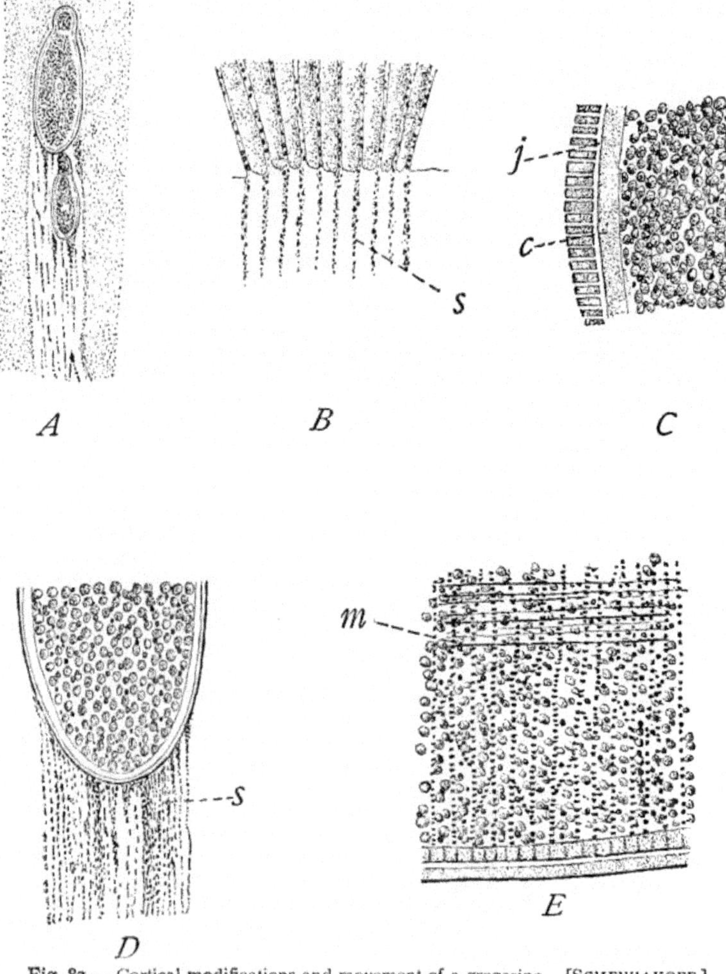

Fig. 82. — Cortical modifications and movement of a gregarine. [Schewiakoff.]

A. Moving gregarine with paths of excreted granules. *B* and *D*. The same, more highly magnified. *C* and *E*. Details of structure. *c*, cortical plasm; *j*, jelly-layer; *m*, myoneme; *s*, secretion.

tioned the peristaltic contraction of certain Gregarinida, or the energetic jerking motion sometimes observed in the same forms. None of these movements, however, brings about a regular translation from place to place, and the Sporozoa are regarded as the most sluggish of the Protozoa. Food-seeking, which in free-living animals is the main occasion for locomotion, is here unnecessary; for the adult animals, placed in the chyle of the host, or in the spaces between cells and tissues, or in the cells themselves, have little occasion for movement, save that which, in young forms, is necessary to reach the host, to maintain their positions, and to prevent displacement. There is, however, in certain forms of Gregarinida, a peculiar gliding motion on the part of the adult organism. This is accomplished without apparent exertion of any kind by the animal, and for a long time was a puzzle to students of the group. Schewiakoff ('94) offered an explanation, based upon actual observation and experiment, and although very improbable at first sight, it is the only one thus far that fits the case (Fig. 82). These observations have been confirmed recently by Siedlecki ('00), who accepts Schewiakoff's interpretation, while Lauterborn also gives a similar interpretation of the movement in diatoms. According to Schewiakoff, the forward, gliding motion is the result of the active secretion of the gelatinous substance from the ectoplasm, which accumulates below the membrane to form a gelatinous layer. The membrane of the cell, as described above, is marked externally by clear longitudinal grooves, and the gelatinous substance after filling these grooves, instead of spreading over the surface of the membrane, flows down and backward in the grooves to the posterior end of the body, where the secretion from different furrows unites to form larger currents, and these, in turn, form still larger streams, which, like a spider's web, solidify upon leaving the body (*D*). Thus, a solid cylinder is formed behind the animal, the posterior end of which fits into the basin-like depression like a cast in its mould. The addition of new jelly by active secretion in the ectoplasm, and the resistance of the solidified portion, causes a forward movement of the animal. The movement, Schewiakoff further observes, is only periodic, for the flowing of the jelly is more rapid than the secretion — a fact which explains the occasional absence of the external gelatinous layer.

E. Reproduction

The most characteristic phenomena connected with the Sporozoa are those of reproduction and development. The many methods occurring in the other forms of Protozoa are here limited to spore-formation, although Labbé describes rather questionable simple divi-

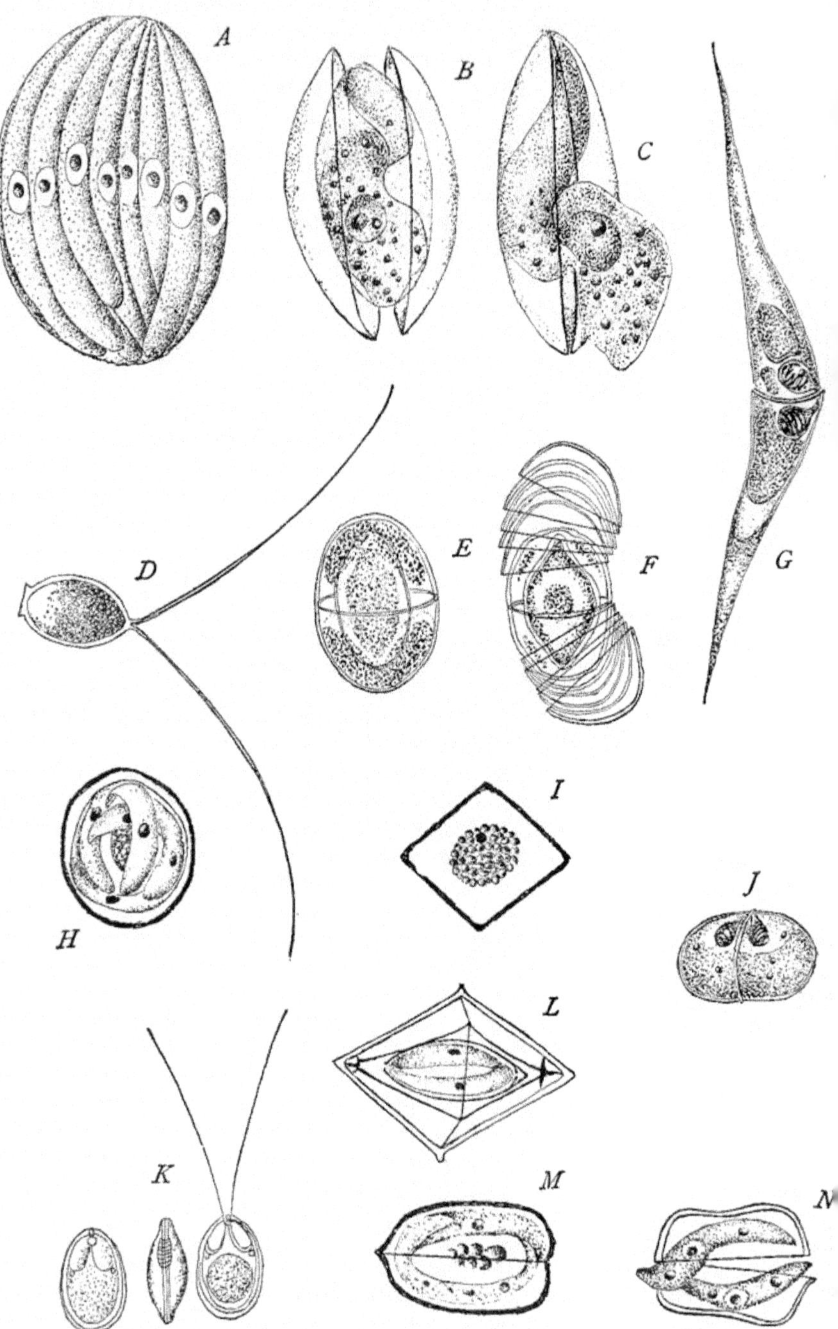

Fig. 83. — Types of spores. [WASIELEWSKY; A. SCHNEIDER; THÉLOHAN, etc.]

A. Eimeria nepæ. *B, C. Barroussia ornata.* *D.* Tailed-spore of Gregarine. *E, F. Ophryocystis Bütschlii.* with multiple epispores. *G. Ceratomyxa sphærulosa.* *H. Klossia helicis.* *I. Crystallospora Thélohani.* *J. Leptotheca agilis.* *K. Myxobolus ellipsoides.* *L. Crystallospora crystalloides.* *M. Goussia clupearum.* *N. Adelea ovata.* Coiled threads are shown in *G* and *J*, the extruded thread in *K*.

sion in Coccidiida (Fig. 6, p. 20). Spore-formation is almost invariably preceded by encystment, an exception being found in the Gymnosporea and Myxosporidiida.

In general, it may be stated that the entire organism takes part in the formation of *archispores* (or *sporoblasts*), each archispore gives rise to spores, and each spore to sporozoites, either directly or indirectly. Each spore, containing from one to many sporozoites, is coated by either a single or a double membrane. When double, the inner membrane is called the *endospore*, and the outer the *epispore* (Fig. 83, *F*). The spores may be of similar or dissimilar size (*macrospores* and *microspores*), they may be ovoid, spherical, biconvex, cylindrical, crystalline, discoid, etc., in form, and may be provided with diverse kinds of appendages, ridges, spines, etc., or with polar capsules containing protrusible filaments (Myxosporidiida). In some cases there is a special apparatus for the dissemination of the spores (*sporoducts*, Fig. 85); in other cases, the spores are liberated by the simple bursting of the outer envelope, or by the rupture of the walls through swelling of a residual protoplasmic mass termed a *pseudocyst*.

The process of spore-formation in the gregarine of an ascidian — *Monocystis ascidiæ* — may be given as an example of a type common to all Sporozoa, although in the several orders the details are variously modified. Two animals come together and form a common cyst (Fig. 84, *A*). The nucleus of each divides by repeated mitoses into a great number of daughter-nuclei, which soon arrange themselves about the periphery (*B*, *C*) like the nuclei of a centrolecithal egg of some Metazoa. A portion of the endoplasm is then budded off about each of the daughter-nuclei, the buds thus formed becoming conjugating gametes. The bulk of the original cells is not used in this process, a considerable portion which Labbé ('96) regards as a reserve store of nutriment[1] remaining unused (*Theilungskorper*, *Cystenrest*, *Réliquat de ségmentation*). During this process the ectoplasm and the membrane in each cell disappear, leaving the gametes and the central residual masses within the cyst (*D*). The gametes now conjugate two by two (*E*) to form the spores (sporocysts). Each of the spores, which from their peculiar shape are known as *pseudonavicellæ*, now in its turn secretes two distinct membranes (epispore and endospore), and within these the nucleus, with its surrounding plasm, divides into eight parts which are disposed quite regularly in the spore (*F*). As in the formation of the archispores, a portion of the plasm is usually left unused (*Sporenrest*, *Restkörperchen*, *Réliquat de différentiation*). Each of these parts is a sporozoite, which, after a developmental period, reproduces an adult gregarine. When mature, the spores or pseudonavicellæ are liberated by the bursting of the outer cyst-walls, brought

[1] See, however, Thélohan, '95.

about either by the simple rupture of the wall or by the swelling of the central mass of useless material. The spores are thus freed, but not the sporozoites; the latter are still confined within their double walls, and cannot be liberated until they are swallowed by some host, where, in the digestive tract, the two coatings are dissolved off by the digestive fluids, and the sporozoites emerge in the form of minute elliptical bits of protoplasm, each containing a nucleus.

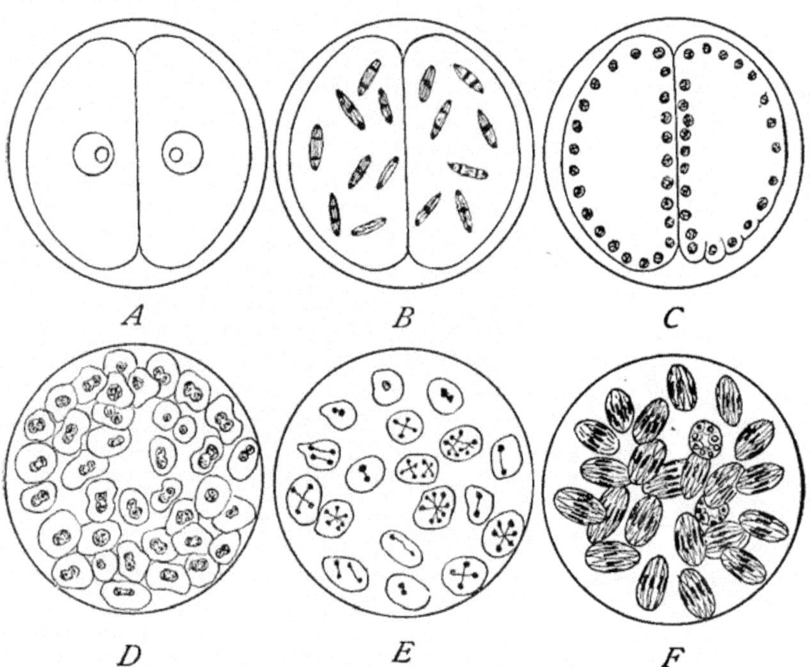

Fig. 84.—Scheme of sporulation in gregarinida.

A. Union of two individuals in a common cyst. *B* and *C.* The formation of gametes of similar size. *D.* Union of the amœboid gametes. *E* and *F.* Formation of sporozoites in the fused gametes.

The process of spore-formation in the many-chambered Gregarinida is more complicated. Thus in *Clepsidrina*, a frequent parasite of insects, the organism when mature throws off the epimerite by which it is attached to an epithelial cell of its own host and, as a sporont, secretes its cysts and undergoes nuclear division as in *Monocystis.* The encysted animal, however, is carried to the exterior with the fæces of the host, and sporulation is outside of the host or *exogenous*, as opposed to the *endogenous sporulation* of *Monocystis*. In these excreted cysts, according to Schneider ('75) and Bütschli ('84), the archispores, instead of, as in *Monocystis*, forming a peripheral layer about a central residual mass, lie in the centre, the unused portion of the original protoplasm forming a thick layer about them. At the same time, a third and very delicate membrane, probably composed of the residual peripheral mass, is formed inside of the cyst and against the second or inner coating. Six to eight radial thickenings can be seen later in this residual portion, and each of these develops a distinct lumen, thus becoming tubular and extending through the residual mass of protoplasm to the new internal membrane. Each tube expands at the extremity into a disc-like cup, while the inner part of the tube is lost in the central mass of spores. In some unexplained way the walls of the primary cyst open, leaving the protoplasm and the spores inclosed only by the third membrane. The tubes already formed then evaginate, and the cylindrical portion of the tube is thrown to the outside. The tubes act as spore-ducts for the inner archispores, each of which contains the definite number of sporozoites (Fig. 85).

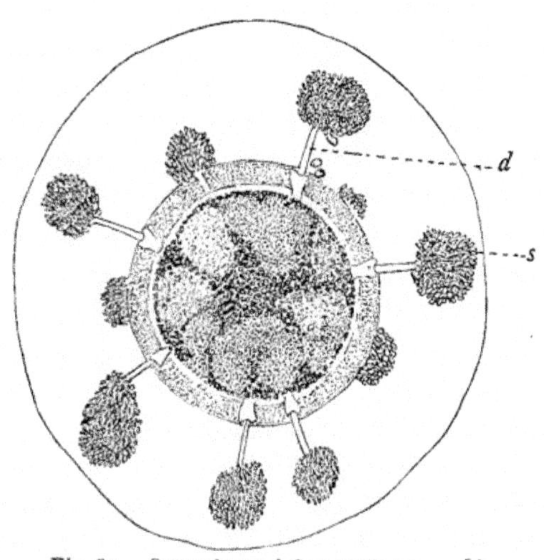

Fig. 85. — Spore-ducts of *Gamocystis tenax*. [A. SCHNEIDER.]
d, spore ducts; *s*, spores in an external gelatinous mantle.

Sporulation of the Coccidiida is strikingly similar to that of the Gregarinida. Here, as a rule, only one membrane (capsule) is formed around the spherical animal; and the nucleus, in addition to division through mitosis, frequently fragments into as many pieces as there are to be archispores (fragmentation). Before the nucleus divides, a certain amount of the chromatin is given off, as in the Gregarinida, to form what Labbé calls the equivalent of the "polar body" of the Metazoa. Again, as in the Gregarinida, the archispores or sporocysts are arranged around the periphery, and a residual mass occupies the centre. The archispores which are liberated by simple rupture of the walls of the cyst, form a definite number of sporozoites, varying from one (*monozoic*) or two (*dizoic*) to many (*polyzoic*). In some forms of Coccidiida either sporozoites or archispores may be formed directly. The number of spores formed is usually small, as in *Coccidium*, where the nucleus divides only twice, producing only four archispores, each of which gives rise to two sporozoites. In other cases (viz. *Pfeifferia* Labbé), a great number of nuclear divisions may take place, and the final daughter-nuclei with their surrounding protoplasm form sporozoites directly and without an intervening archispore stage. A similar direct sporozoite-formation takes place among the Hæmosporidiida, the sporozoites being frequently of two kinds, macrosporozoites and microsporozoites. While not established, it is probable that in all forms this dimorphism in the spores has a sexual significance, the same individual giving rise to only one form. One peculiarity of these sporozoites is that the nucleus is apparently never provided with a nuclear membrane, the chromatin, as in some flagellates, lying freely in the plasm.

Sporulation in the tribe Gymnosporea takes place without the protection of a cyst. The parasite rounds out, but does not secrete a membrane. The nucleus divides into a great number of parts, which migrate to the periphery as in other forms, and there divide. Sporozoites are formed directly without preliminary spore-stages.

An entirely different mode of sporulation occurs in the Myxosporidiida, where the process is somewhat similar to the internal budding of some of the Ciliata. In the genus *Myxobolus*, for example, one of the numerous nuclei of the amœboid form is surrounded by a thickened mass of protoplasm, so that it can be distinguished from the remainder of the animal. The thickened plasm soon forms a mantle about the nucleus, which then divides by mitosis until there are ten or a dozen daughter-nuclei within the specialized protoplasmic region (Thélohan, '95; Gurley, '93). This mass, the sporoblast, which, however, does not quite correspond to the archispores of preceding types, now divides into two equal parts, both of which remain inside of the original protoplasmic mass. Each is an archispore, and each contains three of the ten nuclei. The other four nuclei are left in the free plasm within the membrane and soon degenerate and disappear, corresponding, apparently, to the residual mass of chromatin (polar body) of other forms. Each archispore next divides into three cells (Bütschli, Balbiani for *Myxobolus*), two of which are destined to form peculiar thread-bearing capsules known as the *polar capsules*. The other is much larger and represents the definitive spore. Each sporoblast thus contains one spore, whose nucleus soon divides to form the two nuclei which characterize the young myxospore. The formation of the thread in the polar capsules according to Thélohan ('95) seems to be the same in all species; a vacuole appears in each of the smaller cells of the sporoblast (Fig. 86), then a small knob-like projection grows up from one side of the vacuole, whose outer walls harden until a distinct capsule is formed. The bud of protoplasm within the vacuole now elongates and winds around until a spirally wound filament is the result. The nuclei of the two polar capsules soon degenerate and disappear, leaving only the capsules with their threads, which show a striking similarity to those of the nematocysts of the Cœlenterata. The archispore in many cases develops a bivalve shell, and in this condition can remain for some time within the original spore-forming body (*pan-sporoblast*); or the original membrane may be thrown off, leaving the encapsuled spores suspended freely in the endoplasm of the parent organism. In most cases there is no means of exit for the spores from the body of the host until the latter dies. The archispores thus accumulate until great cysts, sometimes as large as 30 mm. in diameter (Zschokke, '98,

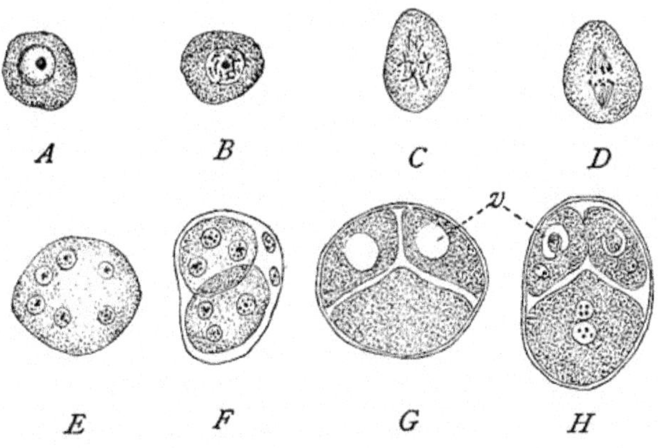

Fig. 86.— *Myxobolus*; capsule-formation. [THELOHAN.]

A–D. Division of the sporoblast nucleus. *F.* The sporoblast is divided into two "sporogenous masses" each containing three nuclei. *G.* Sporogenous mass with protoplasm of the spore and two masses which are destined to develop into capsules and filaments. *H.* The threads are first seen as buds in the vacuole.

Myxobolus bicaudatus), are formed within the tissues of their host. When taken into a new host, the shell of the archispore under suitable excitant, either chemical or physical, soon opens, and the filaments contained within the capsules are thrown out, according to Balbiani, through special apertures, and by the pressure of the capsular walls (Bütschli). These filaments, several times the length of the spore, remain attached, their free ends being swayed about by the currents until they come in contact with and penetrate some cell of the mucous membrane of a new host. The parasite thus anchored retains its position in the lumen until its bivalve shell is thrown off and it can move for itself (Leuckart, Bütschli, Gurley [1]).

In the Myxosporidiida, therefore, sporulation is not the final act of a cell-parasite, but takes place while the animal is performing other

[1] For discussion of various views see Gurley ('93).

normal vegetative functions. It is a case of cellular division of laboi in which possibly some of the multiple nuclei are specially differentiated for reproduction. The number of archispores formed in the pan-sporoblasts varies in the different species and genera. In some cases there is but one (*Chloromyxum*), in others a large number (*Glugea*). The polar capsules, which are particularly characteristic of this type of Sporozoa, also vary in number and in position. They may both be at one end of the spore (anterior), as in *Myxobolus* (Fig. 83, *K*), at the two ends, as in *Myxidium*, or in the centre, as in *Ceratomyxa* (Fig. 83, *G*).

The Sarcosporidiida resemble the Myxosporidiida in forming spores throughout life. The peculiar pouch, which corresponds to the amœboid body of the Myxosporidiida and which may grow to a considerable length (up to 16 mm. in sheep), is filled with masses of nucleated protoplasm which may be called pan-sporoblasts. Those in the centre of the pouch become coated by a membrane and divide into a number of germs or sporozoites known as *Rainey's Corpuscles*, which in some cases appear to have polar thread-bearing capsules similar to those of the Myxosporidiida. The life-history and mode of infection of new hosts is unknown.

Conjugation is a well-authenticated phenomenon in at least three orders: Hæmosporidiida, Gregarinida, and Coccidiida, although the observations have not been numerous enough to warrant further generalizations. Among the Hæmosporidiida, where the intra-cellular parasites frequently leave their cell-hosts, there is a shorter or longer period of free life. During this period two individuals, upon meeting, fuse together, forming one individual (Labbé). The nuclei also fuse, forming a single nucleus. It is an instance of total conjugation, similar to the total fusion in some Monadida, but, unfortunately, the significance of the process and the bearing upon the life-history of the individuals are entirely unknown.

The union of two individuals within a common cyst is not infrequently observed among the Gregarinida, and has been a long-known phenomenon. Two or more individuals may join end to end, protomerite to deutomerite, or side to side, and so form aggregates (Fig. 27, p. 58). If the individuals thus associated happen to be mature at the same time, they may develop a common cyst and so give the appearance of conjugation. Such *pseudoconjugation* frequently leads to the formation of catenoid colonies, where the protomerite of one (*satellite*) becomes attached to the deutomerite of another (*primite*).

We are indebted to Wolters ('91), Siedlecki ('96, '98, '99), and Schaudinn ('96, '99) for more complete accounts of conjugation among the Gregarinida and Coccidiida. According to the former, two gregarines (*Monocystis agilis*) place themselves end to end, but without fusing. The nuclei of the two cells then divide by mitosis, and in each case one of the daughter-nuclei is thrown off as a useless moiety in the same way as a polar globule. The other two daughter-nuclei move toward the partition wall which separates the two individuals, and meet each other in an opening of this wall. They fuse, and this fused mass divides by mitosis, one of the daughter-halves going to each of the conjugants. The nuclei then divide repeatedly, and spores are formed in the usual manner. This method if correctly observed, in contradistinction to pseudoconjugation among the Gregarinida and Hæmosporidiida, is nuclear conjugation as seen in its highest development among the Infusoria; but, unfortunately, there are no observations similar to those of Bütschli, Engel-

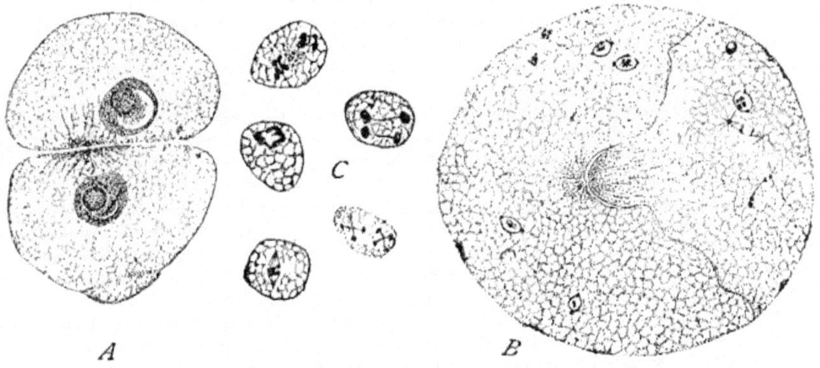

Fig. 87. — *Monocystis ascidiæ* Lankest. [SIEDLECKI.]
A. Fusion of two individuals. *B*. Formation of gametes (*cf*. Fig. 84). *C*. Nuclear division after the fusion of gametes, and sporozoite formation.

mann, Maupas, and others on Infusoria to indicate the significance, and the facts themselves rest upon the observation of a single observer (Wolters). In a closely related form (*Monocystis ascidiæ*), Siedlecki ('99) describes an entirely different process. Here two individuals come together in a single cyst, within which each forms a number of *merozoites* or gametes. The gametes fuse together, and thus affect the conjugation of the two original individuals (Fig. 87).

The greatest advances in our knowledge of the reproduction in Sporozoa, during the last ten years, have been in connection with the Coccidiida, and modern research has shown that the life-history of these forms is bound up with a complicated alternation of generations, the product of the union of sex-cells being the permanent spores by which the infection is carried from one organism to another, while the products of asexual increase lead to auto-infection within the same host.

Up to the last five years the usual description of the life-history of

Coccidiida followed that of Leuckart ('79) in the case of *Coccidium oviformis*, a parasite of the rabbit. According to this view, the adult *Coccidium*, which consists of a globular or oval mononucleate parasite, living in the epithelial cells of the digestive tract and the related organs, encysts and falls into the lumen of the digestive tract, from which it is defecated with the fæces. Inside of the cyst the plasm divides into several parts (in *Coccidium* four), and these parts, after the formation of a firm, resisting membrane, form the permanent spores. Each spore divides into two parts in *Coccidium*, and these two parts constitute the end product of reproduction, according to the older view. Each of the parts forms a germ or sporozoite, which penetrates a new cell-host and develops again to the adult organism.

This cycle, while perfectly logical, left unexplained the immense multitudes of parasites found in the epithelial cells of every *Coccidium*-infected rabbit. The first attempt to explain its wide distribution was made by R. Pfeiffer ('92), who insisted that, in addition to this exogenous spore-formation, there exists an internal reproduction as well, which leads to further infection in the same host, or, as he called it, to *auto-infection*. In contrast with the first method, this was called the *endogenous sporozoite-formation*. This view was based upon the discovery by Pfeiffer of spore-forming cells in the tissues of the host, in addition to those in the lumen of the digestive tract. The majority of investigators along this line have accepted the latter view. There are two notable exceptions, however, one of whom is Labbé, who holds that these smaller forms are only poorly fed individuals and not sporozoites, and explains the undeniable auto-infection through simple division of the parasites.

A large number of papers soon followed, some on *Coccidium*, others on related forms. Mingazzini ('92) followed out the multiple division of the nucleus in the formation of the endogenous sporozoites. Podwyssozki ('94) made the discovery that there are two kinds of these endogenous forms, which were accordingly named *microsporozoites* and *macrosporozoites*. It was Schuberg ('95), however, who first suggested, although he did not confirm the suggestion, that these two forms of sporozoites conjugate and thus lead to sexual reproduction. Labbé strongly opposed the latter view, and held that the larger types of supposed dimorphic spores belong to some unknown species of Coccidiida, and that the smaller forms are degeneration types.

The way was thus prepared for the discovery of conjugation among the cells of the Coccidiida, a discovery made first by Schaudinn and Siedlecki ('97). In two different species, *Coccidium Schneideri* and *Adelea ovata*, it was found that a large cell, an egg, is fertilized by a small one, which has all of the characteristics of a spermatozoön. In the same year Simond worked out again the life-history of the *Coccidium* of the rabbit and described a true copulation between the microsporozoites. Schaudinn regards this, however, as an error, holding that copulation takes place between one of the smaller forms and an enlarged ordinary individual.

Since then, the fact of fertilization, with the resulting formation of sporozoites through spores, has been safely established for a number of species by several different observers, the details alone differing in the several cases. The microsporozoites thus are not true sporozoites, but gametes having a sexual function.

According to these various observations the life-history of the Coccidiida may now be described as follows: the permanent cysts contain spores, each of which contains sporozoites which are taken into the digestive tract with the food. Here the cyst membrane bursts or is dissolved, and the sporozoites are liberated. They penetrate the epithelial cells and grow to the normal size of the adult. They then undergo repeated nuclear division by a process which resembles fragmentation rather than mitosis (Schaudinn), or (possibly) in some cases by binary division also (Labbé), and the nuclear parts wander out to the periphery, where small portions of the cytoplasm form around them and they are pinched off as minute germs which Simond called *merozoites*. These differ in several important respects from the sporozoites, but like them are capable of developing directly into new adult parasites. This process, which Schaudinn calls *schizogony*, leads to the increase of parasites within the host (Fig. 88, *a–c*). During development, some of these merozoites store up reserve nutriment and form large ovoid cells, while others form the mother-cells of the microsporozoites or spermatozoids, without storing up a reserved food supply. The small forms, in some cases, are provided with flagella, which were first made out by Léger and by Wasielewski. Fertilization takes place in a manner almost identical with that of the Metazoa (*d–j*). In some cases a micropyle is formed in the egg through which a spermatozoön can enter, and in all cases after one has entered, a hard membrane corresponding to the vitelline membrane is at once formed (*k*). Complete fusion takes place between the nuclei, and the cleavage nucleus divides by repeated mitosis to form spores.

It is quite probable that the other cases of dimorphism, which have been recorded from time to time, are instances of similar sex-differentiation. The motile forms especially, which numerous observers have recorded, will probably be found to be similar in function. Lavéran ('98), Bosc ('98), Sjöbring ('97), Wasielewsky ('98), Siedlecki ('98), and others have recently described them in different Coccidiida (Fig. 89).

Conjugation in the malaria-causing organism (*Plasmodium malariæ*) is bound up with a change of hosts, thus giving a complicated life-history, which may, and probably does, occur in other kinds of Sporozoa as well, although the phenomenon has been only recently made known.

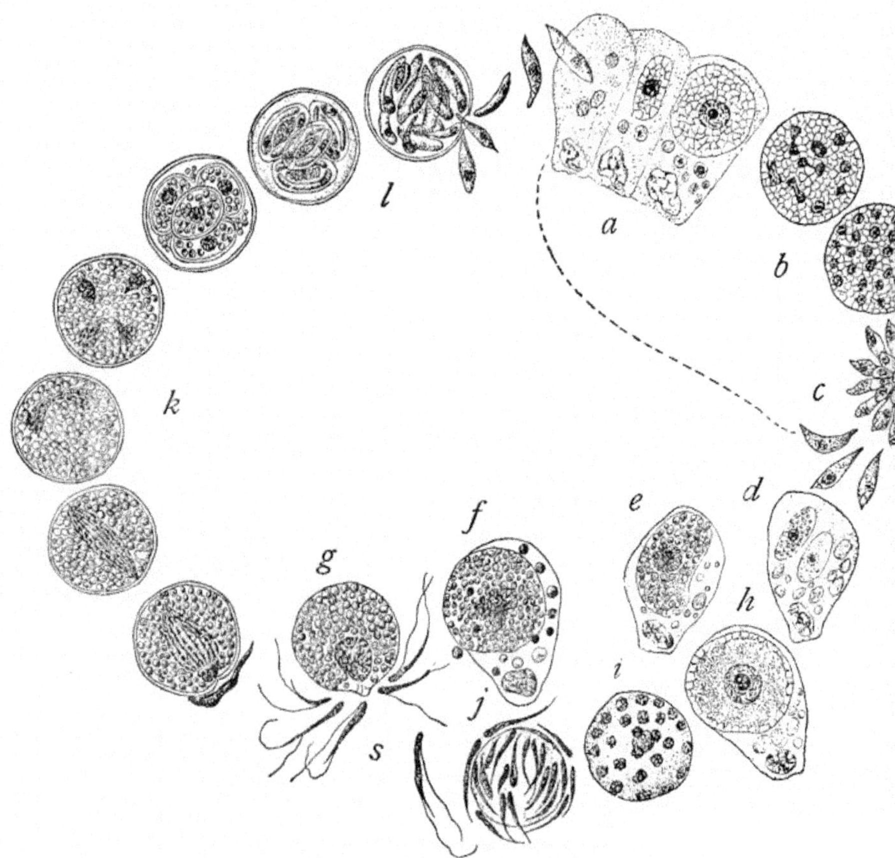

Fig. 88. — Life-history of a *Coccidium*. [SCHAUDINN.]

a, b, c, schizonts and asexual reproduction (schizogony). The merozoites at *c* repeat the cycle or pass on to the following stages. *d, e, f*, development of the female or macrogamete. *h, i, j*, development of the male flagellated gametes; *g*, copulation of the male and female gametes; *k* and *l*, stages in the formation of the four spores and sporozoites.

The sporozoön which is now positively known to be the cause of "malarial disease," lives in the human blood under various forms, which may possibly be distinct species differing from one another in the number of spores produced and in the pathogenic effects. Several varieties at least have been described and specially named on account of minute differences, but it is probable that these can be reduced to three principal types: *Plasmodium malariæ*, *P. vivax*, and *Laverania malariæ*. These all agree in having an intra-corpuscular,

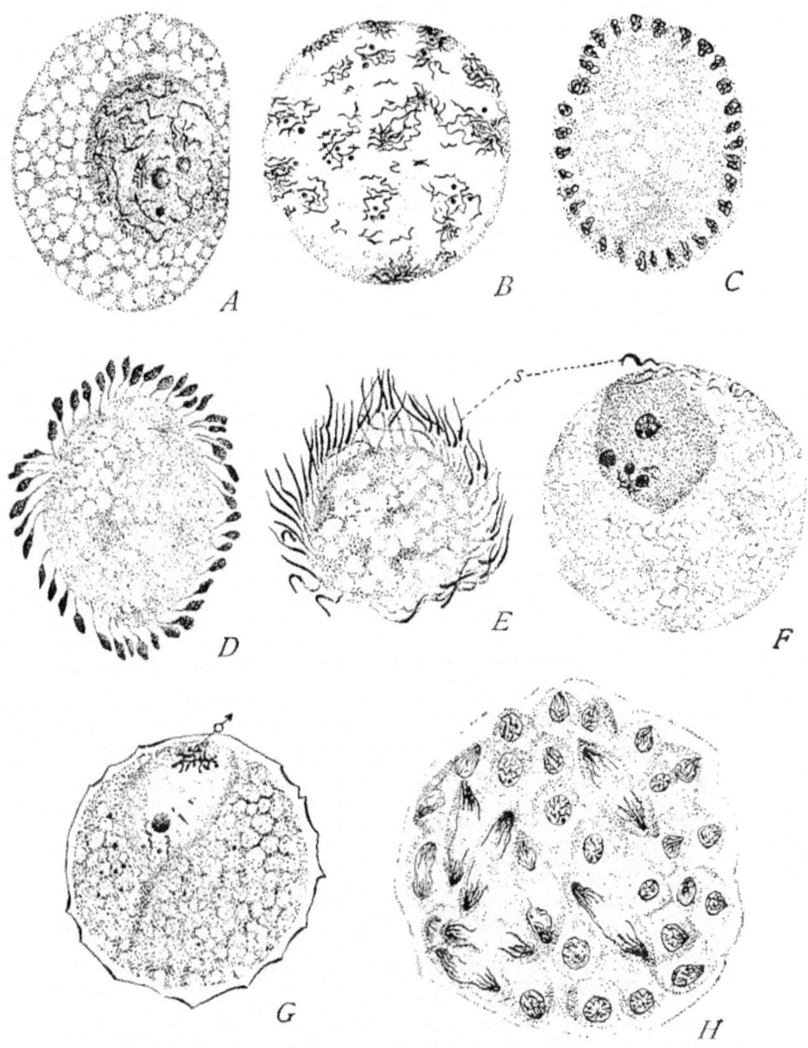

Fig. 89. — Conjugation and sporulation in *Klossia helicina*, Labbé. [SIEDLECKI.]
A-E. Formation of microgametes. *F.* Conjugation. *G.* Union of the male and female nuclei. *H.* Formation of spores.

amœboid stage during which the body-plasm becomes stored with melanin granules or metamorphosed hæmoglobin. They agree also in having an intra-corpuscular spore-forming stage (*schizogony*, Fig. 90, *F*, *G*, *L*, *M*), the spores bringing about auto-infection by penetrating new blood-corpuscles and there repeating the cycle; and under certain conditions they all produce flagellated bodies or spores (*Polymitus* form, *R*). They differ in the number of spores that are formed, although in the same variety the number appears to be inconstant, so that mere difference in number cannot be considered a good specific character. A much more satisfactory means of distinguishing them lies in the regular periodicity of spore-formation, which accompanies well-marked morbid symptoms in the patient. Thus, in *Plasmodium vivax*, spore-formation occurs every forty-eight hours (approximately); in *P. malariæ*, every seventy-two hours. The pyrexial attacks in all cases consist of similar symptoms, a stage of fever following one of chill and followed by a stage of perspiring. Spore-formation is the signal for a chill in the patient. The pigment granules (melanin) become aggregated in the centre of the cell, while the protoplasm breaks up into spores about them. The blood-corpuscle which contains the parasite then disintegrates, and the spores and melanin are liberated. It is supposed that this disruption of the corpuscle, by liberating a toxin (melanin) created and stored up by the parasite, is the direct cause of the attack. If this hypothesis, which is certainly based upon considerable evidence, should be verified by future investigation, the malaria-organism will be the only protozoön known to produce poisonous growth-products. Quinine in the system is apparently fatal to the parasite by preventing its growth and sporulation.

The spores of the malaria organism are not covered by a protective coating as in the majority of Telosporidia, and are, therefore, unsuited for an exposed life outside of their host. It was early recognized, however, that there must be an extra-corporeal period in the life-history of the parasite in order that the species should be perpetuated. That there actually is such a period is shown by the spread of the disease throughout a community.[1] Nevertheless, the whereabouts of the organism and its form during the extra-corporeal existence have remained a mystery until within the last few years. The key to the puzzle has been given by the flagellated body or *Polymitus* form. The blood when examined fresh from a malaria patient shows the ordinary form of the parasite, but after a short period of exposure to the air (10 to 30 minutes, Manson, '98), the parasite develops long flagelliform processes, which vibrate with great vigor and not infrequently break away from the body of the cell to swim about like *Spirilla* in the plasm. Danilewsky ('91) believed this stage to be an independent flagellated parasite of a special nature, and he named it *Polymitus*. Lavéran, Metsch-

[1] In Italy it is estimated that 2,000,000 people are ill every year with malaria (Santori, 1900).

nikoff, Manson, and others regarded it as an essential developmental stage of the malaria organism, and believed that the free-

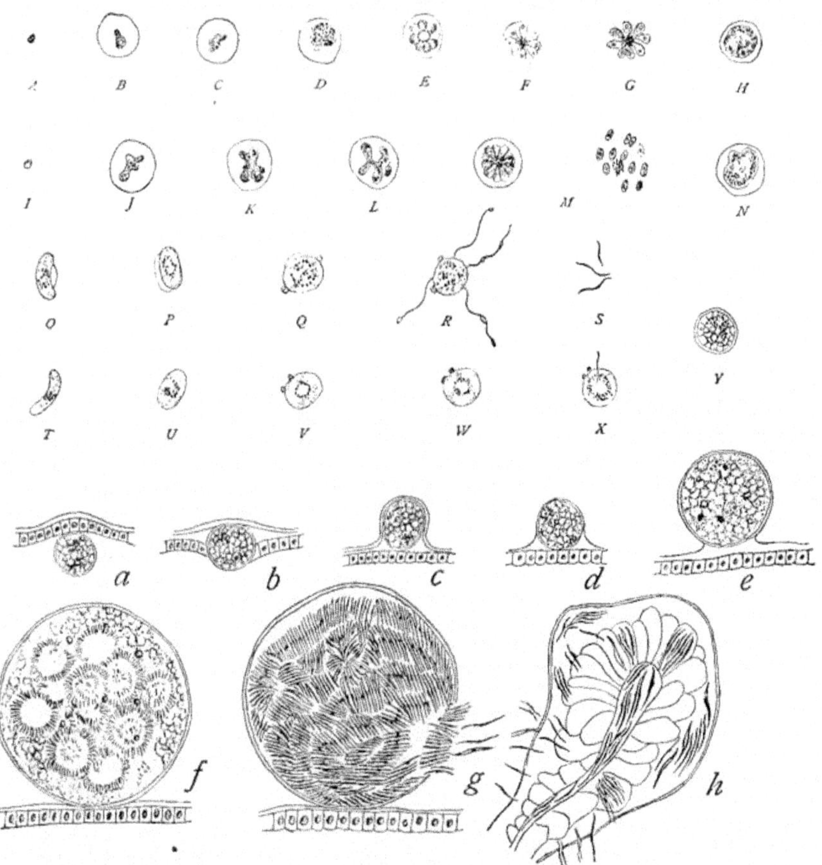

Fig. 90. — Life-history of Malaria, causing Sporozoa. [Ross and Fielding-Ould.]

A–F. Stages in the development of merozoites. *A–I.* The sporozoite. *B, C, D,* and *J, K, L.* The growing sporozoite in blood-corpuscles. *E.* Asexual reproduction (schizogony). *F, G,* and *M.* Liberation of merozoites and melanin granules. *O–W.* Stages in the development of sexual individuals. *R. Polymitus* form. *S.* Fully developed microgametes. *T–W.* Development of the female individual (macrogamete). *X.* Fertilization of a macrogamete by a microgamete. *Y.* The fertilized cell *copula*, with its vitelline membrane. *a–e.* The copula in the stomach of the mosquito (*Anopheles sp.*). *b.* The copula penetrating the epithelium which lines the stomach of *Anopheles*. *b, c, d, e.* Growth of the copula in the body-cavity of *Anopheles*. The small sphærules at *V, W,* and *X* are supposed to be analogous to polar bodies of metazoan eggs. *f.* Sporulation in the body-cavity of *Anopheles*. *g.* Liberation of the sporozoites. *h.* Salivary gland (in section), with sporozoites in the lumen, in the cells, and penetrating the membrane.

swimming detached processes are germs which reproduce the adult. Many others (Grassi, Felletti, Celli, Sanfelice, Sacharoff, Labbé, etc.), however, regarded these forms as degenerating parasites induced by the abnormal conditions of exposure and not present, normally, in the blood. Manson ('96) was one of the first to call attention to the fact that the formation of these so-called degeneration forms, which occurs only at the time when the blood is exposed to the air, is evidence of the beginning of extra-corporeal life. He suggested a theory that the *Polymitus* forms are flagellated spores, the extra-corporeal homologue of the intra-corporeal spores. He further suggested that, since the parasite is normally incased within a blood-corpuscle, it is unable to leave the host by its own efforts and must be removed by some blood-eating animal, probably a suctorial insect, such as a mosquito, common in swampy, malarial regions. At the same time ('96) Lavéran, in France, proposed an identical hypothesis. Subsequent investigation has given the complete confirmation of this hypothesis. The work of Major Ross, in India, of Koch, Grassi, and others, elsewhere, has established beyond a doubt, that extra-corporeal life of the parasite is spent in the mosquito, and that the disease is spread by these insects through inoculation. About eight or ten days after drawing blood from a malaria patient, the insects are able to transmit the germs to new hosts by inoculation through the proboscis. After a number of experiments, Ross and Grassi found that certain genera of mosquitoes, *e.g. Culex sp.*, are incapable of fostering the human parasite, while all species of the genus *Anopheles* are particularly susceptible.

The history of the parasite in the mosquito has been variously interpreted. Manson ('96, '98), on *a priori* grounds, suggested that the *Polymitus* form is developed after the blood is taken from the host into the colder digestive tract of the insect, the change of medium acting upon the parasite in the same way that the air does. Here, he argued, it penetrates an epithelial cell and repeats the life-history of an ordinary form, sporulating and increasing by auto-infection. MacCallum ('97), however, described a true conjugation between a *Polymitus* form and a free pigmented parasite in a very similar organism (*Halteridium* Labbé), which is parasitic in the blood of the American crow. In this case the impregnated *Halteridium* slowly changes form, becoming elongated and more or less worm-like, and moves about in the blood-plasm. Thus in this case the *Polymitus* form corresponds to a spermatozoön, and the ordinary individual, as in *Coccidium*, to an egg. Ross and Grassi, finally, have demonstrated the same relation in *Plasmodium malariæ*; the *Polymitus* form fuses with an ordinary individual in the intestine of the mosquito, and as in *Halteridium*, the copula becomes a motile individual which, after a short period, penetrates the epithelial cells lining the digestive tract (Fig. 90, *a–c*). Here it resembles one of the Coccidiida, growing at the expense of the cell-host and finally sporulating. The spores do not form protective coatings, but divide at once into sporozoites (*f*, *g*).

These make their way into the body cavity, or lymph spaces, of the mosquito, and ultimately find their way to the salivary glands, from which they may be deposited, together with the salivary fluid, in the blood of man (*h*). The life-history of a malaria organism thus involves a complete change of hosts, one phase being in the warm blood of man, the other a Coccidia-like stage in the Insecta, a group which above all others is noted for the frequency and number of sporozoan parasites. It is not improbable that a similar change of hosts occurs in the parasites belonging to this same group of Hæmosporidiida, which have been observed in birds (Lavéran, Labbé, '94); in reptiles (Labbé, '94; Langmann, '99); and amphibia (Labbé, '94; Langmann, '99); also it is possible that many of the uncertain forms, such as *Serumsporidium* of the Crustacea (Pfeiffer, '95) or *Lymphosporidium* of the trout (Calkins, '99), have a similar complicated life-history.

Apart from their pathogenic effects in man, the Sporozoa are frequently a pest in the lower animals. The Sarcosporidiida have already been mentioned as producing morbid symptoms resembling Trichinosis, in the domestic animals often leading to death. The Myxosporidiida occasion great loss to fish culturists by causing ulcers which ultimately result in the death of the fish, and to silkworm culturists on account of costly and extensive epidemics produced by them among the silkworms. These organisms (*Glugea bombycis*) were so disastrous to the silk industry during the years 1854–1867, that a loss was estimated of at least 1,000,000,000 francs (about $190,000,000). In regard to this epidemic Huxley ('70) writes: "In the years following 1853 this malady broke out with such extreme violence that, in 1858, the silk crop was reduced to a third of the amount which it had reached in 1853; and, up till within the last year or two, it has never attained half the yield of 1853. This means not only that the great number of people engaged in silk growing are some 30 millions sterling poorer than they might have been; it means not only that high prices have had to be paid for imported silkworm eggs, and that, after investing his money in them, in paying for mulberry leaves and for attendance, the cultivator has constantly seen his silkworms perish and himself plunged in ruin; but it means that the looms of Lyons have lacked employment, and that, for years, enforced idleness and misery have been the portion of a vast population which, in former days, was industrious and well to do."

The caterpillars, although infested by the parasites which were frequently so numerous that all of the organs of the body swarmed with them, were nevertheless able to produce the moth. The latter, though stunted and undeveloped, could lay eggs which themselves contained spores of the organism, and these spread the disease.

The disease was checked only by careful examination of the food of the caterpillar, and by microscopic examination of all eggs and rejection of the infected ones.

No remedy is known for the many other diseases due to Sporozoa, especially among domestic animals, or fresh-water fish, and careful prophylactic measures analogous to those employed in stopping the silkworm epidemic may be the only means of checking them. Such measures have already been successfully applied to prevent the spread of malaria, and the experiments which are now going on in all parts of the world justify the hope that this disease will be ultimately stamped out.

F. Inter-relationships of the Sporozoa

The Sporozoa, modified beyond doubt by adaptation to a parasitic mode of life, have ever been a puzzle to systematists. Kölliker ('48) early suggested that they are single cells, and included them in his Protozoa. Stein ('48) agreed with him as to their primitive structure, but was loath to regard them as single animal cells, and compromised by calling them *Symphyta*, a group of the Protozoa. Another view, developed by Henle ('45) and Bruch ('50) and taken up by Leydig ('51) and Leuckart ('52), was based upon the superficial resemblance of the Gregarinida to Nematode worms. It found little support, however, against Kölliker's view. Still another theory of the origin of the Sporozoa has been held by those who, following Gabriel ('75, '80), regard these forms as plants, placing them with the Mycetozoa, among the Fungi. Bütschli at first[1] favored the view that the Sporozoa are derived from the Rhizopoda, basing his belief upon the method of reproduction, general morphology, and physiology. Later, however,[2] he considered their relationship to the Flagellidia as much more close, not to the simplest forms, but to the higher types with a well-differentiated cuticle. The flagellum and mouth parts, he assumed, became lost with gradual adaptation to the intra-cellular mode of life, while the methods of reproduction became specialized in response to the requirements of a new environment. This view is strengthened by the close agreement in finer structures of the Gregarinida and the Flagellidia, especially as regards the differentiations of the cuticle and the presence of muscular elements. Their movements, too, recall those of the Flagellidia, especially certain species of *Astasia*, where, in the non-flagellated condition, the plasm moves forward by a peculiar peristalsis, while the secretion of a jelly from the sub-cuticular or cortical plasm is identical in the two groups. The nuclei show perhaps a closer resemblance to those of the Rhizopoda

[1] ('83), p. 479.

[2] ('84), p. 807.

than to the Flagellidia, but the conjugation processes are much more like those of the Flagellidia. Haeckel follows Bütschli in regarding the Sporozoa in this light, and derives them from the Phytoflagellida through adaptation, first to a saprophytic and then to a parasitic mode of life. Wasielewsky also favors the flagellate origin, basing his opinion, however, upon the uncertain ground of flagellated swarm-stages of certain Sporozoa as well as upon the general resemblance to the Astasiidæ. In general, however, it must be admitted that there is very little support for any one of these theories, and all attempts to trace the origin of the Sporozoa upon the mere basis of their present degenerate condition are highly speculative.

CLASSIFICATION

CLASS III. **SPOROZOA.** The Sporozoa are Protozoa which are never provided with flagella or cilia in the adult state. They are always endoparasites in cells, tissues, or cavities of other animals, and food is taken in by osmosis. Reproduction is always by spore-formation, and germs (*sporozoites*) are produced either directly from the parent, or indirectly through spores.

Subclass I. **TELOSPORIDIA.** Sporozoa in which spore-formation ends the individual life, the entire cell then forming spores.

Order 1. **GREGARINIDA.** Telosporidia possessing a distinct membrane, with myonemes during adult life, locomotion being accomplished mainly by their contraction. The young stages alone (*cephalonts*) are intra-cellular parasites, the adults (*sporonts*) being found in the digestive tract or the body cavities. Sporulation takes place after or without conjugation, but within a cyst which is never formed while the parasite is intra-cellular.

Suborder 1. **CEPHALINA.** Gregarinida possessing an organ for attachment (*epimerite*), and with or without septa dividing the cell into chambers.

Tribe 1. **Gymnosporea.** The adults are solitary or associated; the sporozoites are formed directly from the adult without encystment.

Family 1. **Aggregatidæ.** Colonies consisting of two or more individuals. Several residual protoplasmic masses are found during sporulation in each cyst. Genera: *Aggregata* Frenzel ('85).

Family 2. **Porosporidæ.** The individuals are usually solitary. The sporozoites are arranged in groups around a central residual mass. Genera: *Porospora* A. Schn. ('75).

Tribe 2. **Angiosporea.** Cephalina with well-developed spores, which are provided with spore-membranes (*epispores* and *endospores*).

Family 1. **Didymophyidæ.** Chain-forming aggregates, two individuals being so closely joined as to appear like one with three chambers. Genera: *Didymophyes* Stein ('48).

Family 2. **Gregarinidæ.** The individuals are solitary or associated. The epimerite is simple and regular. The cysts may or may not have spore-ducts. Genera: *Gregarina* Dufour ('28); *Gamocystis* A. Schn. ('75); *Hirmocystis* Léger ('92); *Hyalospora* A. Schn. ('75); *Euspora* A. Schn. ('75); *Sphærocystis* Léger ('92); *Cnemidiophora* A. Schn. ('82); *Stenophora* Labbé ('99).

Family 3. **Dactylophoridæ.** The epimerite is asymmetrical and irregular. Genera: *Rhopalonia* Léger ('93); *Echinomera* Labbé ('99); *Trichorhynchus* A. Schn. ('82); *Pterocephalus* A. Schn. ('87); *Dactylophorus* Balbiani ('89).

Family 4. **Actinocephalidæ.** The individuals are always single. The epimerite is simple or lobed and symmetrical. The cysts open by simple dehiscence. The spores are boat-shaped, bi-conical, or cylindro-conical. Genera: *Sciadiophora* Labbé ('99); *Anthorhynchus* Labbé ('99); *Pileocephalus* A. Schn. ('75); *Amphoroides* Labbé ('99); *Discorhynchus* Labbé ('99); *Stictospora* Léger ('93); *Schneideria* Léger ('92); *Asterophora* Léger ('92); *Stephanophora* Léger ('92); *Bothriopsis* A. Schn. ('75); *Coleorhynchus* Labbé ('99); *Actinocephalus* Stein ('48); *Pyxinia* Hammerschmidt ('38); *Légeria* Labbé ('99); *Phialoides* Labbé ('99); *Beloides* Labbé ('99).

Family 5. **Acanthosporidæ.** Solitary. The spores are provided with equatorial or polar spines. Genera: *Corycella* Léger ('92); *Acanthospora* Léger ('92); *Ancyrophora* Léger ('92); *Cometoides* Labbé ('99).

Family 6. **Menosporidæ.** Solitary. The epimerite is on a long neck. The spores are crescent-shaped. Genera: *Menospora* Léger ('92); *Hoplorhynchus* Carus ('63).

Family 7. **Stylorhynchidæ.** The spores are formed in chains, and the cysts have a double envelope. Genera: *Lophocephalus* Labbé ('99); *Cystocephalus* A. Schn. ('86); *Oocephalus* A. Schn. ('86); *Sphærorhynchus* Labbé ('99); *Stylorhynchus* Stein ('48).

Family 8. **Doliocystidæ.** Cephalina without septa dividing the cell into *protomerite* and *deutomerite*, but consisting of a single chamber with epimerite. Genera: *Doliocystis* Léger ('93).

Suborder 2. **ACEPHALINA.** Gregarinida consisting of a single chamber, and without epimerite. They are parasites in the body cavity or cavities of the various organs of different animals. Genera: *Monocystis* Stein ('48); *Zygocystis* Stein ('48); *Zygosoma* Labbé ('99); *Pterospora* Racovitza and Labbé ('96); *Cystobia* Mingazzini ('91); *Lithocystis* Giard ('76); *Ceratospora* Léger ('92); *Urospora* A. Schn. ('75); *Gonospora* A. Schn. ('75); *Syncystis* A. Schn. ('86).

Order 2. **COCCIDIIDA.** Telosporidia having a spherical or oval form, without a free and motile adult stage, and never amœboid. Sporulation takes place within cysts formed while the organism is an intra-cellular parasite.

Family 1. **Disporocystidæ.** The cell forms two sporocysts, each sporocyst forming two or four sporozoites. Genera: *Cyclospora* A. Schn. ('81), with two sporozoites; *Isospora* A. Schn. ('81), and *Diplospora* Labbé ('93), with four or more sporozoites.

Family 2. **Tetrasporocystidæ.** Each organism forms four sporocysts, each of which produces two sporozoites. Genera: *Coccidium* Leuckart ('79) (including *Goussia* Labbé); *Crystallospora* Labbé ('96).

Family 3. **Polysporocystidæ.** Each organism produces an indefinite number of sporocysts, each of which produces one sporozoite,—[*Barrouxia* A. Schn. ('85), *Diaspora* Léger ('99)], two sporozoites,—[*Adelea* A. Schn. ('75), and some species of *Hyaloklossia* Labbé ('96)], three sporozoites,—[*Benedenia* A. Schn. ('75), or four, Genus *Klossia* A. Schn. ('75)].

Order 3. **HÆMOSPORIDIIDA.** Sporozoa of small size living in the blood-corpuscles or plasm of vertebrates. The adult form is mobile, and in some cases is provided with myonemes. They reproduce by endogenous or asexual spore-formation while in the host, and by exogenous spore-formation after conjugation. Genera: *Lankesterella* Labbé ('99); *Caryolysus* Labbé ('94); *Hæmogregarina* Danilewsky ('85); *Caryophagus* Steinhaus ('89); *Halteridium* Labbé ('94); *Hæmoproteus* Kruse ('90); *Plasmodium* Marchiafava & Celli ('85); *Laverania* Grassi & Feletti ('92); *Cytamœba* Labbé ('94).

Subclass II. **NEOSPORIDIA.** Sporozoa which form sporocysts throughout life; the entire cell is not used in the formation of spores.

Order 1. **MYXOSPORIDIIDA**. Neosporidia of amœboid or spherical shape; multinuclear. The initial free stage is passed in the cavities of the organs, or in the tissues of the host. In sporulation a definite or an indefinite number of sporoblasts is formed, each of which gives rise to one or several spores; the latter are provided with one or several polar capsules, which contain coiled threads like a nematocyst. Each spore gives rise to one amœboid sporozoite.

Suborder 1. **PHÆNOCYSTINA**. Spores with polar capsules distinctly visible when fresh.

Family 1. **Myxidiidæ**. Myxosporidiida forming two or more spores at the same time. Spores variable in form inclosing two polar capsules. Genera: *Sphærospora* Thélohan ('92); *Leptotheca* Thél. ('95); *Ceratomyxa* Thél. ('92); *Myxidium* Bütschli ('82); *Sphæromyxa* Thél. ('92); *Cystodiscus* Lutz ('89): *Myxosoma* Thél. ('92).

Family 2. **Chloromyxidæ**. The spore has four polar capsules. Genera: *Chloromyxum* Mingazzini ('90).

Family 3. **Myxobolidæ**. Adult stages very rare, ordinarily found encysted in the tissues; usually polysporous. The spores have one or two polar capsules. Genera: *Myxobolus* Bütschli ('82); *Henneguya* Thél. ('92).

Suborder 2. **MICROSPORIDIINA**. Myxosporidiida in which the spores have but one polar capsule, which is invisible in the fresh state without the use of reagents.

Family 1. *Nosematidæ*. With a bivalve spore. Genera: *Nosema* Nägeli ('57) (*Glugea* Thélohan, '92); *Plistophora* Gurley ('93); *Thélohania* Henneguy ('92).

Order 2. **SARCOSPORIDIIDA**. Sporozoa in which the initial stage is passed in muscle-cells of vertebrates. The form is usually elongate, tubular or oval, or sometimes spherical. It forms cysts with a double membrane, in which are formed kidney-shaped or falciform sporozoites, or else spores (?), provided with a polar capsule and projectile thread. Genera: *Sarcocystis* Lankester ('82).

SPOROZOA INCERTÆ SEDIS

Amœbosporidia. Sporozoa possessing an amœboid body, and reproducing either by division or by spore-formation after conjugation. Genera: *Ophryocystis* A. Schn. ('84).

Serumsporidia. Sporozoa which reproduce by division (?) or by spore-formation, the sporozoites being minute oval or spherical bodies They are found in the cavities or cœlomic fluids of Invertebrates and Vertebrates. Genera: *Serumsporidium* L. Pfeiffer ('95); *Blanchardina* Labbé ('99); *Lymphosporidium* Calkins (1900).

SPECIAL BIBLIOGRAPHY V

Gurley, R. R. — The Myxosporidia, or Psorosperms of Fishes, and the Epidemics produced by them. *Bull. U. S. Fish Comm.* XI., 1893.

Labbé, A. — Recherches zoologiques, cytologiques et biologiques sur les Coccidies. *Arch. d. zool. expér. et gén.* (3) IV., pp. 517–654, 1896.

Labbé, A. — Sporozoa. In Das Tierreich, *Berlin*, 1899.

Légér, L. — Recherches sur les Gregarines. *Tablettes zoologiques*, III., pp. 1–182, 1892.

Leuckart R. — Die Parasiten des Menschen. *Leipzig*, 1879.

Ross, D. — On Some Peculiar Pigmented Cells found in Two Mosquitoes fed on Malarial Blood. *Brit. Med. and Surg. Jour.*, 1897, pp. 1786–1788.

Schneider, A. — Sur les psorospermies oviformes ou Coccidées, espèces nouvelles ou peu connues. *Arch. d. zool. expér. et gén.* (1) IX., pp. 387–404, 1881.

Siedlecki, M. — Étude cytologique et cycle évolutif de la coccidie de la seiche. *Ann. d. l'Inst. Pasteur*, XII., 1898.

Siedlecki, M. — Ueber die geschlechliche Vermehrung der Monocystis ascidiæ R. Lank. *Bull. d. l'Acad. d. Sci. d. Cracovie*, 1899.

Thélohan, P. — Recherches sur les Myxosporidies. *Bull. Sci. de la France et de la Belgique*, XXVI, pp. 101–394, 1895.

Wasielewsky, Von. — Sporozoenkunde, ein Leitfaden für Aerzte Tierärzte und Zoologen. *Jena*, 1896.

CHAPTER VI

THE INFUSORIA

"Die Infusionsthiere gehören in den Kreis der Protozoen. Innerhalb desselben bilden sie eine eigene und zwar die am höchsten stehende Klasse." — STEIN.[1]

As was long since clearly recognized by Stein, the Infusoria are the most highly differentiated of all Protozoa and often attain a degree of complexity which is perhaps greater than in any other cells. Their form varies considerably in the several divisions, but all are characterized by certain structural features by which they can be distinguished at a glance. All are provided with cilia which may be retained throughout life (*Ciliata*), or may be replaced in the adult phases by suctorial tentacles, cilia being present only during the embryonic phases (*Suctoria*); they possess mouth parts which are adapted for swallowing, for simple ingestion, for sucking, or which may be entirely degenerate through parasitism; and they are provided with two kinds of nuclei, known as *macronuclei* and *micronuclei*. They reproduce by simple division and by budding, or rarely by spore-formation.

I. THE CILIATA

Among the Ciliata the arrangement of the cilia upon the body affords a character which was first used by Stein ('59), and is still retained as a means of distinguishing the subdivisions of this group. In the first and probably the most primitive type, *Holotrichida*, the cilia are arranged uniformly over the entire body of the animal and show no regional differentiations (Fig. 91, *A*). In the second type, *Heterotrichida*, the cilia are uniform over the main portion of the body, while a specialized set fused into a curved series of firm vibratory plates, or *membranelles*, are found in an *adoral zone* about the mouth (*B*). In the third type, *Hypotrichida*, the body is flattened dorso-ventrally and the dorsal side is entirely free from cilia, while on the ventral side the cilia are frequently fused together into stiff seta-like organs, the *cirri*, and as in the Heterotrichida, they may form a curved line of membranelles around the mouth (*C*). Finally, in the *Peritrichida*, the highest type of this class, the cilia are reduced to one or two bands or girdles in addition to the adoral zone (*D*).

Although, with the exception of the motile organs, no single item of structure is found here which is not occasionally met with in other

[1] ('59), p. 54.

classes of Protozoa, yet in no other class are they all present in a single cell. Each of the different elements thus brought together has a definite function to play in the life of the organism, and intra-cellular division of labor is developed to a high degree.

Leading an active life and forced to seek food in all sorts of places, from the clearest waters to the internal fluids of various hosts, the Ciliata have acquired a very great diversity of form. The simplest and probably the most primitive forms are mónaxonic, the mouth being anterior and the anus posterior (Fig. 91, *A*). Symmetry, however, is the exception and asymmetry the rule, the latter condition arising by the gradual shifting of the mouth to a more or less well-

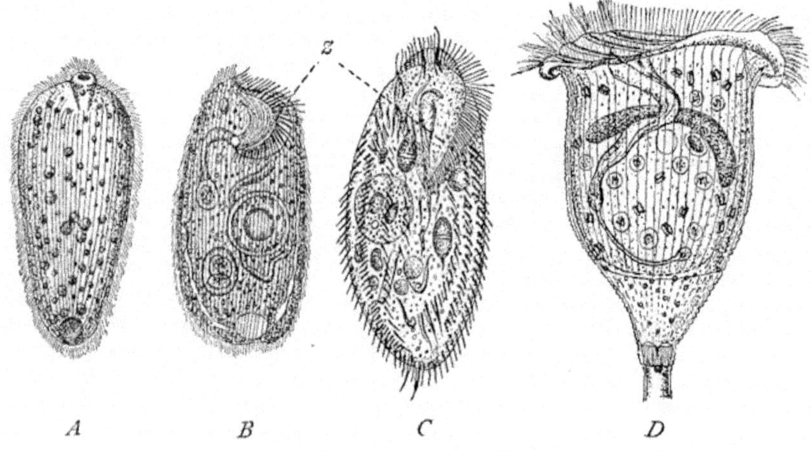

Fig. 91. — Types of Ciliata. [BÜTSCHLI.]

A. Prorodon teres Ehr.; an holotrichous form. *B. Climacostomum virens* Ehr.; an heterotrichous form. *C. Pleurotricha grandis* St.; an hypotrichous form. *D. Vorticella umbellaria*; a peritrichous form. *z*, adoral zone.

defined ventral side, while the anus becomes more or less dorsal. The functional anterior end may thus be either ventral, or superior to the mouth, when the latter becomes sub-terminal. The simple monaxonic ground-type is subject to other minor variations among the Holotrichida, which point the way toward the more striking deviations among the Heterotrichida and the Hypotrichida. A frequent modification is the anterior prolongation of what might be considered the upper lip, as in *Dileptus* or *Lionotus* (Fig. 92), where bilaterality and asymmetry are well established. The mouth becomes more and more ventral in the family Trachelinidæ (Holotrichida), while in the order Hypotrichida it is always ventral and the original monaxonic structure is replaced by dorso-ventral differentiation and complete asymmetry.

A. Protoplasmic Structure

As in all other Protozoa, the endoplasm consists of alveoli of varying size and arrangement, the network being built up of plasm of greater or less density. Within the endoplasm there is a constant streaming of the granules, which varies greatly in the different species and even in the same individual under different conditions. In some

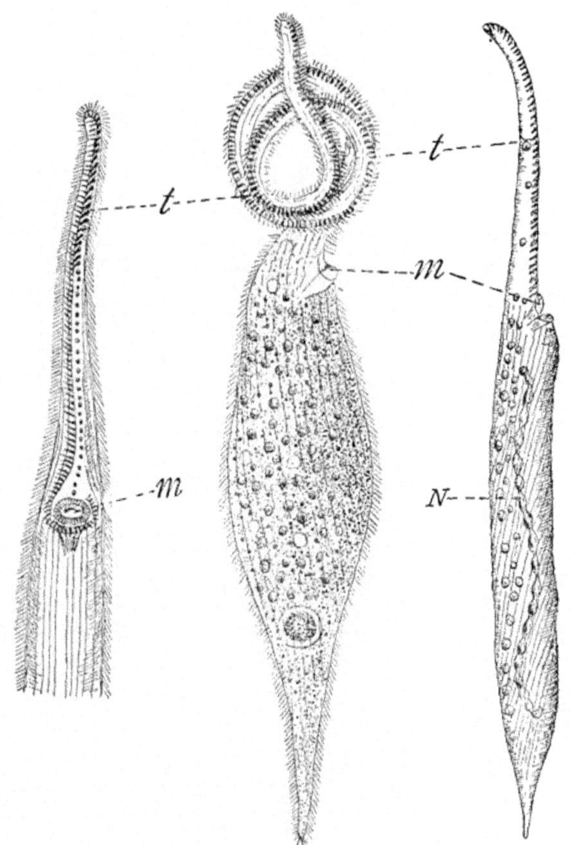

Fig. 92. — *Dileptus anser* O. F. M. [Bütschli.]
m, mouth; *N*, macronucleus; *t*, trichocysts on the tentacle-like end.

cases the course of the streaming recalls cyclosis in some plant-cells, and frequently follows a well-defined and unvarying route (*Colpoda cucullus*, or *Paramœcium bursaria*). In all cases the current seems to start from the mouth, passing backward around the cell either to the right or left, then forward and back to the mouth (Bütschli). In this stream, circulating through the body of the animal, are carried food-products in various stages of digestion and assimilation, as well as excretory products in the form of granules similar in all respects to those found in other Protozoa. Engelmann ('83) found that in one of the Vorticellidæ (*Vorticella campanula*) the animal is colored by diffuse green pigment, which he took to be chlorophyl, and he further showed that oxygen is generated and that the animal can assimilate like a plant, but that it is not limited to this kind of nutrition, since it also takes in solid food through the mouth. Other colored particles which are found in various kinds of Ciliata, especially in those forms which subsist upon plant food, are presumably due to the coloring matter contained in the food. Schewiakoff ('89) has shown that the colored balls which appear in some cases are merely fluid drops colored with the pigment contained in *Oscillaria* and other vegetable cells. In many cases, also, green algal cells live as symbionts within the endoplasm. Le Dantec ('92) found that these cells (*Zoochlorella*) are apparently taken in as food and become inclosed within gastric vacuoles, the fluids of which have no effect upon them. Soon the vacuoles disappear and the algæ are left free in the plasm, where they live and multiply. There are also various fats and excretory products either crystalline or granular in form. The crystals, according to Schewiakoff ('93), are granules of calcium phosphate.[1] Among the pigmented inclusions of the cell must be noted the so-called "eye-spots" or stigmata, which, in some instances, are accompanied by lenticular differentiations of various kinds. These pigmented spots are, as a rule, mere heaps of granules colored either red, brown, black, or orange, and are probably the same in function as the similar products in Mastigophora.

The protoplasm becomes more dense toward the periphery, and, as in the Sarcodina, it finally becomes too compact for the granules to penetrate. This outer portion is, therefore, comparable to the ectoplasm of the less differentiated Protozoa. The importance of this layer is seen in the fact that nearly all of the organs which characterize the Ciliata, including the myonemes, cilia, membranes and membranelles, the trichocysts, nematocysts, and the complex membranes and tests, are modifications of, or are produced by, the ectoplasm. In some forms the thickened plasm immediately adjoining the endoplasm is distinctly marked off from the more external portions, forming a continuous layer around the entire body. This layer, which is in reality neither ectoplasm nor endoplasm, but intermediate between the two, is called the *cortical plasm* (*Rindenparenchyma*, Stein), and is characterized by the reduced size and number of its vacuoles, by the absence of granules and streaming motion, and

[1] Cf. p. 286, *infra*.

by its fixity in the cell. It is occasionally thickened to form the denser ends of the body, as in the tail of *Stentor*. In some cases, also, processes from the cortical plasm invade the endoplasm to surround the nucleus and hold it in a fixed position in the body (*Dasytricha*). The cortical plasm, furthermore, is the seat of the peculiar and characteristic offensive and defensive trichocysts. In some forms, probably representing the primitive condition, these are distributed about the body (*Paramæcium*), but, even more than the cilia, they have been subject to reduction in most parts of the body until, in the majority of forms, they are restricted to a limited area, while in the Hypotrichida and Peritrichida they occur only sporadically. In *Prorodon* (Holotrichida), the trichocysts are found in the anterior end only, and in the family *Trachelinidæ* they are found only on the ventral side, while in *Trachelius*, *Dileptus*, etc., they are on the ventral side of the anterior process (Fig. 92). In *Lionotus*, they are reduced to a single line along the ventral side of the anterior process.

The trichocysts are so minute that their finer structure has not been definitely made out, although a few different types have been studied (Fig. 12, *C*, p. 39). Rod-like forms have been seen in *Loxophyllum*, *Lionotus*, and *Strombidium*, and spindle forms in *Paramæcium*, *Frontonia* and *Nassula*. When protruded from the body they are, for the most part, apparently of the same size and shape as when within the ectoplasm. Occasionally when protruded, however, they have small hooks or swellings on the end (Maupas). They vary in size from three to twelve microns when within the body, but when protruded they measure from thirty to sixty microns. The cause of the protrusion is unknown; certain reagents act as irritants and cause them to explode and throw out the long threads. Their function, too, is purely conjectural, although it is generally supposed that they serve as defensive weapons. In some cases they appear to serve as weapons of offence as well, especially in those ciliates where they are limited in number to a comparatively few large ones. According to Maupas ('83), these Infusoria chase their prey and launch their trichocyst darts, which penetrate the outer coating of the victim and paralyze it, possibly through the action of some noxious fluid. Forms much larger than the hunters are frequently brought down in this way, to be swallowed either whole or piecemeal. The attack is not necessarily fatal, for the larger forms frequently revive.[1]

The position of the trichocysts is primarily in the cortical plasm, but they are rarely entirely immersed, being much more frequently suspended in the streaming endoplasm. They are occasionally drawn

[1] See *ante*, p. 50.

out from the cortical plasm and are then carried about in the stream.

In addition to the trichocysts, some of the Ciliata carry still more effective weapons in the form of *nematocysts*. In *Vorticella umbellaria* (Clap. & Lach.), Engelmann described from twelve to twenty pairs of capsules, each of which contained a coiled thread which, as in the Cœlenterata, could be thrown out upon irritation.

The cortical plasm may be considered the inner portion of the ectoplasm, the outer portion of which forms the covering of the animal, the membrane, cuticle, or pellicle. The latter is extremely variable in thickness and in complexity. It is apparently homologous with the external membrane of ordinary animal cells, and, according to Bütschli, is formed by condensation of the protoplasmic ground substance, and, as Stein ('59) first maintained, is in no sense a secretion.[1] Bütschli ('88), Schuberg ('87), and many others regarded it as the agglutination of the outer thickened lamellæ of the external alveoli into a continuous membrane. In forms where the cortical plasm is absent, as in Hypotrichida, the cuticle, or, as Bütschli prefers to call it, the *pellicle*, forms a thin coating to the cell and lies directly upon the endoplasm. In many cases the outer protoplasm either becomes changed into, or else secretes, an external casing or house, which may be either loose or tight-fitting. This covering may be of jelly (*e.g. Ophrydium*), or of chitin (*e.g. Folliculina*), or of a horny product without any mineral elements (*e.g. Coleps*). In *Coleps hirtus* (Ehr.) the horn-like covering is tight-fitting, and composed of separate pieces, which form four girdles about the body (Fig. 93). Each girdle is composed of separate pieces, each of which is straight on one edge and serrated upon the other in such a

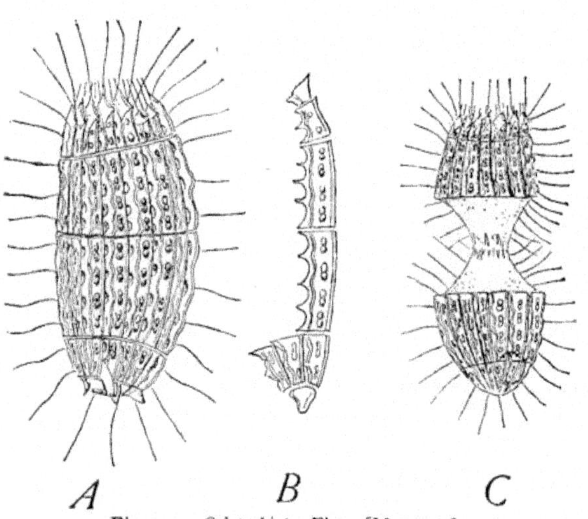

Fig. 93.— *Coleps hirtus* Ehr. [MAUPAS]
A. Side. *B*. One of the 15 rows of plates composing the test. *C*. Division-phase.

[1] Cf. Stein ('59), p. 56.

manner that, when they are put together, the serrations slightly overlap the straight edge of the next adjacent piece, thus leaving openings for the protrusion of the cilia. The lower end is covered by separate pieces, which open centrally for excretion. On the anterior end, the mouth, crowned by a ring of large cilia, is protected by a set of plates with teeth-like projections, which act somewhat like the similarly arranged teeth of a sea-urchin.

In a great many cases the outer body wall is marked by striations of various kinds showing the lines of insertion of the cilia. A typical example is shown in the membrane of *Holophrya*, one of the Holotrichida (Fig. 91, *A*, and 97, *g*). Here the cilia are inserted in regular lines which run from the anterior to the posterior end. In other cases, as in *Lembadion* (Holotrichida), the cilia are inserted on minute papillæ, which lie in rows upon the cuticle with more or less distinct furrows between them, thus forming secondary, but very distinct, markings in addition to the primary lines formed by the insertion points of the cilia. That the striation is due to the ciliation can be easily seen in cases where the cilia are absent from one portion of the body and present in others, as in many of the Holotrichida. The rows are not always straight, as in *Holophrya*, but are variously changed through the alteration of the axial relations. The most frequent variation from the primitive condition is the spiral arrangement (*e.g. Lacrymaria coronata*, Fig. 94), where the course of the cilia has become changed by the alteration in the position of the mouth. A very curious type of striation is seen in *Dasytricha ruminantum* (Schuberg, '88) (Holotrichida), where the striations do not converge at the mouth as they do in the majority of forms, but in a line above it. The exception is significant, however, as showing the line of the shifting of the mouth, the path being marked by the meeting points of the converging striæ (Fig. 95).

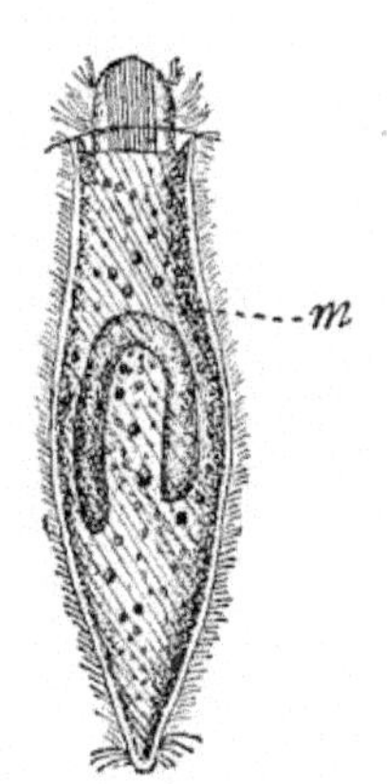

Fig. 94.—*Lacrymaria coronata* Cl. and Lach. [BUTSCHLI.] *m*, spirally wound rows of cilia.

The external markings were early recognized by Ehrenberg, who interpreted them as the insertion points of the cilia as described above. Stein, however, held that the markings are invariably due to the presence of myonemes which form the insertion base of the cilia. Both observers were right in part, for striations in some of the Heterotrichida and Peritrichida are due to the presence of myonemes, but in the Holotrichida, where, with one or two exceptions (*Holophrya*, *Prorodon*), myonemes do not occur, the markings are unmistakably due to the cilia.

The myonemes are ectoplasmic differentiations which are contractile in nature and are formed, according to Bütschli and Schewiakoff, from the walls of the alveoli which make up the sub-cuticular layer of the membrane. Although probably arising in the peripheral alveolar region, these threads occasionally become separated from this position and are then found in the cortical plasm or even in the endoplasm. Myonemes are most highly differentiated and are best known in the Vorticellidæ, where the sudden contraction of the bell, or the instantaneous rolling-up of the stalk, are due to their action. While the most conspicuous myonemes in *Vorticella* run from the centre of the disk to the very base of the stalk, Entz ('91) has described additional fibrils which have a similar but less important function. According

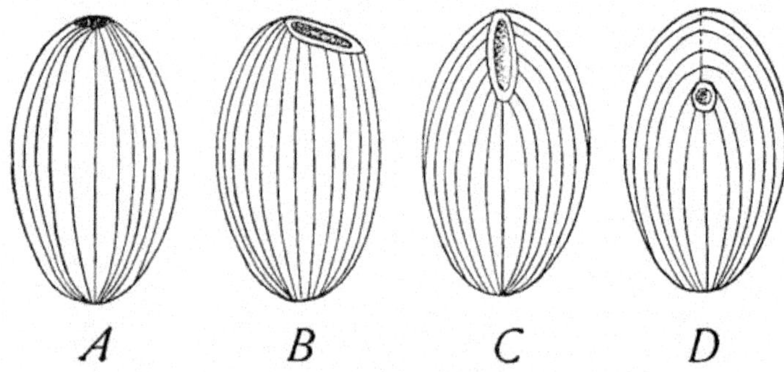

Fig. 95.—Supposed change of position of the mouth in Ciliata. [BÜTSCHLI.]
A. Original position (as in *Holophrya*). *B*. The mouth has become elongated (as in *Enchelys* or *Spathidium*). *C*. Similar stage from the ventral side. *D*. The mouth has become closed behind, leaving the opening away from the body extremity. The markings on the membrane now meet in the line represented by the original mouth-slit (*e.g.* in *Glaucoma*).

to this observer there are two sets of myonemes, one internal, the other external. Each set includes two groups of myonemes, one circular in its course, the other longitudinal. The external layer, observed by Lachmann ('56) and Stein ('59, '67) but denied by many, is formed of a large, single fibre composed of fibrillæ, which winds spirally about the bell from the junction of the peduncle to the centre of the disk. It is this myoneme, Entz maintains, which gives the annulate appearance to the bell, and, like a muscle-fibre, it is characterized by fine transverse striations. A second circular set is formed by another single fibre, which, however, is confined to the peristome disk, and is located deep in the ectoplasm (Fig. 91, *D*). This fibre takes only a few spiral turns at the base of the elevated disk and around the edge of the collar, and functions as a sphincter-muscle to close over the disk. Two sets of longitudinal myonemes complete the muscular system. Of these, the external set, lying between the two circular myonemes, consists of fine fibres running from the peduncle to the disk where their course is radial. The largest and most important of all of the myonemes are those forming the fourth set. These are longitudinal muscle-fibres of considerable thickness running from the centre of the disk radially toward the periphery, then continuing down the sides of the bell as far as the ciliary girdle (*Wimperring*), where they leave the wall of the body and come together to form the thick muscle-strand of the stalk. The latter highly contractile organ consists of a wall and of the central, contractile strands which are bathed with a fluid contained within the walls of the stalk. The wall itself, according to Entz, but contrary to Bütschli, is a continuation of the living wall of the bell, in which membrane and underlying mus-

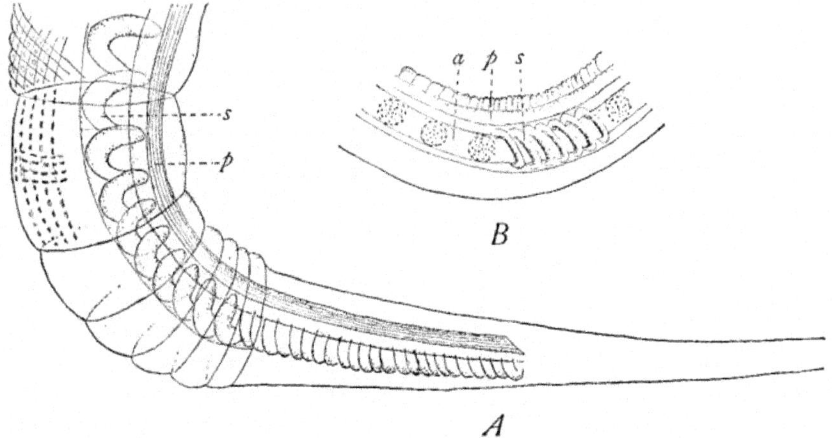

Fig. 96. — *Zoothamnium arbuscula* Ehr. [ENTZ.]

A. Lower portion of main trunk. *B.* One of the branches of the main trunk. *a*, axoneme; *p*, spasmoneme; *s*, spironeme.

cular structures can be distinguished, as in the main portion of the body. Bütschli, on a less substantial basis, described the stalk as a secretion similar to the stalks of the Mastigophora and Sarcodina, and chitinous in composition. The main strand within the stalk is formed by the collection of the strands of the inner longitudinal myonemes, and is covered by a delicate sheath which separates it from the fluid or gelatinous matter surrounding it. Bütschli regards this sheath as a continuous coat from the alveolar layer of the bell. The strand has three threads which Entz calls *spasmoneme*, *spironeme*, and *axoneme* (Fig. 96). The fibres of the first run to the base of the stalk. The other two are closely connected, and both are made up of microsomes, which Entz described as nucleus-like granules (*karyophans*) surrounded by an ovoid matrix (*cytophan*). These granules, so conspicuous in the stalks of *Vorticella*, evidently correspond to the *Elementar-Granula* (Greeff, '71) or cyto-microsomes. Entz figured them as arranged in rows like a string of beads. Without going deeply into the subject, which is far from settled, it will suffice here to state that two views are now held as to the seat of contractility in the stalks of *Vorticella*. One set of observers hold that the outer membrane of the stalk is the contractile portion, and that the contained thread merely counteracts the force of the membrane, which tends to contract and roll up the stalk. In other words, the myonemes of the spasmoneme are regarded as elastic and not as contractile fibrils, at rest when the stalk is coiled, active when the bell is extended (Cohn, '62; Metschnikoff, '63; Rouget,

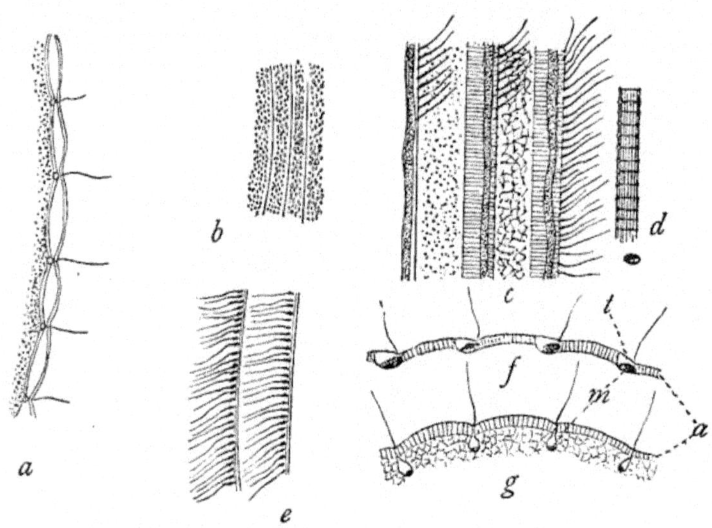

Fig. 97.—Myonemes and cilia. [METSCHNIKOFF, BÜTSCHLI, and JOHNSON.]
a, b, d, e. Cuticle and myonemes of *Stentor cæruleus*. *d.* More highly magnified piece of myoneme. *g.* Optical section through the body wall of *Holophrya discolor*.

'61; Schaaffhausen, '68; Entz, '91). Entz described the membrane of the stalk and the spasmoneme as antagonistic elements. The former, which stretches out while at rest and contracts when irritated, opposes the latter, which acts in the reversed manner. The axoneme he regarded as a sort of nerve-centre. The opponents of this view hold that the rolling of the stalk is accomplished by the contraction of the muscle-like spasmoneme (Stein '67, Clap. & Lachmann '58, Engelmann '76, Bütschli '88, etc.).

The myonemes lie in minute canals according to Bütschli ('88) and Schewiakoff ('89); in direct contact with the plasm according to Johnson ('93) and Entz ('91), and they probably vary in position in different forms. The structure and position of a myoneme in the ectoplasm can be more easily seen from the accompanying figure than from a description (Fig. 97).

Other contractile elements are occasionally found: the most remarkable, perhaps, is the peculiar muscular band which surrounds the peristome of *Bursaria truncatella.* This highly differentiated muscular organ, which functions as a sphincter, is, like the myonemes, derived from the alveolar layer immediately below the pellicle.

The ectoplasm appears throughout to be the seat of motion. Not only are the contractile myonemes differentiations of this important layer, but the cilia and all of their modifications are likewise derived from it. The cilia themselves appear to be mere prolongations of the alveolar layer. They are minute, probably of similar diameter throughout, and except for regional differentiation in the vicinity of the mouth, are of uniform length. As a rule, they are inserted upon minute elevations or papillæ on the cuticle and appear to be connected by minute fibrils with the myonemes. The finer structure of the cilia has not been satisfactorily made out, but the present results tend to the view that they are simple, firm threads without differentiations. Unlike flagella, they act in unison, and their motion is that of a paddle rather than a lash, as in flagella. Jensen ('93) has figured the absolute lifting power of the ciliary apparatus of *Paramæcium* at 0.00158 milligrammes, or nine times the weight of the animal. The cilia are grouped together in various ways, forming more or less complex motile organs. These are rarely seen in Holotrichida, but in the other orders they may be pointed aggregates (cirri), plate-like vibratile organs (membranelles), or broad, undulating membranes. All of these modifications are found in the Hypotrichida, where the motile apparatus is especially characteristic. The arched dorsal surface is without cilia, but occasionally holds a varying number of bristles which have, possibly, a sensory function. On the flat, ventral side of the most primitive forms of this order, the cilia are very generally distributed (*Peritromus*, Fig. 113, *B*), but in the more differentiated forms they are reduced in number, and modified into cirri, membranes, and membranelles. In many forms they may be entirely absent, the only motile apparatus being the membranelles on the ventral side about the mouth. These form the adoral zone, which stretches from the mouth forward on the left side of the peristome, and as far as the dorsal anterior region. In some cases a row of cilia stretches along the floor of the peristome parallel with the membranelles, a single cilium opposite each membranelle. The right border of the peristome (Fig. 98) is continued into a vibratile membrane, and close to the left of this and running parallel with it is another row of cilia (præoral cilia *poc*). In the centre of the peristome is a second undulating membrane, the endoral membrane (*em*), which passes downward and into the pharynx, and this, also, is sometimes accompanied by a row of cilia even into the pharynx (endoral cilia *eo*).

Cirri, membranelles, and membranes are each striated, and when treated with certain reagents (*e.g.* gold chloride, Maupas), can be reduced to separate fibres which are similar to cilia. The simplest of these aggregations are the cirri. These bundles of threads are usually pointed, and either curved or straight, forming the *Griffeln und Haken* of Ehrenberg. A simple condition is seen in the tail-like process of *Urocentrum*, which has a distinctly fibrillar structure and can be readily reduced to a brush of very fine hairs (Fig. 99). Although the striated appearance and the reduction to the component fibrils make it probable that cirri arise by the fusion of cilia, an objection is met in the fact that their development, after division, shows no such origin. On the contrary, they arise from the ectoplasm as cirri and not as cilia. This objection, however, seems hardly sufficient to counterbalance the evidence in favor of the concrescence theory, evidence which is strengthened by the position of the cirri along the lines of the cilia-markings.

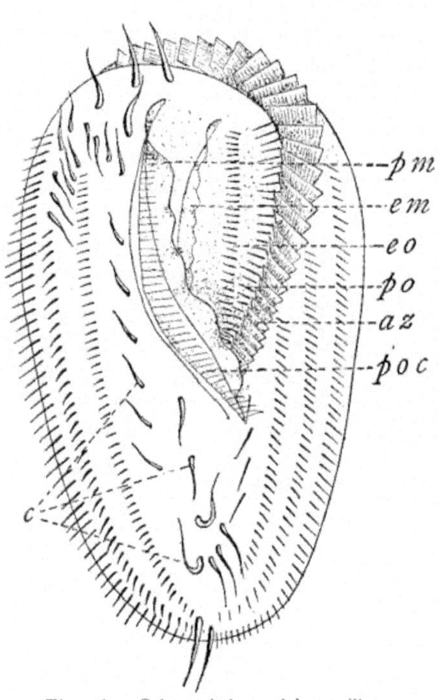

Fig. 98. — Schematic hypotrichous ciliate. *az*, adoral zone; *c*, ventral cirri; *em*, endoral membrane; *eo*, endoral cilia; *pm*, præoral membrane; *po*, paroral cilia; *poc*, præoral cilia.

The membranelles are flat plates of striated appearance usually in the form of triangles, squares, or parallelograms. Each membranelle is inserted in a furrow below which is a basal stripe of thickened protoplasm continuous with the longitudinal ciliary markings (Heterotrichida). Like the cirri, they can be readily reduced to component filaments resembling cilia, and there is, therefore, every reason to suppose that the membranelles which form such a characteristic differential for all orders save the Holotrichida, are merely the differentiated portions of the ciliary rows.[1] The basal stripes of the membranelles, which are spirally arranged upon the peristome, are in turn inserted, in some cases at least, in a thick fibrous strand which

[1] Johnson ('93) alone regards the membranelles in *Stentor* as endoplasmic in origin.

runs around the peristome connecting the series, and possibly forming a nervous organ (Delage, '96; Moore, '93).

The undulating membranes, finally, which are almost always confined to the oral region, and like the membranelles chiefly concerned with food-taking, have probably a similar origin, although the connection with the cilia is less apparent. They are frequently, as in the Hypotrichida, placed deep in the vestibule, but in many forms they are confined to the pharynx itself, as in many of the Holotrichida.

In addition to cilia, membranelles, and membranes, the ectoplasm has other modifications, such as pseudopodia (*e.g. Stentor*) and tentacles. The pseudopodia are used for anchoring the animal, and are produced at the posterior end by the so-called foot-disk (Johnson, '93). The cortical plasm gives rise to these processes, and also to the peculiar tentacle-like appendages found in some forms. In *Actinobolus* (Fig. 100) these pseudopodial tentacles are particularly well known through the complete study made by Entz ('82). Here the threads pass out between the cilia and not infrequently reach a length of twice or even three times the body diameter. The threads are of nearly uniform thickness, with blunt or slightly knobbed ends (Entz). These tentacles, while occasionally stiff and unyielding, can be shortened or lengthened, or drawn into the body in a manner surprisingly suggestive of pseudopodia, while the protective and offensive function is shown by the presence of trichocysts at their extremities. Similar tentacles are found in *Mesodinium* and *Ileonema* (see Fig. 115).

Fig. 99. — *Urocentrum turbo* O. F. M. [BUTSCHLI.]

While the ectoplasm is devoted to the functions of motion and irritability, the endoplasm is charged with digestion and reproduction. Thus the membranelles and membranes are important in creating the current which brings the food particles; the trichocysts are occasionally developed as food-killing organs, and these, with the mouth, vestibule, and pharynx, are ectoplasmic in origin.

While all these special modifications are developed for the purpose of food-getting, the endoplasm, with its digestive processes, shows but little advance, so far as can be made out, over the already complicated endoplasm in the less highly organized forms. Similar food substances are treated in similar gastric vacuoles, and the products of assimilation are carried about in the plasm by similar cyclosis, while indigestible remains are excreted in the same way. The Protozoa thus offer in the most striking manner an example of how species may have originated through structural adaptations of the parts (ectoplasm) that are in direct contact with the environment. The mouth parts, which are functionally the beginnings of the digestive system, are formed by the invagination of a limited portion of

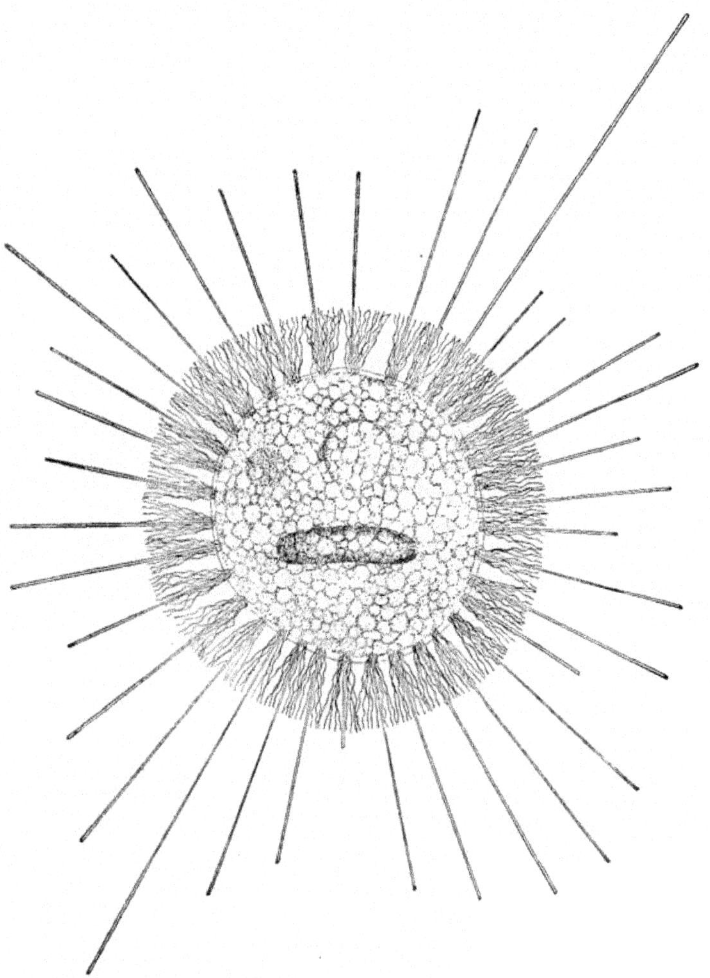

Fig. 100. — *Actinobolus radians* St.

the ectoplasm in a manner analogous to the formation of the stomodæum of Metazoa. The peristome is the beginning of the mouth depression, becoming more and more deeply depressed as the mouth region is reached. It is not present in all forms, the mouth in its original position probably being anterior and terminal in the monaxonic body, and leading by a mere passage into the endoplasm below. Mouthless forms are known, but these have degenerated through parasitism and are not primitive (Opalinidæ). In the majority of forms the mouth is displaced from the original terminal position and has become ventral and central. Bütschli maintains that this change in the position of the mouth is brought about by the gradual shifting from the anterior end, as shown by the meeting point of the lines of ciliary markings. As previously indicated, the original course of these lines is from the anterior to the posterior end, but in numerous transitional forms in which the mouth has a more or less ventral position, the markings become curved to agree with the changed position, while the course which the mouth is supposed to have taken is shown by the converging lines (Fig. 95).

In almost all cases the mouth is not in direct communication with the endoplasm, but is separated from the latter by a longer or shorter pharynx, œsophagus, or gullet, which frequently bears cilia, membranes, or membranelles. The œsophagus is likewise an ectoplasmic invagination, as is also a second œsophageal apparatus, found in some forms (Vorticellidæ), where the mouth leads into a comparatively large ciliated or membraned space, known as the *vestibule* (Fig. 101, *C*, *D*), and this leads into the œsophagus or gullet proper, which, in turn, communicates with the endoplasm. This space begins as a wide tube and gradually narrows down to a more or less narrow aperture or constriction at the œsophagus. The anus and the contractile vacuole, in some forms, open to the exterior through the vestibule. In some of the Holotrichida, the region about the pharynx is strengthened by accessory apparatus developed in the cortical layer, which in this region is greatly thickened and which in some cases contains secretions in the form of bars arranged in a peculiar basket structure (*A*, *B*).

The membranelles which surround the mouth are usually in motion, as are the membranes and cilia which extend into the vestibule or œsophagus. Even while the animal is lying quiet, the membranelles continue their active vibrations, keeping a constant current of water toward the mouth. This current brings a supply of bacteria, diatoms, algæ of various kinds, rotifers of small size, or parts of animals undergoing disintegration, flagellates, etc. A distinction can be made here between herbivorous and carnivorous forms, although the differences can hardly extend to structural adaptations, unless it be, perhaps, in some carnivorous forms, where special weapons of offence (the trichocysts) are found. Probably all forms are more or less omnivorous and make little or no selection of food. The food particles are thrown by the current of the membranelles into the peristomial depression and thence into the vestibule or œsophagus, until they come in contact with the endoplasm at the base of the latter. Here they are readily absorbed by the endoplasm, in which, together with a small amount of water, they are confined in a small gastric vacuole. The vacuole enlarges by the constant addition of new material, until it is caught up in the current of the endoplasm and dragged away. In this improvised "stomach" it is slowly digested, a new drop being formed in the meantime at the mouth opening. If food is abundant, the animal may become filled with these gastric vacuoles. The liquid of the vacuole is, at first, simply water, like the surrounding medium, but gradually becomes acid through osmosis in the plasm, and the digestible substances are

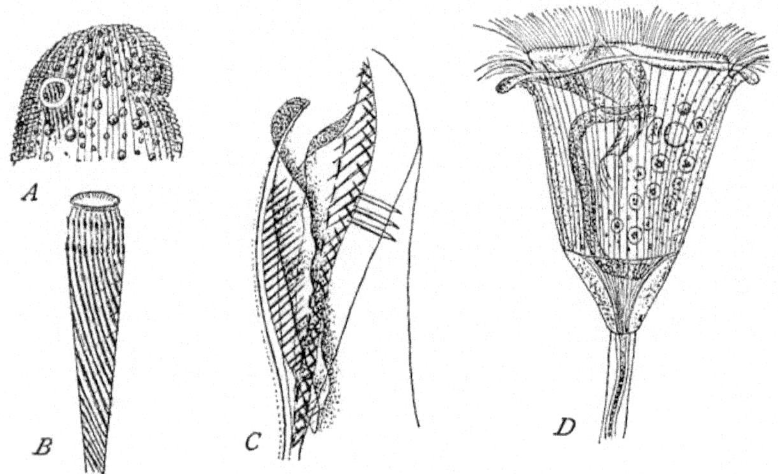

Fig. 101.—Buccal apparatus. [BÜTSCHLI.]
A, B. Nassula aurea Ehr. *A.* From the ventral side. *B.* The buccal apparatus strongly magnified. *C. Urostyla grandis* Ehr. *D. Vorticella nebulifera* O. F. M.

slowly dissolved, the residue being cast to the exterior through the anus.

Unlike the mouth, the anus, as a rule, is a simple opening in the outer wall (Maupas, '83; Bütschli, '88), although Stein ('67) described an anal tube in certain forms (*Nyctotherus*). In the Heterotrichida it is sub-terminal in position. In the Hypotrichida it is never terminal, but usually dorsal, and toward the left edge. In the Peritrichida, especially in the Vorticellidæ, the anus opens into the vestibule.

B. Contractile Vacuoles

A certain amount of water is taken in with the food through the mouth, and at the same time (as in those cases where the mouth is absent) by absorption through the body wall, and it is the function of the contractile vacuole to get rid of the surplus. This organ is variously complicated by the development of a more or less extensive series of canals, which empty in a common excretory vacuole. Always situated in the cortical plasm, the contractile vacuoles are fixed in position and communicate with the exterior at systole by a permanent aperture, which, however, becomes covered internally during filling or diastole. They vary in number from one to a hundred, or even more, and are absent, apparently, in only one form (*Opalina*), although Vejdovsky ('92) describes contractile vacuoles in a closely allied form, *Monodontophrya longissima*, while even in *Opalina* the reminiscence of the vacuole is seen in the remnants of the feeding canals (Delage, '96). In its simplest form the vacuole is single and terminal, a condition which may be found in each of the four orders. When there are more than one, they are grouped around the original vacuole in a terminal position, or arranged along one or more lines upon the dorsal side. In *Discophrya* and *Hoplitophrya* (Holotrichida) there is no regular vesicle, but a long contracting canal which runs the length of the body. *Spirostomum* (Heterotrichida) has a terminal vesicle, with one long feeding canal, and from this the canal system is developed in a variety of ways. Thus there is a vesicle with two feeding canals in *Climacostomum* (Fig. 91, *B*), one terminal vesicle and four feeding canals in *Urocentrum*. In *Stentor* there is a single vesicle near the peristome, with two feeding canals, one of which runs to the end of the body, while the other runs around the peristome edge. Fabre-Dumergue ('89) holds that canals, for the most part invisible, are present in all ciliates. This is certainly true in *Frontonia*, where there are one or two vesicles on one side and an immense number of feeding canals, which anastomose and branch to form a complicated network, involving the entire body. In some forms the vesicle communicates with the exterior directly, but it may be complicated by the formation of ducts or reservoirs. In the holotrichous form, *Lembadion*, the vesicle lies

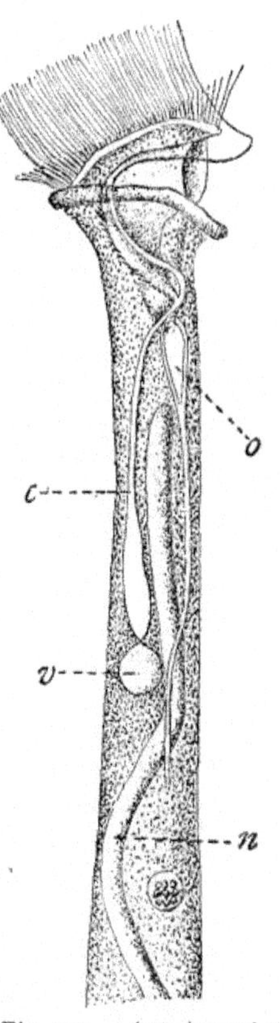

Fig. 102. — Anterior end of *Ophrydium eichhornii* Ehr. [WRZESNIOWSKY.]

c, the reservoir of the vacuole (*v*) emptying through a long canal into the vestibule; *o*, the œsophagus; *n*, the nucleus.

dorsally in the middle of the body, but is connected with the terminal aperture by a long canal. In the Peritrichida there are one or two vesicles, which empty by contraction into special reservoirs, and these, in turn, empty into the œsophagus (Fig. 102).

According to Delage ('96) the contraction of the vesicle is brought about by the contractility of the surrounding cortical plasm. Rhumbler ('98) has shown, however, that the contained water may so affect the plasm that it becomes differentiated like ectoplasm, and this gives some substantial basis for the view that a special membrane surrounds the vacuole. There is no evidence, however, to show that this modified protoplasm acts like a sphincter. Bütschli holds that the vesicle contracts through a mechanical force exerted upon the thin plasmic layer between the vesicle and the opening of the excretory pore by the pressure of the filling vacuole and the turgor of the cell. At the completion of the diastole the pressure becomes too great for the lamella, and the latter is ruptured, allowing the contained fluid to pass to the exterior.

C. The Nucleus

The nuclei of the Infusoria show some of the most striking structural characteristics connected with the Protozoa. Here there is a differentiation of the nuclear material into two forms, a larger *macronucleus*, and a very much smaller *micronucleus*. With the single exception of *Polykrikos* (Dinoflagellidia), this differentiation of the nuclei is found nowhere outside of the present group. The functions of the two kinds of nuclei are supposed to be respectively vegetative and reproductive (Bütschli), but this distinction is, perhaps, too sweeping. Julin ('93) held that the macronucleus stands not only for nutrition, movement, sensation, and regeneration, but for asexual division as well, in fact is a "somatic nucleus," while the micronucleus functions only as a sexual nucleus. There may be one or many of each kind in each cell. The macronucleus, which is invariably present, recalls the nucleus of tissue cells. It is usually single, and, lying in the endoplasm, it may be carried about with the flowing granules, or maintained in a permanent position in the cortical plasm, or by processes from this plasm (*e.g. Isotricha*). Its form is quite variable and has little significance for systematic work, for in the same species under certain conditions it may even become amœboid (Loeb & Hardesty, '95). The usual form is spherical, but it may be elongated into an oval, or into a flattened rod which may be curved or straight, or it may be divided into small pieces resembling a string of beads, connected by a membrane (Fig. 103). It is always provided with a membrane (Maupas), but the chromatin contained within it is variously distributed. The vesicular structure, in which the nuclear substance is so distributed as to leave more or less space filled with "nuclear sap," is almost never seen, the macronucleus appearing solid and completely filled with chromatin. Bütschli described the finer structure as almost invariably alveolar, the meshes corresponding to those of the surrounding plasm. The entire network stains deeply with the nuclear dyes, but at certain stages, especially during division, distinct fine lines can be made out connecting the

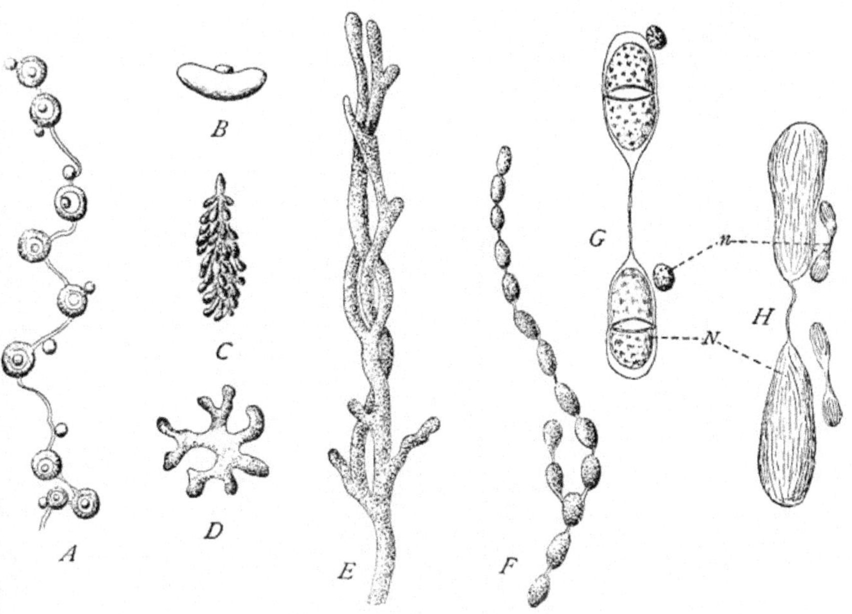

Fig. 103.—Types of macronuclei. [SAVILLE KENT.]
A. Macro- and micronucleus of *Loxodes rostrum* O. F. M. *B.* Of *Nyctotherus cordiformis* Leidy. *C.* Macronucleus of *Plagiotoma lumbrici* Duj. *D. Dendrosoma radians* S. K., a young nucleus. *E. Dendrosoma radians* S. K. *F. Stentor polymorphus* Ehr. *G. Stylonychia mytilus* O. F. M. *H.* The same in division. *N,* the macronucleus; *n,* the micronucleus.

chromatin granules, and corresponding to the linin reticulum of most nuclei. In some cases (*e.g. Loxophyllum*) a permanent spireme is present, as in the nucleus of cells from the salivary gland of *Chironomus* larvæ, and is transversely striated, indicating disks which Balbiani ('90) thought are alternately chromatin and linin (Fig. 104). In many cases there are internal modifications of the nuclear material forming so-called "nucleoli," although there is a possibility that these structures are similar to the intranuclear bodies found in Mastigophora.

In many macronuclei a peculiar division of the organ is made by a split, which the Germans call a *Kernspalt* and the French a *fente*. While this peculiar feature of the nucleus has not been explained, what may be an important light has been thrown upon it by Balbiani ('95), who described the appearance as due to the presence of two materials within the nuclei, one of which is chromatin, the other, achromatic material or archoplasm. This interpretation, however, cannot be accepted as final.

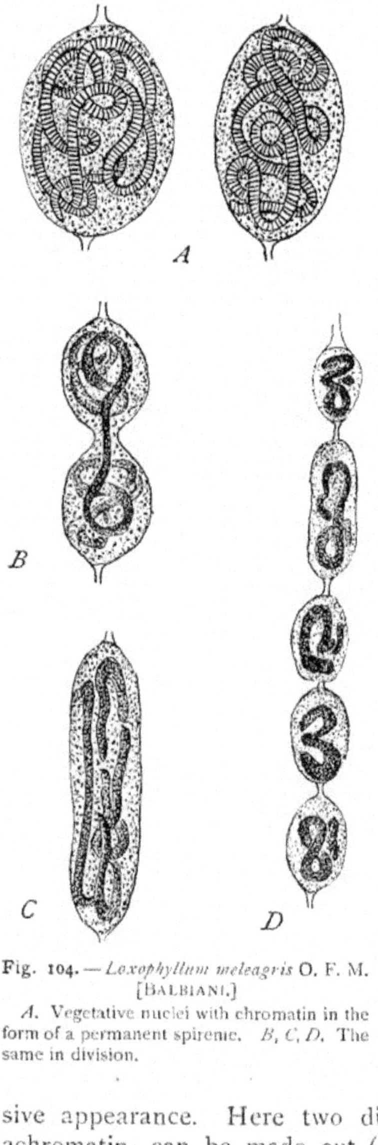

While the macronuclei are, as a rule, single in number, the micronuclei are often multiple. Probably all Ciliata have at least one micronucleus, although the small size and the extreme difficulty in staining sometimes render it hard to find. In one case at least (*Opalina ranarum*) there is only one kind of nucleus. The number of micronuclei is usually greater where the macronucleus is elongate, and especially where it is beaded (*Stentor*, *Spirostomum*, etc.). As a rule the micronuclei are closely attached to the membrane of the macronucleus, occupying a minute indentation in the latter, but in some cases they are well separated. In form they are round, ellipsoidal, or spindle-shaped, but the form varies with the nuclear activity, and does not mean much in itself. Their longest axis measures from 1 μ to 10 μ, and like the macronuclei, they are covered with a distinct membrane, while the chromatin is usually massed at some part of the nucleus. In certain cases the appearance is like that of the macronucleus with the chromatin in the form of a densely packed reticulum, giving to it a massive appearance. Here two distinct portions, the chromatin and achromatin, can be made out (Fig. 105).

Fig. 104.—*Loxophyllum meleagris* O. F. M. [BALBIANI.]
A. Vegetative nuclei with chromatin in the form of a permanent spireme. *B*, *C*, *D*. The same in division.

Division of the nuclei takes place by mitosis in the micronuclei, and, as a rule, by amitosis in the macronuclei. The latter is the simpler; in many spherical or elliptical nuclei the structure merely draws out and segments into two equal parts. It is more or less complicated, however, in different macronuclei, until well-developed mitosis replaces simple division (*e.g. Spirochona*). There is reason to regard the simple division of the larger nuclei as the mere degeneration of mitosis, by a process in which the various stages have gradually disappeared until only the preliminary stages of such division are to be found. These preliminary stages are seen in the transformation of the reticulum of chromatin into thread-like masses, which recall the spireme of

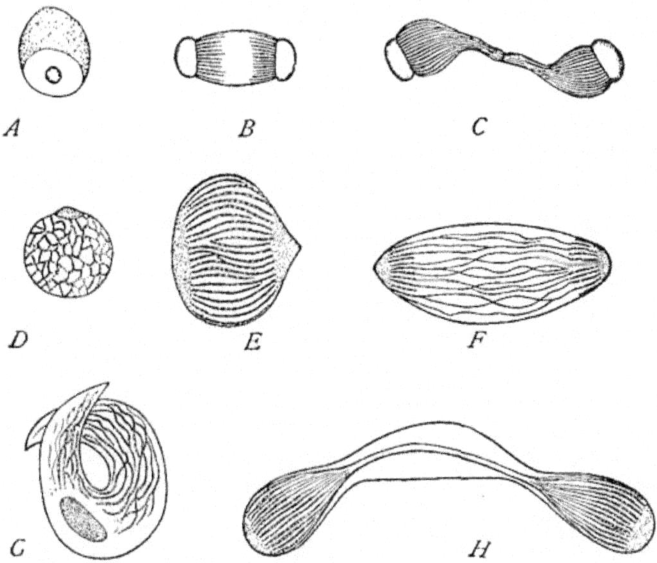

Fig. 105. — Mitotic division of the nuclei of *Spirochona* and *Paramœcium*. [R. HERTWIG.] *A–C*. Macronucleus of *Spirochona* with well-developed pole plates. *D–H*. Different stages in division of the micronucleus of *Paramœcium*.

higher cells; the threads are then divided across into equal parts. In *Spirochona*, however, the process of division is strikingly similar to mitosis in Metazoa (Fig. 105, *B*, *C*).

Division of the micronuclei is accompanied by the formation of polar attraction-spheres, and by the rearrangement of chromatin into chromosomes. Before division, the micronucleus swells to nearly twice its size during resting stages (Hertwig, '77), while the granular chromatin begins to collect in lines — at first equally thick, but later concentrated in the equatorial region (Fig. 105, *D–H*). Division then takes place through the centre.

D. Encystment

The phenomenon of encystment may be seen in the Ciliata as in all other groups of the Protozoa. It occurs when the animal is in danger of drying, in some cases before division, in others, for the purpose of digesting a full meal. The cilia are drawn in, the mouth and peristome disappear, the contained body-granules are voided, and a gelatinous secretion is poured out from the ectoplasm. The secretion soon hardens, becoming chitinous. The vacuole continues to pulsate for some time, and the secretion forms a liquid layer about the animal under the cyst. The cysts are variously diversified with spines and processes of different kinds, and are occasionally multiple, the spaces between the cysts being filled with water (Fig. 17, *B*, *C*, *F*, p. 47).

E. Reproduction

Reproduction among the Ciliata takes place almost exclusively by simple division or fission. It is practically the same for all forms, the variations being of minor importance. The nuclei first divide, new mouth parts are developed in the posterior half, and then the cell divides. The first indication of division in *Stentor*, for example, is a rift in one side of the animal below the adoral zone. This rift rapidly develops motile organs (membranelles), and acquires the full length of the lower daughter-individual. The nuclei in the meantime divide, and the original animal draws out, leaving a slender foot for the upper or anterior cell, and a swollen portion for the pharyngeal region of the second individual. The new adoral zone is completely formed before actual division; the steps in the process are shown in the accompanying diagram from Johnson ('93) (Fig. 106). The new contractile vacuole, according to him, arises *de novo* in *S. cæruleus*, and by a dilatation of the longitudinal canal in *S. rœselii*. The new vacuole thus formed remains in connection with the longitudinal canal, the upper part of which becomes drawn out with the torsion of the adoral zone to form the much-discussed ring-canal discovered by Lachmann.

In some forms, as in *Spirochona*, division simulates budding, unequal division giving rise to mother- and daughter-cells.

When division takes place within the cyst, the various mouth parts may or may not first be absorbed, but in all cases the vacuole still continues to pulsate. Here, as a rule, division is double, resulting in broods of four which escape as embryos, and gradually grow into the parent form. This condition closely simulates spore-formation, which results when, as in *Opalina*, the number of divisions within the cyst reaches three, four, five, or six.

Nowhere among the Protozoa has the process of conjugation been so thoroughly studied in connection with the life-history of the organism as in the Ciliata. Worked out first by Bütschli and Engelmann in 1876, it has since been carefully studied by numerous observers, and the conditions preliminary to conjugation, during, and subsequent to it have been made known in a great variety of forms representing all divisions of the class. Bütschli and Engelmann early recognized that conjugation is necessary for continued life activity of the organism, and came to the conclusion, which has been fully confirmed by subsequent observers, that conjugation is a process of rejuvenation or a renewal of vitality, the need of which is shown by the reduced size

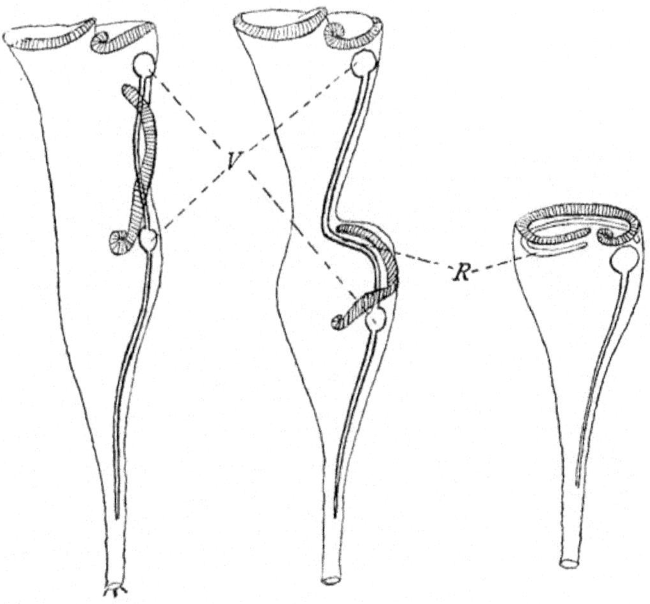

Fig. 106. — Diagrams to illustrate the division of *Stentor rœselii*. [JOHNSON.] *V*, the vacuole; *R*, the ring canal.

and general degenerate condition of the organism prior to conjugation. Maupas ('89), in his classical work on conjugation among Infusoria, found that the number of generations which may be formed from one conjugating period to the next varies with the species, but is usually between three hundred and four hundred and fifty. He found, furthermore, that certain conditions are necessary for conjugation. These conditions are: (1) maturity of the organisms, *i.e.* forms which have just conjugated will not again conjugate until after a certain number of generations; (2) partial lack of food, *i.e.* if plenty of food is present, conjugation will not take place even though the individuals are well along in degeneration; (3) diverse ancestry, *i.e.* the conjugants must come from different ancestral conjugants.

The two conjugants fuse either temporarily or permanently, and the external structures, such as the membranelles, are absorbed. If the fusion is temporary, as in the majority of forms, the two ectoplasms fuse at or near the mouth parts, and a protoplasmic bridge is formed between the two organisms. The two organisms then become sluggish, and rest for a considerable time upon the bottom without movement of any kind. They ultimately separate and begin to divide.

The micronuclei play the most important part in conjugation. Each divides two or more times to form four or more daughter-nuclei, some of which degenerate, while one divides again, one half to fuse with a similarly derived nucleus of the other organism, while the other half remains as the receptive nucleus, or the *female pronucleus*. The two conjugating nuclei cannot be distinguished from those which degenerate, but are apparently only those which lie nearest the bridge joining the two organisms. Meanwhile the macronucleus undergoes complete degeneration, breaking up into a number of pieces, which are gradually absorbed by the protoplasm. The new macronucleus is formed by the enlargement of a daughter-micronucleus derived from the fusion nucleus. Hoyer ('99), however, asserts that in *Colpidium colpoda* it forms by the union of two daughter-nuclei.

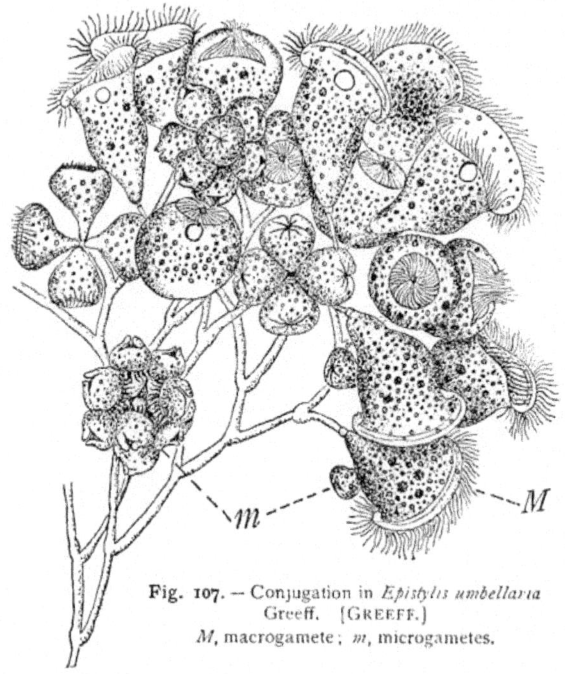

Fig. 107. — Conjugation in *Epistylis umbellaria* Greeff. [GREEFF.]
M, macrogamete; *m*, microgametes.

When there are two or more micronuclei in each conjugant, the process is repeated for each of them, although it is not known whether this holds when, as in *Stentor*, the number reaches sixty or seventy.

In a few cases (Vorticellidæ) the conjugants are of diverse size.

The larger form or *macrogamete* is usually a normal-sized individual, although in some cases it is somewhat larger than the ordinary cells (*Zoothamnium*). The microgametes, on the other hand, are considerably smaller, and from four to eight are formed by each cell. These never develop a stalk, but leave the parent colony, and swim about by means of the ring of cilia around the lower pole. They finally come in contact with the macrogamete and fuse with it, the union taking place at the lower end of the attached organism, and near the insertion point of the stalk (Fig. 107).

II. THE SUCTORIA

The Suctoria differ decidedly from the Ciliata, from which they have undoubtedly sprung. With the exception of *Hypocoma* (Fig. 115, *C*), which remains ciliated throughout life, the Suctoria possess cilia only during the embryonic stages. They are, for the most part, sedentary forms, and grow upon a chitinous peduncle, which is attached at the lower end to some foreign object. The upper end of the peduncle is hollowed out into a bowl, within which the animal lies. Owing to its attached mode of life, and to the equal pressure on all sides, the general form of the animal is spherical or radially symmetrical. In some cases there is a well-defined membrane, but the various students of the group are not agreed as to its structure. It is never striated, as in the Ciliata, and there is no cortical plasm. The endoplasm shows no differentiations other than the usual food granules or assimilation products common to all Protozoa.

An essential point of difference from the ciliate structure is the presence of *tentacles*, which, in the majority of Suctoria, are the only motile appendages of the adult. In many respects they are similar to the tentacles of *Actinobolus*, *Ileonema*, and *Mesodinium* (Fig. 115), but differ in the very important fact that they are hollow, while the extremities bear the mouth openings.

There are two general types of tentacles: one, according to Bütschli, captures prey, while the other devours it. Of the latter forms, there are also two types. One is long and broad, and, like a thorn, pointed at the extremity; the other is nearly uniform in diameter and flattened at the top, or hollowed out into a cup-like sucking organ (Fig. 108). These are distinguished as the *styliform* and *capitate* tentacles (Delage). Both sets of tentacles are hollow, and their lumena open at the ends. There is a difference of opinion, however, in regard to their inner structure and function. Bütschli holds that some of them are solid, and others hollow. He maintains that, in the solid forms, the internal portion is formed of endoplasm, which is continuous with the inner plasm of the cell. Delage claims that they are all hollow. The function of the endoplasm, according to Bütschli, differs in the styliform and the capitate tentacles. In the latter, the prey is retained by the sucking disk at the extremity while the endoplasm within the tube meantime works up and down like a pump-piston, and a vacuum being thus formed, the cuticle of the prey is burst, and the fluid endoplasm flows down the tentacle canal to the endoplasm of the captor, where it is digested.

Such an explanation of the action of these tentacles is regarded by most observers as extremely doubtful. Delage finds no motion in the endoplasm during feeding, save in the rhythmic pulsations of the contractile vacuole, an organoid which Eismond ('90) believed is the cause of the suction. The excretion of water from the vacuole, he argued, creates a semi-vacuum in the protoplasm, and the pressure

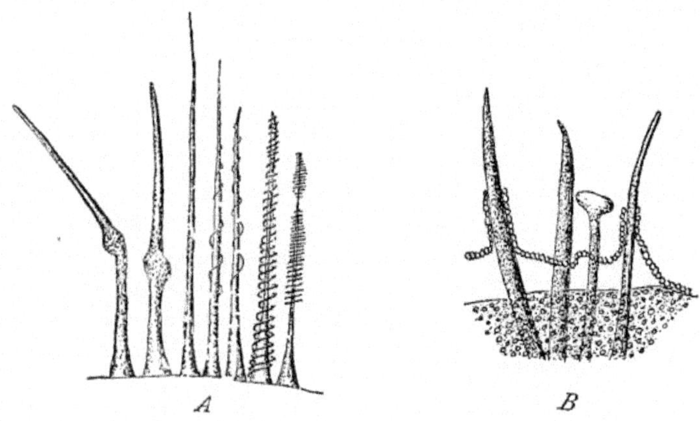

Fig. 108. — Tentacles of Suctoria. [R. HERTWIG.]
A. Different types of styliform or piercing tentacles. *B.* Capitate and piercing tentacles.

from without forces food or loose particles, etc., through the tentacle openings to the endoplasm. This explanation, although somewhat fanciful, is certainly as plausible as the principle of the pump, but the matter must remain for the present as one of the many unsolved problems connected with this group. In some cases (*Trichophrya angulata*) the tentacles are apparently unnecessary for food-taking, as Dangeard ('90) found that particles are occasionally engulfed, as in the Rhizopoda, at any point of the naked body.

In the styliform tentacles, on the other hand, the sharp points pierce the membrane of the prey, while the endoplasm contained within the tentacle possibly flows into the prey, whose endoplasm is digested *in situ*. In some cases they appear to have a paralyzing effect upon other forms, and ciliates coming in contact with them have been seen to stop their movements as though stunned.

In all cases the tentacles are remarkably like pseudopodia, and may change their form and their position, and may even be entirely withdrawn into the body, to reappear, possibly, at some other place. Some forms have the power of withdrawing their tentacles and developing cilia, which may be retained for a longer or shorter

Fig. 109. — *Dendrosoma radians* Ehr. [SAVILLE KENT.]
n, nucleus.

period. In some cases the tentacles distinctly originate in the endoplasm, and penetrate the membrane (Hertwig, '76; Ishikawa, '96).

The nuclei, like those of the Ciliata, are of two kinds, macro- and micronuclei. The former are little different from the macronuclei of the more generalized Ciliata, while very little is known about the latter. In the colony-form *Dendrosoma*, where the many branches suggest a hydroid colony, the macronucleus extends through all the branches and trunks, penetrating the entire system, like the cœnosarc of a hydroid (Fig. 109).

The contractile vacuole never becomes so complicated as in some Ciliata, but consists, usually, of a single vesicle, which may be surrounded by a circle of small vacuoles emptying into it. In some cases there is a short excretory duct leading from the vesicle to the excretory pore in the membrane, which, as in the Ciliata, is a permanent opening.

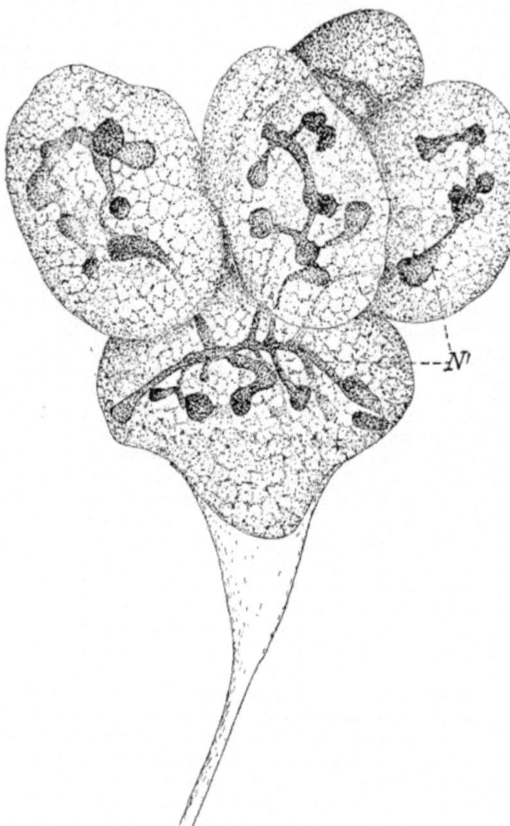

Fig. 110. – Exogenous budding in *Ephelota Bütschliana* Ishi. *N*, nucleus.

These animals encyst only for protection, never, apparently, for reproduction. As in the Ciliata, the process consists of the secretion of a chitinous mantle about the cell, the tentacles being withdrawn into the body.

Reproduction is almost invariably by simple division, which may be either equal or partial (budding). In the simplest cases the upper portion of the cell is constricted off, and moves away from the lower portion, which remains upon its stalk (*Podophrya, Spirophrya, Urnula*, etc.). The detached part develops cilia and, after a longer or shorter free-swimming period, settles down, loses its cilia, and secretes a stalk. Partial division, or budding, may be either *endogenous* or *exogenous*. The latter is the simpler; an individual prepares as for division, but instead of dividing into two equal portions, a number of papillæ appear at the outer surface, each becomes a bud, receiving a portion of the nucleus (Fig. 110). Endogenous division arises by the invagination of such a budding area, while the walls surrounding it grow together above the developing buds, which, when ripe, break through the birth-opening left in the covering membrane (Fig. 111). In some cases the buds are multiple, again single, and a number may develop at the same time within the brood-sac (*Acineta*, *Ophryodendron*).

The embryos thus formed are variously ciliated in the different genera. In some they are holotrichous, in others hypotrichous, and in others peritrichous (*c*, *d*).

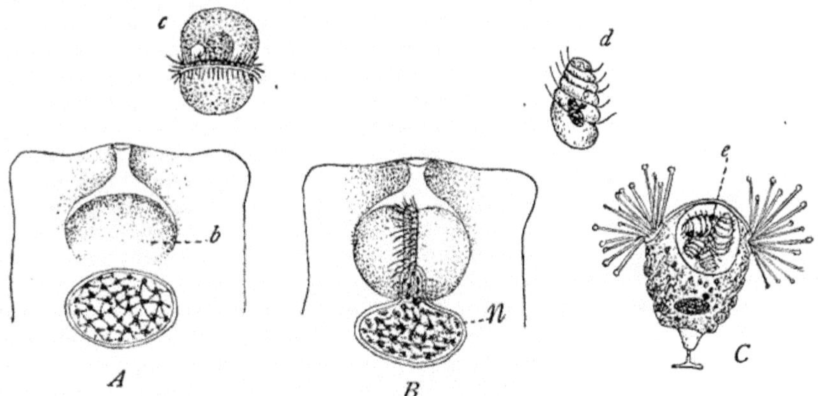

Fig. 111. — Endogenous budding in Suctoria. [BÜTSCHLI.]
A–B. Two stages in the formation of the bud in *Tokophrya quadripartita* Cl. and Lach. *c*. The swarm-spore liberated. *C*. Buds in *Acineta tuberosa* Ehr. *d*. A swarm-spore liberated.

Conjugation occurs here as in the Ciliata, but the process rests upon the single observations of Maupas, who shows, however, that it differs in no essential features from that already described.

III. INTER-RELATIONS OF THE INFUSORIA

In searching for the origin of the Ciliata, the naturalists of thirty years ago had an apparent advantage, in that the supposed ciliated girdle of the Dinoflagellidia offered a direct transition to the peritrichous Ciliata, which, accordingly, were regarded as coming from the flagellate stem at a comparatively late date. Unfortunately for the theory, however, it was ascertained by Bütschli ('85) and others that the girdle of cilia is only a vibrating flagellum in the transverse groove. In other directions the search for the origin of these forms has been almost equally vain. The singularly conservative structure which the ciliate body presents leaves but little clue to their ancestry. The universal presence of macro- and micronuclei is paralleled by only one other known case, the almost universal reproduction by transverse division is met with elsewhere but rarely. The sole possibility which presents itself is that the infusorian stem was derived from the flagellate at a very early period, and that the side branch became progressively differentiated until the well-marked characteristics of to-day distinguish the Infusoria as an entirely independent group. The first forms to diverge from the flagellate stem may have been like the type described by Cienkowsky, under the generic name of *Multicilia* (Fig. 112, *A*), a form with a number of long flagella. It is thought by Bütschli that the Ciliata might have been derived from such generalized forms by progressive increase, with shortening of the motile elements, until cilia were the

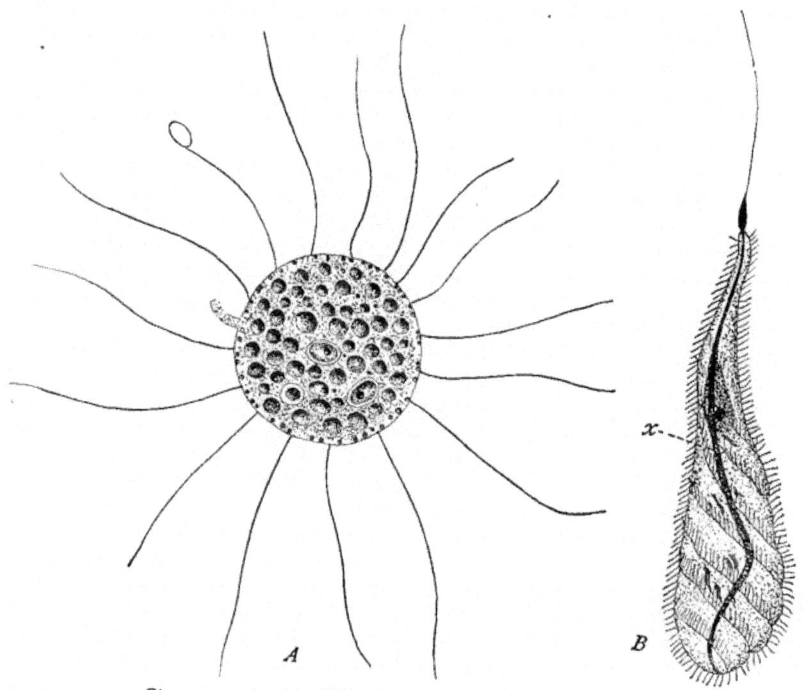

Fig. 112.—*A. Multicilia lacustris* Lauterb. [LAUTERBORN.] *B. Pyrsonympha vertens* Leidy. [PORTER.] *x*, the vibrating band in the inner plasm.

outcome. There is no close connection, however, between cilia and flagella, such as exists between the flagella and the pseudopodia. Other forms, more or less similar to *Multicilia*, have been described by various observers, so that the hypothesis of Bütschli is not without warrant. Among these forms are *Grassia*, *Trichonympha*, *Leidyonella*, *Lophomonas*, *Pyrsonympha*, etc., which are placed by some among the Flagellidia (Delage), by others among the Ciliata (Bütschli). Another point of view has been based upon the relations of the Ciliata to the Suctoria, and through them to the Sarcodina

(Entz, Maupas). This view will be more appropriately examined in connection with the Suctoria.

The Holotrichida appear to be the most generalized of the entire group of Infusoria, but a few forms among them have a slight regional differentiation of cilia suggesting the characteristics of the Heterotrichida (*Lembus*, *Pleuronema*, *Ophryoglena*, etc.). In fact, there appears to be no sharp line between the two divisions, although the presence of an adoral band of cilia in the Heterotrichida is a sufficient differen-

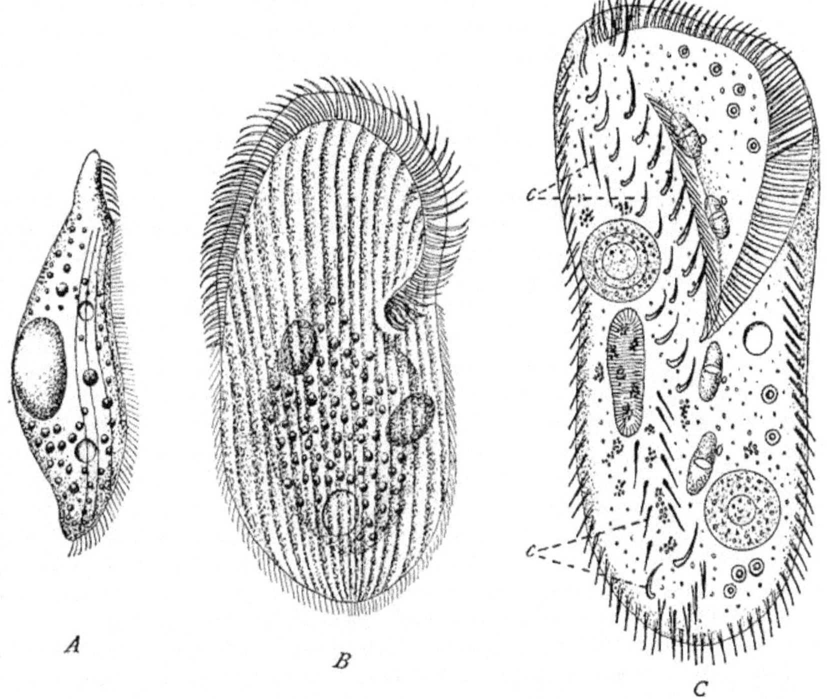

Fig. 113. — Illustrating Bütschli's hypothesis of the origin of the Hypotrichida. [BÜTSCHLI.]
A. Stephanopogon colpoda Entz. *B. Peritromus emmæ* St. *C. Onychodromus grandis* St. *c*, cirri.

tial. In some forms the uniform coating of cilia is broken in certain regions, giving characteristic girdled forms, which are included as a separate order apart from the Holotrichida by some writers (Haeckel). In the Holotrichida, also, there are a few forms which show a distinct tendency toward bilateral symmetry, due primarily to a bending of the body, and followed by a reduction of the cilia upon the arched side (*Stephanopogon*, Fig. 113, *A*). Bütschli derives the Hypotrichida from the Heterotrichida by the supposition of the loss of cilia upon the arched dorsal side and incomplete closure of the adoral ring of cilia, which are here fused to form the characteristic membranelles; the mouth, as in Heterotrichida, remaining on the ventral side. In the most generalized forms, such as *Peritromus* or *Oxytricha* (*B*), the cilia are well distributed over the ventral surface, but in most of the other Hypotrichida they are reduced, and many are obliterated or fused into characteristic cirri. The cirri in the Oxytrichinæ are primitively arranged in six rows, but in the various genera the number becomes reduced, and frequently only isolated cirri mark the original position of the row (*C*).

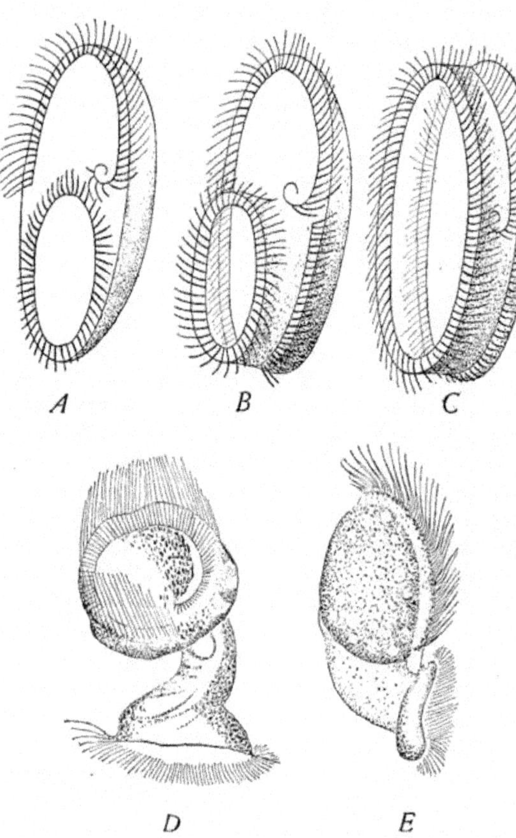

Fig. 114. — Illustrating Bütschli's view of the origin of the Vorticellidæ. [BÜTSCHLI.]

The *Trichodina* form *C* is supposed to have arisen from the *Lichnophora*-like form *A* by the outgrowth of the lower ciliated area, first forming an intermediate stage *B*. This ring of cilia becomes lost in the Vorticellidæ, appearing only when the individuals are free-swimming. *D*, *E*. Side and front of *Lichnophora cohnii* Clap.

The Peritrichida, finally, show the most far-reaching deviations from the holotrichous type, from which they are probably derived through the Heterotrichida and the Hypotrichida. In all members of this group the adoral zone is continued into a spiral, which may have as many as five complete turns (*Campanella*). The chief interest concerning this adoral zone is that in some forms the spiral is turned to the left, similar to that of all of the other groups of Ciliata, while in other forms, belonging to the great family of the Vorticellidæ, the spiral is turned to the right. The sinistral type of the Peritrichida originates, according to Bütschli, from an hypotrichous form, becoming attached at the posterior half of the ventral surface, with loss of the posterior girdle of cilia, and elevation of the anterior region bearing the adoral zone, which, as in the other groups of Ciliata, is turned toward the left. The key to the other group of Peritrichida is seen, Bütschli maintains, in the family Lichnophoridæ, where the individuals closely resemble hypotrichous forms (Fig. 114), being oval, flattened ventrally, and arched dorsally. The cilia, as in the Hypotrichida, are limited to the ventral surface, and an adoral zone is present, which runs from the mouth near the middle of the left body edge, entirely around the anterior region of the body, to form an incomplete arc which terminates in the line of the mouth, but on the right side. Another closed ring of cilia is present in the posterior half of the ventral surface. The anterior and posterior rings of cilia are separated in *Lichnophora* by a stretch of plasm in such a way as quite to divide that surface into a posterior and an anterior division (*E*). The posterior part becomes modified to form an attaching organ upon which the animal creeps about upon its host; the anterior region at the same time is elevated, and held in a position at right angles to the plane of attachment, the apparent stalk which supports it being in reality the intermediate plasm between the posterior and the anterior regions of the ventral surface (*D*). Thus the curious anomaly arises of an animal whose anterior and posterior ends represent parts of the same ventral surface.

Bütschli derives the entire family of the Vorticellidæ from this primitive type, through forms like the Urceolarinæ, where the attaching disk, primarily, is not so far removed from the peristome, nor so stalk-like, as it is in the present-day Lichnophoridæ (*D*, *E*). He argues that the *Vorticella*-type is derived from the Urceolaria-type by the attaching part of the ventral surface, *i.e.* the posterior part being carried outward from the remainder of the ventral surface, and thus borne upon a platform so that the two portions of the same surface are no longer in the same plane (*B*). The anterior ring of cilia is then supposed to have grown around the base of the elevated portion until the original adoral zone of cilia now forms a ring about the entire ventral surface. The new arm of this line of cilia grows on past the mouth-opening and forms a spiral, which, looked at from the ventral side, turns to the left, as in all other Ciliata seen from the same surface. Looked at from the other side, however, *i.e.* dorsally, the spiral turns to the right (*C*). This condition is practically represented by the genus *Trichodina*, which moves about on the skin of various Invertebrata by means of the ciliated or attachment disk, in reality the posterior part of the original primitive ventral surface; while the other portion is now carried dorsally and parallel to the attachment disk, the mouth being on the left side of this anterior part. In the Vorticellidæ this posterior or attaching part becomes drawn out into the long contractile stalk, while the ciliated condition, as represented by *Trichodina*, is again brought about in *Vorticella*, when the latter breaks away from its stalk, develops a ciliated band in the posterior region, and swims freely about. The ciliated band is homologous with the posterior ring of *Lichnophora* and the attaching disk of *Trichodina*, while the adoral zone of cilia conforms to the typical left-handed spiral of the remaining ciliates, when looked at from the same morphological point of view. In *Gerda*, the peristomial region has degenerated, while the ciliated disk remains as the organ of locomotion.

Returning now to the origin of the Ciliata as a group, quite another view has been maintained by a number of observers, the essential point of which is that the Ciliata are connected with the Sarcodina through the Suctoria, the tentacles in the latter being regarded as modified pseudopodia. This view was apparently first suggested by Stein ('54) when he included the Heliozoa and the simpler forms of Suctoria in the genus *Actinophrys*. The assumption was taken up seriously by Maupas ('81), who held that through the Suctoria, the Ciliata were derived from the Sarcodina, and Pénard ('90) accepted the same view in regarding *Actinolophus capitatus* as a connecting link between the two groups. Claparède and Lachmann were the first to deny the connection of Ciliata and Rhizopoda, but made the even more improbable assertion that the Suctoria are derived from the Flagellidia through forms like *Syncrypta volvox*. The close relation of the Suctoria and the Ciliata was brought into prominence through Stein's famous, though erroneous, Acineta-theory, in which the Suctoria were supposed to be young forms of Ciliata. The connection between the two was, however, first put on a substantial basis by the discovery of the ciliated embryos of the Suctoria, a connection which was early accepted by students of the Protozoa, and which was greatly emphasized by the discovery that, like the Ciliata, the Suctoria have macro- and micronuclei.

At the present time it is almost universally held that the Suctoria are offshoots of the Ciliata, although the opposite view is maintained by some observers, who, with Entz ('79, '82), consider the Ciliata as permanent forms of the ciliated embryos of Suctoria. Entz himself regards the matter as insoluble, and believes that the evidence is about equally balanced. A number of cases certainly gives strength to Entz's position, for many of the Enchelinidæ, in addition to their cilia, have distinct tentacular processes (*Ileonema*, *Mesodinium*, *Actinobolus*, etc., Fig. 115).

Actinobolus, discovered by Stein, and more recently examined by Entz, has long tentacle-like threads evenly distributed about and between the cilia (Fig. 100). They can be lengthened or shortened or entirely drawn in by the animal. In *Mesodinium*, there are only four of these tentacles, which are arranged about the mouth (Fig. 115, *B*). *Ileonema* has only one (*A*). These processes were considered so important from the phylogenetic standpoint that Mereschowsky ('82) formed a special group, the *Suctociliata*, for their reception. Neither Entz, nor Stein, nor Mereschowsky, however, regarded the tentacles as food-taking organs like the tentacles of the Suctoria; the former, at best, could assign to them no other function than that of assisting in the capture of aliments. Maupas regarded them simply as pseudopodia, and upon them as a basis formulated his view connecting the Ciliata with the Sarcodina. Bütschli strongly opposed Entz's view as to the origin of Suctoria and Ciliata, and believed that there is no

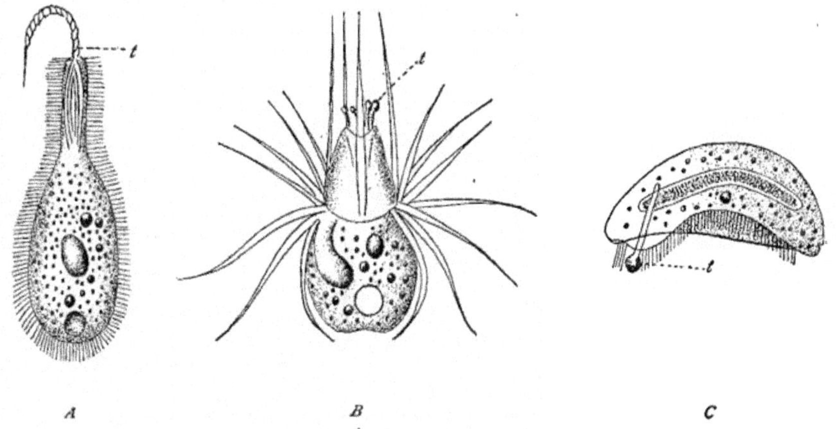

Fig. 115. — Ciliata with tentacles.
A. Ileonema dispar Stokes. [STOKES.] *B. Mesodinium pulex* Clap. and Lach. [ENTZ.] *C. Hypocoma parasitica* Grub. [ENGELMANN.] *t*, tentacles.

direct connection between the tentacles of the two groups, but regarded them as independent adaptations. The hypothesis advanced by Bütschli is that primitive forms of Suctoria (such as *Hypocoma* (*C*), which has but one suctorial tentacle, and which retains its cilia throughout life, the cilia being upon the ventral side only, as in hypotrichous forms of Ciliata) were derived from hypotrichous ciliates by the mouth portion becoming progressively drawn out into a tentacle. Haeckel ('96), adopting Bütschli's view, compared the simple, single, and terminal mouth-tube of a primitive suctorian with the long, proboscis-like oral region of certain holotrichous ciliates, such as *Lacrymaria olar* or *L. phœnicopterus*. In the closely allied forms, *Didinium* and *Mesodinium*, the oral tube is not ciliated and is contractile, so that when food is taken in, the tube widens into more or less of a disk similar to many suctorian tentacles. In *Mesodinium*, this tube is not only retractile, but is also surrounded by four tentacle-like processes which simulate some kinds of tentacles in the Suctoria. As the majority of the larvæ of the Suctoria are ciliated in girdles, Haeckel holds that this division of the Holotrichida represents the nearest allies of the Suctoria, and that the loss of cilia in the adult is already foreshadowed by the regional loss of cilia in these girdled forms.

The entire matter, finally, of the origin of Infusoria from more generalized forms of Protozoa remains unsolved; the various hypotheses are interesting possibilities, but no more can be said for them. This problem, like that of the origin of the Protozoa, may never come nearer settlement; for, without the assistance of palæontological and embryological evidence, which in other great groups of the animal kingdom are of inestimable value in tracing ancestors, the possibility of tracing their origin is reduced to a minimum.

CLASSIFICATION

CLASS V. **INFUSORIA.** Protozoa in which the motor apparatus is in the form of cilia, either simple or united into membranes, membranelles, or cirri. The cilia may be permanent or limited to the embryonic stages. With two kinds of nuclei, macronucleus and micronucleus. Reproduction is effected by simple transverse division or by budding. Nutrition is holozoic or parasitic.

Subclass I. **CILIATA.** Infusoria provided with cilia during adult as well as embryonic life. Reproduction is brought about typically by simple transverse division. Mouth and anus are usually present. The contractile vacuole is often connected with a complicated canal system.

Order 1. **HOLOTRICHIDA.** Ciliata in which the cilia are similar and distributed all over the body, with, however, a tendency to lengthen in the vicinity of the mouth. Trichocysts are always present, either distributed about the body or limited to a special region.

Suborder 1. **GYMNOSTOMINA.** Holotrichida without an undulating membrane about the mouth, which remains closed except during food-taking intervals.

Family 1. **Enchelinidæ.** The mouth is always terminal or sub-terminal, and is usually round or oval in outline. Food-taking is usually a process of swallowing. Genera: *Holophrya* Ehr. ('31); *Urotricha* Clap. & Lach. ('58); *Enchelys* Hill (1752), Ehr. ('38); *Spathidium* Duj. ('41); *Chænia* Quennerstadt ('68); *Prorodon* Ehr. ('33); *Dinophrya* Bütschli ('88); *Lacrymaria* Ehr. ('30); *Trachelocerca* Ehr. ('33); *Actinobolus* Stein ('67); *Ileonema* Stokes ('84); *Plagiopogon* Stein ('59); *Coleps* Nitsch ('27); *Tiarina* Bergh ('79); *Stephanopogon* Entz ('84); *Didinium* Stein ('59); *Mesodinium* Stein ('62); *Bütschlia* Schuberg ('86).

Family 2. **Trachelinidæ.** The body is distinctly bilateral or asymmetrical, with one side, the dorsal, slightly arched. The mouth may be terminal or sub-terminal, or the entire mouth-region may be drawn out into a long proboscis. An œsophagus or gullet may or may not be present; when present, it is usually supported by a specialized framework. Genera: *Amphileptus* Ehr. ('30); *Lionotus* Wrzesniowski ('70); *Loxophyllum* Duj. ('41); *Trachelius* Schrank ('03); *Dileptus* Duj. ('41); *Loxodes* Ehr. ('30).

Family 3. **Chlamydodontidæ.** The general form is oval or kidney-shaped. The mouth is almost always in the posterior region. The pharynx is supported by a rod-apparatus or a smooth, firm tube.

Subfamily 1. *Nassulinæ.* Ciliation is complete. Genera: *Nassula* Ehr. ('33).

Subfamily 2. *Chilodontinæ.* The body is generally flattened, and the cilia are stronger on the dorsal side, or are confined to that region. Genera: *Orthodon* Gruber ('84); *Chilodon* Ehr. ('33); *Chlamydodon* Ehr. ('35); *Opisthodon* Stein ('59); *Phascolodon* Stein ('57); *Scaphidiodon* Stein ('57).

Subfamily 3. *Erviliinæ.* The cilia are confined to the ventral surface or to a portion of it. The posterior end invariably possesses a movable style arising from the posterior ventral surface. Genera: *Ægyria* Clap. & Lach. ('58); *Onychodactylus* Entz. ('84); *Trochilia* Duj. ('41); *Dysteria* Huxley ('57).

Suborder 2. **TRICHOSTOMINA.** In addition to the general coating of cilia there is an undulating membrane or membranes at the edge of the mouth or in the pharynx. The mouth is always open.

Family 1. **Chiliferidæ.** The mouth is in the anterior half of the body or close to the middle The pharynx when present is short. The so-called "peristome area" leading to the mouth is absent or only slightly developed. Genera: *Leucophrys* Ehr. ('30); *Glaucoma* Ehr. ('30); *Dallasia* Stokes ('86); *Frontonia* Ehr. ('38); *Ophryoglena* Ehr. ('31); *Colpidium* Stein ('60); *Chasmatostoma* Engelmann ('62); *Uronema* Duj. ('41); *Urozona* Schewiakoff (Bütschli) ('88); *Loxocephalus* Kent ('81); *Colpoda* Müller (1773).

Family 2. **Urocentridæ.** The mouth, with a long, tubular pharynx, is in the centre of the ventral side. The cilia are confined to two broad zones around the body at each end. Genera: *Urocentrum* Nitsch ('27).

Family 3. **Microthoracidæ.** Small asymmetrical forms, with the mouth invariably in the hinder portion. The cilia are always more or less dispersed, sometimes limited to the oral region. There may be one or two undulating membranes. Genera: *Cinetochilum* Perty ('49); *Microthorax* Engelmann ('62); *Ptychostomum* Stein ('60); *Ancistrum* Maupas ('83); *Drepanomonas* Fresenius ('58).

Family 4. **Paramœcidæ.** The mouth is sometimes in the anterior, sometimes in the posterior, half of the body, and is accompanied by a large, triangular "peristome area" running from the left anterior edge of the body to the mouth. Genera: *Paramœcium* Stein ('60).

Family 5. **Pleuronemidæ.** The mouth is at the end of a long peristome which runs along the ventral side: the body is dorso-ventrally or laterally compressed. The entire left edge of the peristome is provided with an undulating membrane which occasionally runs around the posterior end of the peristome to form a pocket leading to the mouth. The right edge of the peristome is provided with a less developed membrane. There may or may not be a well-developed pharynx. Genera: *Lembadion* Perty ('49); *Pleuronema* Duj ('41); *Cyclidium* Ehr. ('38), a sub-genus of the preceding; *Calyptotricha* Phillips ('82); *Lembus* Cohn ('66).

Family 6. **Isotrichidæ.** The body is more or less plastic, but not contractile. The cuticle is thick and provided with evenly distributed cilia. The mouth is posterior and accompanied by a distinct pharynx. They are parasites in the digestive tract of ruminants. Genera: *Isotricha* Stein ('59); *Dasytricha* Schuberg ('88).

Family 7. **Opalinidæ.** The form is oval, and the body may be short or drawn out to resemble a worm. They are characterized mainly by the absence of mouth and pharynx. Genera: *Anoplophrya* Stein ('60); *Hoplitophrya* Stein ('60); *Discophrya* Stein ('60); *Opalinopsis* Fœttinger ('81); *Opalina* Purkinje and Valentin ('35); *Monodontophrya* Vejdowsky ('92).

Order 2. **HETEROTRICHIDA.** Ciliata characterized by the possession of a uniform covering of cilia and an *adoral zone*, consisting of short cilia fused together into membranelles.

Suborder 1. **POLYTRICHINA.** Heterotrichous ciliates provided with a uniform coating of cilia.

Family 1. **Plagiotomidæ.** The peristome is a narrow furrow, which begins, as a rule, close to the anterior end, and runs backward along the ventral side to the mouth, which is usually placed between the middle of the body and the posterior end. A well-developed adoral zone stretches along the left side of the peristome, and it is usually straight. Genera: *Conchophthirus* Stein ('61); *Plagiotoma* Duj. ('41); *Nyctotherus* Leidy ('49), a sub-genus; *Blepharisma* Perty ('49); *Metopus* Clap. & Lach. ('58); *Spirostomum* Ehr. ('35).

Family 2. **Bursaridæ.** The body is usually short and pocket-like, but may be elongate. The chief characteristic is the peristome, which is not a furrow, but a broad triangular area, deeply insunk, and ending in a point at the mouth. The adoral zone is usually confined to the left peristome edge, or it may cross over to the right anterior edge. Genera: *Balantidium* Stein ('67); *Balantidiopsis* Bütschli ('88); *Condylostoma* Duj. ('41); *Bursaria* O. F. Müller (1773); *Thylakidium* Schewiakoff ('92).

Family 3. **Stentoridæ.** The peristome is relatively short and limited to the front end of the animal, so that its plane is nearly at right angles to that of the longitudinal axis of the body. The adoral zone of cilia either passes entirely around the peristome edge, or ends at the right-hand edge. The surface of the peristome is spirally striated and provided with cilia. Undulating membranes are absent. Genera: *Climacostomum* Stein ('59); *Stentor* Oken ('15); *Folliculina* Lamarck ('16). Genera *incertæ sedis*: *Cœnomorpha* (*Gyrocorys* Stein) Perty ('52); *Maryna* Gruber ('79).

Suborder 2. **OLIGOTRICHINA.** Heterotrichous ciliates characterized by the reduced cilia, which are limited to certain localized areas.

Family 1. **Lieberkühnidæ.** This name was given by Bütschli for certain little-known forms, which were at first considered young Stentors.

Family 2. **Halteriidæ.** The peristome has no cilia, and only a few scattered ones can be found on the ventral and dorsal surfaces. Genera: *Strombidium* Clap. & Lach. ('58); *Halteria* Duj. ('41).

Family 3. **Tintinnidæ.** The body is attached by a stalk to a theca. Inside of the adoral zone of membranelles is a ring of cilia (par-oral cilia). Genera: *Tintinnus* Fol. ('89); *Tintinnidium* Kent ('81); *Tintinnopsis* Stein ('67); *Codonella* Haeckel ('73); *Dictyocysta* Ehr. ('54).

Family 4. **Ophryoscolecidæ.** Heterotrichous ciliates characterized by a thick cuticle and deep funnel-like peristome. The posterior end is provided with distinct spine-like processes, while the terminal anus is provided with a well-defined anal tube. Genera: *Ophryocolex* Stein ('59); *Entodinium* Stein ('59); *Diplodinium* Schuberg ('88).

Order 3. **HYPOTRICHIDA.** Ciliata in which the cilia are limited to the ventral surface of a dorso-ventrally flattened body; they are frequently fused to form larger appendages, the cirri, and an adoral zone of membranelles. The dorsal surface is frequently provided with bristles. A pharynx may be absent or but slightly developed.

Family 1. **Peritromidæ.** The peristome is but slightly marked off from the remaining frontal area. The cilia on the ventral surface are uniform in size and arrangement, and are not differentiated into cirri. Genera: *Peritromus* Stein ('62).

Family 2. **Oxytrichidæ**. The peristome is not always distinctly marked off from the frontal area. In the most primitive forms the ciliation on the ventral surface is similar to that of the preceding family. Almost invariably in these primitive forms some of the anterior and some of the posterior cilia are fused into large and more powerful appendages, the cirri, which are distinguished as the *frontal* and *anal* cirri, respectively. In the majority of forms all of the cilia are thus differentiated; strong marginal cirri are formed in perfect rows, and ventral cirri in imperfect rows. In addition to the adoral zone of membranelles, there is an undulating membrane on the right side of the peristome, and, in some cases, a row of cilia between the membrane and the adoral zone. These are the par-oral cilia, and they form the par-oral zone. Genera: *Trichogaster* Sterki ('78); *Urostyla* Ehr. ('30); *Kerona* Ehr. ('38); *Epiclintes* Stein ('62); *Stichotricha* Perty ('49); *Strongylidium* Sterki ('78); *Amphisia* Sterki ('78); *Uroleptus* Stein ('59); *Sparotricha* Entz ('79); *Onychodromus* Stein ('59); *Pleurotricha* Stein ('59); *Gastrostyla* Engelmann ('62); *Gonostomum* Sterki ('78); *Urosoma* Kowalewsky ('82); *Oxytricha* Ehr. ('30); *Stylonychia* Stein ('59); *Actinotricha* Cohn ('66); *Balladina* Kowalewsky ('82); *Psilotricha* Stein ('59); *Tetrastyla* Schewiakoff ('92); *Holosticha* Wrzesniowski ('77).

Family 3. **Euplotidæ**. Hypotrichous ciliates, which are characterized mainly by the considerable reduction of the cilia, as well as the frontal, marginal, and ventral cirri; the anal cirri, on the other hand, are always present. The macronucleus is band-formed. Genera: *Euplotes* Stein ('59); *Certesia* Fabre-Dumergue ('85); *Diophrys* Duj. ('41); *Uronychia* Stein ('57); *Aspidisca* Ehr. ('30).

Order 4. **PERITRICHIDA**. Ciliata usually of cylindrical or cup-like form, in which the cilia are reduced, as a rule, to those which form the adoral zone, but secondary rings of cilia may be present.

Family 1. **Spirochonidæ**. Peritrichous ciliates in which the peristome is drawn out into a curious funnel-like process, either simple or rolled. They are parasitic forms in which reproduction by budding is characteristic. Genera: *Spirochona* Stein ('51); *Kentrochona* Römpel ('94); *Kentrochonopsis* Doflein ('97).

Family 2. **Lichnophoridæ**. In addition to the adoral zone, there is a secondary circlet of cilia around the opposite end. The adoral zone is a left-wound spiral. A single genus, *Lichnophora*, Claparède ('67), which is parasitic on various marine arthropods.

Family 3. **Vorticellidæ**. Attached or unattached forms of peritrichous ciliates, in which the adoral zone, seen from above, forms a right-wound spiral (dexiotropic). A secondary circlet of cilia around the under end may be present either permanently or periodically.

Subfamily 1. *Urceolarinæ*. Vorticellidæ having a permanent secondary circlet of cilia which incloses an adhesive disk, and without a peristome fold. Genera: *Trichodina* Stein ('54); *Cyclochæta* Jackson ('75); *Trichodinopsis* Clap. & Lach. ('58).

Subfamily 2. *Vorticellidinæ*. Peritrichous forms without a permanent secondary circlet of cilia, and provided with a peristome fold which can be contracted sphincter-like to inclose the peristome. Genera: *Scyphidia* Lachmann ('56); *Gerda* Clap. & Lach. ('58); *Astylozoön* Engelmann ('62); *Vorticella* Ehr. ('38); *Carchesium* Ehr. ('30); *Zoothamnium* Stein ('54); *Glossatella* Bütschli ('88); *Epistylis* Ehr. ('30); *Rhabdostyla* Kent ('82); *Opercularia* Stein ('54); *Ophrydium* Ehr. ('38); *Cothurnia* Clap. & Lach. ('58); *Vaginicola* Clap. & Lach. ('58); *Lagenophrys* Stein ('51).

Subclass II. **SUCTORIA**. Infusoria having no cilia during the adult stages, but provided with them during the embryonic period. In a few cases the cilia are retained. They have tentacles of various kinds, some adapted for sucking, some for piercing.

Family 1. **Hypocomidæ**. These are unattached forms of Suctoria with a permanently ciliated ventral surface, and with one suctorial tentacle. Reproduction is effected by cross-division. A single genus, *Hypocoma* Gruber ('84).

Family 2. **Urnulidæ**. A family of small attached forms, with or without a cup or theca; with one or two, rarely more, simple tentacles. Swarm-spores holotrichous. Genera: *Rhyncheta* Zenker ('66); *Urnula* Clap. & Lach. ('58).

Family 3. **Metacinetidæ**. Thecate forms; the base of the cup is drawn out into a long stalk, and the walls are perforated for the exit of the tentacles. A single genus, *Metacineta* Bütschli ('88).

Family 4. **Podophryidæ**. Stalked or unstalked forms of more or less globular shape. The tentacles are numerous and distributed about the entire surface or limited to the apical region; some of them are knobbed, others pointed and have a prehensile function. Genera: *Sphærophrya* Clap & Lach. ('58); *Endosphæra* Engelmann ('76); *Podophrya* Ehr. ('38); *Ephelota* Str. Wright ('58); *Podocyathus* Kent ('81).

Family 5. **Acinetidæ**. The individuals are naked and stalked, or thecate and stalked or unstalked. The tentacles are numerous, usually knobbed and all alike. Reproduction is effected by inner or *endogenous* budding, which may be simple or multiple. The swarm-spores are usually peritrichous, but may be holotrichous or hypotrichous. Genera: *Tokophrya* Bütschli ('88); *Acineta* Ehr. ('33); *Solenophrya* Clap. & Lach. ('58); *Suctorella* Frenzel ('91).

Family 6. **Dendrosomidæ**. Suctoria without stalks or theca. The tentacles are numerous, all alike, and knobbed and grouped in distinct tufts; they may be simple or branched. Reproduction by endogenous division; the swarm-spores are peritrichous. Genera: *Trichophrya* Clap. & Lach. ('58); *Dendrosoma* Ehr. ('38); *Staurophrya* Zacharias ('93).

Family 7. **Dendrocometidæ**. Sessile Suctoria resting upon the entire basal surface or upon a portion of it raised as a stalk. The numerous tentacles are short and knobbed, and distributed over the entire apical surface or localized upon branched arms. Spore-formation is endogenous; the swarm-spores peritrichous. Genera: *Dendrocometes* Stein ('67); *Stylocometes* Stein ('67).

Family 8. **Ophryodendridæ**. Stalked or sessile forms possessing numerous long, rarely knobbed tentacles, which are supported upon proboscis-like processes of the apical side. Reproduction is brought about by endogenous budding. The swarm-spores are peritrichous. Genera: *Ophryodendron* Clap. & Lach. ('58).

SPECIAL BIBLIOGRAPHY VI

Bütschli, O. — Protozoa, Infusoria. In Bronn's Klassen und Ordnungen des Thierreichs. *Leipzig*, 1886–88.

Entz, G. — Protistenstudien. *Budapesth*, 1888.

Kent, W. Saville. — A Manual of the Infusoria. *London*, 1881.

Schewiakoff, W. — Beiträge zur Kenntniss der Holotrichen Ciliaten. *Bibliotheca Zoologicà*, Heft 5, 1889.

Stein, Fr. — Der Organismus der Infusionsthiere. *Leipzig*, 1863 and 1867.

CHAPTER VII

SEXUAL PHENOMENA IN THE PROTOZOA

"Die Bedeutung des Konjugationsaktes ist ein Verjüngung der ihn begehenden Tiere. Durch diese Verjüngung erscheinen uns die aus der Konjugation hervorgehenden Individuen sehr geeignet, zu den Stammvätern einer Reihe durch Theilung sich fortpflanzender Generationen zu werden, im Laufe welcher allmählich ein Sinken der Lebensenergie sich einstellt. Letzter Umstand findet seinen Ausdruck darin, das die Grösse der Individuen mehr und mehr sinkt sodass schliesslich eine minimalgrösse erreicht wird, worauf eine neue Konjugationsepoche eintritt." — BÜTSCHLI.[1]

THE power of the animal or plant to reproduce its kind from a portion of its own body is bound up, in the higher forms of life, with sexual processes and, in its more familiar forms, accordingly, is characterized as *sexual reproduction.* It involves the union of two cells having quite different characteristics; the spermatozoön or male cell, being minute and active, the ovum or female cell, larger and quiescent. Reproduction of the individual without sexual processes is, however, possible, even in the higher organism, as we daily witness in plant "cuttings," and as is almost equally well known in the higher forms of invertebrates such as insects, where, without fertilization, an ovum may develop into an adult form. In the lower forms of Invertebrata, such as the worms and the Cœlenterata, so-called "asexual reproduction" by division or by budding is widespread, and in the Protozoa this method of reproduction is the usual form.

The phenomena of parthenogenesis, or development from the egg without fertilization, and reproduction by simple division or by budding as seen in the Protozoa and the lower Metazoa, have recently led to the pertinent query: With what right do we distinguish the phenomena of reproduction as "sexual" and "asexual"? (Hertwig, '99). In two recent publications R. Hertwig ('98, '99) has discussed this question in a very interesting and convincing manner, and he maintains that, in order to speak of "sexual" reproduction, it must be shown that in the Protozoa, for example, fertilization has some direct effect upon division or that a certain specific form of division results from fertilization, neither of which, he says, is true in the majority of cases.

All observers are apparently agreed that *asexual reproduction* as seen in the processes of binary fission and spore-formation is a result of growth, usually expressed by the statement that increase

[1] ('76), p. 421.

in volume continually tends to outrun that of surface, for the former increases as the cube of the diameter, the latter only as the square. The mass of living protoplasm must, therefore, increase more rapidly than the surface which serves to keep it alive, and the isolated cell tends to come first to a physiological standstill, and second to a period of decline, since the surface of nutrition, respiration, and excretion is incommensurate with the bulk of protoplasm. Cell-division, therefore, which Spencer characterized as an indication of the limit of growth, becomes an apparent necessity.

Growth, or the preponderance of constructive processes, leads thus indirectly to increase of surface by division of the cell, but in conjugation or fertilization the very opposite phenomenon occurs, *viz.* the increase in bulk of the cell with a consequent relative decrease in surface because of the union of two cells. Geddes and Thompson ('90), like Spencer, emphasize the connection between division and the advent of preponderating katabolism within the cells, and in their interesting work on the Evolution of Sex, interpret male and female organisms in terms of relative metabolism, the male being regarded as relatively katabolic, the female as anabolic. These authors interpret fertilization as a "katabolic stimulus to an anabolic cell, and on the other side, of course, as an anabolic renewal to a katabolic cell, as well as the union of opposed hereditary characteristics." [1]

If, as Minot ('79) suggested, every newly formed organism be regarded as having a certain initial potential energy which is gradually used up in its life-activities to be restored by conjugation, then the union of two cells may be interpreted as a renewal of vigor or a "rejuvenescence" (Maupas), a view of fertilization first expressed by Bütschli ('76), Engelmann ('76), and Minot ('77, '79), and apparently confirmed later by Maupas ('88, '89) through his admirable observations on the life-cycle of Infusoria. O. Hertwig ('76) also, in connection with fertilization of the metazoan egg, arrived at a similar dynamic view of fertilization, maintaining that protoplasm gradually tends toward a state of stable equilibrium expressed by decreased activity, etc., and that it is restored to a more unstable or more labile condition by conjugation or fertilization. The force of these views as to the need of conjugation for different species of Infusoria, at least, can hardly be questioned; for, as repeatedly stated in the preceding chapters, reproduction by simple division may go on for a certain number of generations, but cannot continue indefinitely, unless at certain intervals, which Maupas has shown to be more or less definite, two individuals unite in conjugation. This union, in some wholly unexplained way, imparts to each of the conjugants

[1] *Loc. cit.*, p. 232.

a renewed vitality, or in Bütschli's words ('76) a renewal of youth (*Verjüngung*), expressed by increased activity in movements and reproduction. Conjugation thus, as R. Hertwig insists, is not the beginning of a series of reproductive acts, but occurs at or near the end of such a series. Maupas's results seem to offer conclusive evidence that the absence of conjugation involves a cumulative degenerative process which ultimately ends in death. The phenomena of so-called sexual reproduction and of sex-differentiation have, in all probability, grown out of this apparently fundamental requirement of living protoplasm, namely, the *periodic union of two cells;* and I believe with Bütschli, Engelmann, Maupas, Hertwig, and many others, that it cannot in itself be regarded as primarily a reproductive act. None of the facts that have been determined show that the morphological distinction of the sexes is a primary attribute or property of living organisms, nor do any of the dynamic views of fertilization afford an explanation of sex-differences any more than does the statement of Geddes and Thompson cited above. Even though the views of these authors be accepted, we should still have to admit that no explanation of fertilization can be wholly satisfactory unless it is equally applicable to forms which cannot be distinguished as more anabolic or katabolic, *i.e.* to the conjugation of equal-sized Infusoria, Mastigophora, or Sporozoa, as well as the union of differentiated male and female cells. If then we define sex as *the condition by which single-celled or many-celled organisms are differentiated into male and female*, we must admit that the origin of sex is only a part of the problem, for fertilization occurs between individuals in which there is no apparent sex-difference, as well as between those possessing it.

In the present chapter I have brought together some of the evidence which bears upon these several points. A number of phenomena which accompany reproduction in the higher animals, and which have attracted so much attention among biologists of all times, are seen in simpler forms in Protozoa. Among these the phenomena of sex-differentiation, of maturation or preparation for fertilization, and fertilization itself are of paramount interest.

The explanation of sex-differentiation is not as yet made more easy by the study of Protozoa, and here as in higher animals it must remain entirely hypothetical until future research throws more light upon the problem. The conditions accompanying conjugation, however, have been carefully studied and analyzed in relation to other vital functions of the cell, and the evidence thus acquired gives a clue, I believe, to an ultimate explanation of fertilization. An important underlying principle was first made out by Bütschli ('76) and Engelmann ('76) upon degenerating Infusoria, and was expressed by Minot three years later in the sentence, "the exhaustion of the rejuvenating power (*i.e.* the exhaustion of the initial potential of vitality) becomes the stimulus for the formation of the sexual products."[1]

A. Phenomena of Conjugation

With our present knowledge it is impossible to say that conjugation is absent in any group of Protozoa, and until the life-cycle of every genus is fully known, the conservative but logical view which Bütschli expressed in regard to flagellates, is the most acceptable. He says: "I, personally, am inclined to the view that the significance of this process in the life of these organisms is so general and deep-reaching that the failure to observe it in certain groups up to the present time is no reason for considering it absent."[2] Nevertheless, the observations which have been recorded show the greatest diversity in the process, the variations passing from extremely simple fusion, which only by stretching the meaning of the term can be called sexual, to the highly differentiated male and female organisms where, as in the Metazoa, fertilization is followed by cleavage.

Stages in the development of so-called sexual reproduction may be considered as follows: —

1. The permanent or temporary union of similar adult cells (*Isogamy*) (Sarcodina, Sporozoa, Mastigophora, Ciliata).
2. The union of individuals apparently similar in all respects save size (*Anisogamy*) (Sarcodina, Mastigophora, Ciliata).
3. The union of reduced individuals. Swarm-spores (*Isogamy* or *Anisogamy*) (Sarcodina, Mastigophora).
4. The union of specialized individuals — male and female cells (Spermatozoa and eggs) (Sporozoa, Flagellidia).

1. *The permanent or temporary union of similar adult individuals* (*Isogamy*).

Thanks to the unbroken observations of Dallinger and Drysdale, the conjugation and full life-history of some of the lowest forms of Protozoa (monads) have been made out. All of the forms examined reproduce by simple division for a few days, and then conjugate. In *Cercomonas longicauda* Duj. (*typica* Kent), one of the Monadida, reproduction by ordinary fission continues for two to four days, when the offspring, without losing their flagella, become amœboid and conjugate two by two (Fig. 116). The union begins with the fusion of the pseudopodial processes, and, as it progresses, the flagella are withdrawn. The nuclei finally unite (*D*), and the product of the union, or the *zygote*, forms a thin-skinned globular cyst (*E*). After a

[1] ('79), p. 199. [2] ('83), p. 778.

short resting period the cyst breaks and an innumerable quantity of fine spores pour out (*F*).

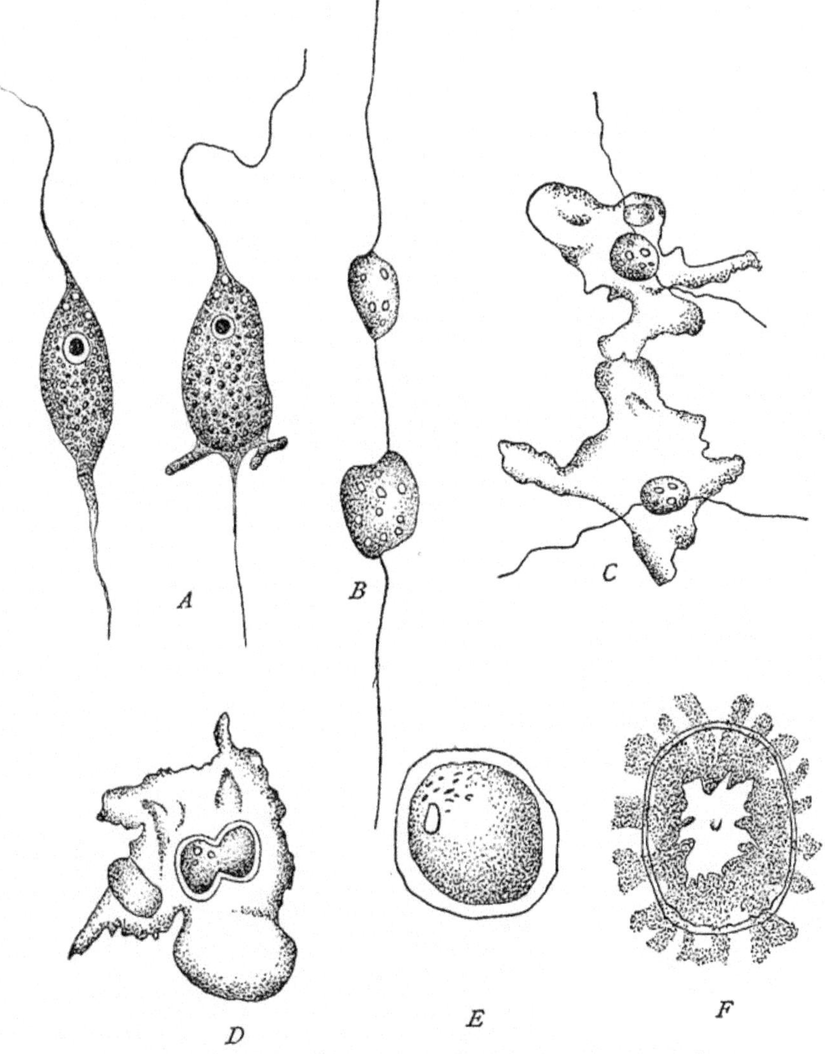

Fig. 116.—Conjugation of *Cercomonas*. [DALLINGER and DRYSDALE.]

The vegetative cells increase by transverse division (*A, B*). When sexually mature they become amœboid, and then fuse (*C, D*). The result is a zygote (*E*), which ultimately bursts and liberates masses of spores (*F*).

In *Tetramitus*, one of the Polymastigida, a similar period of reproduction by longitudinal division finally results in amœboid forms which conjugate, forming cysts and spores (Fig. 117). The process of conjugation is somewhat modified in certain *Bodos* of the group Heteromastigida, where, after a period of binary fission, as in *Cercomonas*, the individuals become amœboid, two or three fusing while in

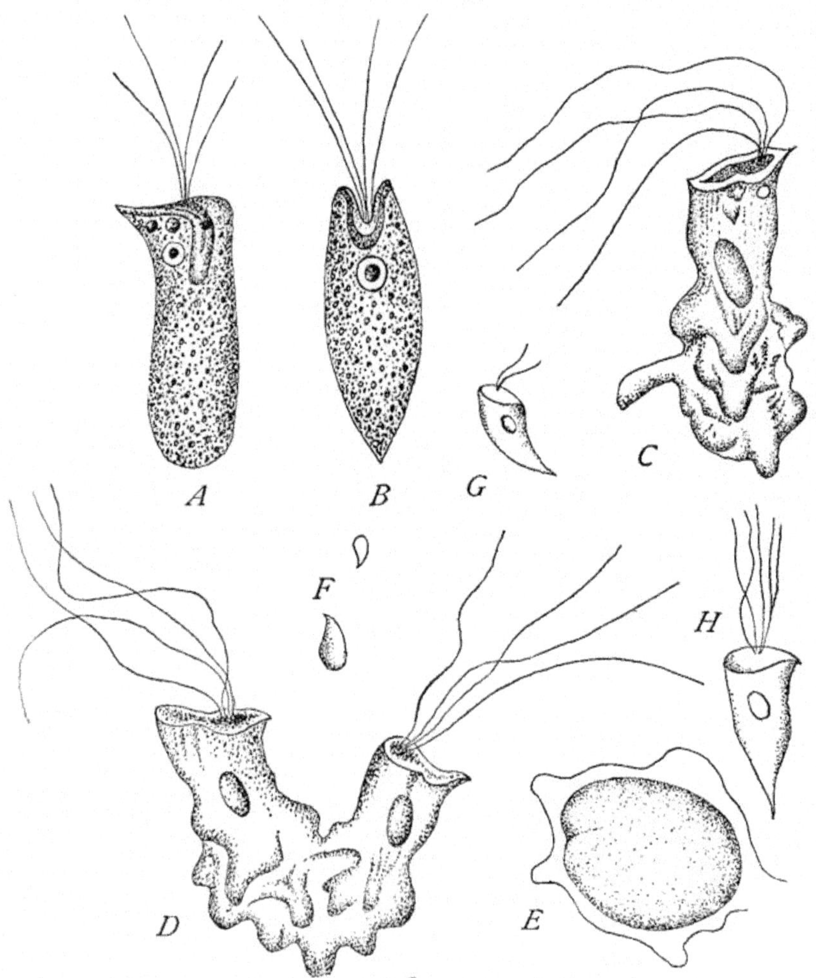

Fig. 117. — Conjugation in *Tetramitus rostratus* Perty. [STEIN.]
A, B. Individuals from the front and side. *C.* Amœboid form. *D.* Conjugation of amœboid forms. *E.* Cyst. *F, G, H.* Development of the spores.

this condition to form a common mass. After a resting period, the encysted mass breaks up into an immense number of small individuals, or spores, differing from those in *Cercomonas*, in that each one is similar to the parent organism. This union, however, appears to be purely facultative, for the same process of encystment and spore-formation may take place in an isolated individual, and in this case, at least, reproduction cannot be dependent upon sexual union or conjugation, although it does not signify that conjugation is not necessary for the continued power of reproducing. Kent ('81), who has confirmed Cienkowsky's ('65) observations upon conjugation of *Bodo augustatus*, states that a difference exists in the number of spores that are formed in the fusion and in the solitary cysts, only four spores arising from the latter.

A significant feature in the conjugation of these forms is that the individuals lose their customary outline and become amœboid prior to

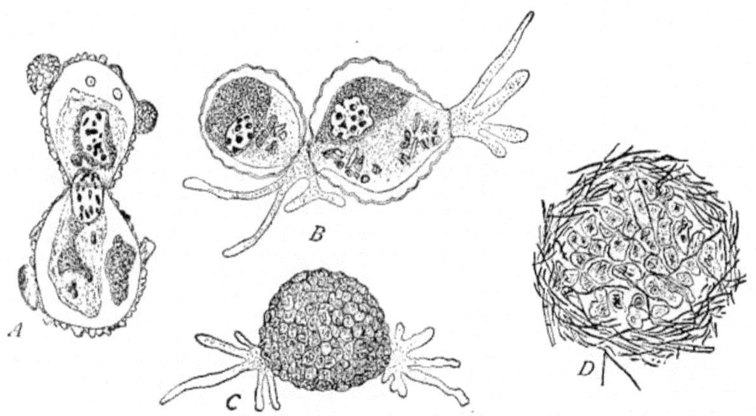

Fig. 118. — Conjugation in Rhizopoda. [RHUMBLER.]
A, B, C. *Difflugia lobostoma* Duj. *D.* Aggregated condition of *Amœba verrucosa* Ehr.

fusion, thus showing that some change has taken place in the consistency of the plasm.

A great number of observations have been made among the Rhizopoda upon so-called conjugation phenomena between similar individuals, the process varying in complexity from simple contiguity to the more complicated fusion of cell-bodies. While many of the earlier observations probably dealt with division phases rather than with conjugating individuals, a possibility of error first pointed out by Claparède and Lachmann ('56), conjugation of shelled forms has been safely established through the observations of Bütschli ('74), Jickeli ('84), Blockmann ('88), Pénard ('90), Rhumbler ('98), and others, and of unshelled forms by Schultze (*Gromia*, '75), Holman ('86), and Kühne ('64).

The phenomena of *cytotropy*, or the mutual attraction of two or more cells, among the Sarcodina at least, if not in all forms, are probably closely connected with conjugation and may possibly be the first stages in the development of sexual reproduction.[1] It is certainly reasonable to argue that the mutual attraction of two previously separated blastomeres of the frog's egg (Roux, '94), or the reunion of an amputated pseudopodium with the main body of *Difflugia* (Verworn, '88; Rhumbler, '98), the union of numerous naked *Amœba verrucosa* into a common aggregate (Rhumbler, '98), and the union of two conjugating individuals, are all phenomena of the same order. The phenomenon in *Amœba* appears to have no bearing upon the function of reproduction, for, according to Holman's and Kühne's accounts, conjugation is not followed by reproduction, while true conjugation phases may yet be found in other stages of the life-history of *Amœba* (Fig. 118, *D*). The discovery by Schaudinn of swarm-spores in the allied form *Paramœba*, and of the union of swarm-spores in the more or less closely allied rhizopod *Hyalopus*, makes it not at all improbable that the same thing may occur in *Amœba verrucosa*.

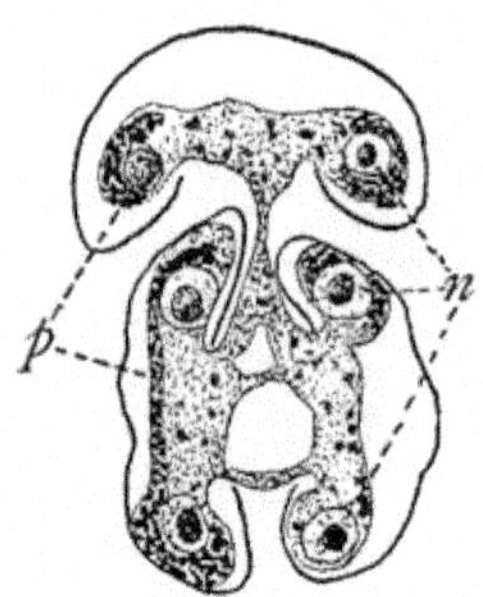

Fig. 119.—Conjugation in *Arcella vulgaris* Ehr. *p.* Perinuclear plasm. *n.* Nuclei.

Thus cytotropy, leading first to contiguity, may result in *plastogamy*, or the fusion of cell-plasms. The protoplasm, however, must be in the proper plastic condition for such a union. Some forms, such as the Mycetozoa and some Heliozoa, are apparently always in this condition, and contact results in fusion. In many such cases plastogamy leads to nothing further, nuclear fusion (*karyogamy*) not occurring. Many instances of such union are found among the Sarcodina, and up to the present time they have been seen nowhere else. These unions take place not only in viscous forms such as the plasmodia of Mycetozoa, *Actinophrys* or *Actinosphærium*, but also in the denser types of Rhizopoda. Thus, in *Arcella*, two or more individuals may fuse together (Fig. 119), and in the shelled rhizopod *Difflugia lobostoma*, Rhumbler ('98) has shown that two, three, or four individuals may be found in plastogamic union in from six to ten per cent of all cases. He also made the interesting and significant observation that this union cannot be induced at will, and concluded that plastogamy takes place here only under certain conditions of the plasm.[2]

Such plastogamic union in the cases cited has apparently no effect upon the united organisms; both Johnson and Schaudinn found no changes in the nuclei in *Actinophrys* and *Actinosphærium*, and

[1] Cf. Cuénot ('97); Rhumbler ('98).

[2] Cf. union during the amœboid stage of *Cercomonas* and other flagellates.

Rhumbler reached similar results in the fused *Difflugias*. The latter, however, calls attention to the fact that two organisms thus united are subject to the interchange of substances through osmosis, and he maintains that such an interchange must take place between them. This interchange may even extend to the substances of the nuclei, which are constantly renewed from, and given off to, the cytoplasm.

Although fusion of the nuclei of the forms just mentioned does not take place, the stimulus of the cytoplasmic interchange in some similar cases is apparently sufficient to bring about reproduction. Thus, in certain Reticulariida (*Patellina corrugata* and *Discorbina globularis*, Schaudinn, '95), two, three, four, or even five individuals may fuse and form embryos without a previous nuclear union. In these cases plastogamy alone is apparently a sufficient stimulus for reproduction.

It is by no means a fanciful assumption to postulate the union of the two nuclei, or karyogamy, through conditions similar to those which lead to the union of two organisms through cytotropy, *i.e.* mutual attraction and consequent fusion when the nuclear plasm is in the right condition, a condition defined by Dangeard ('99) as "sexual hunger."

Almost all cases of karyogamy are complicated by nuclear processes analogous to maturation in Metazoa, a few doubtful cases among the Mastigophora and Rhizopoda alone indicating that fusion may take place without a preliminary loss of a portion of the nucleus. Thus Jickeli ('84) and Rhumbler ('98) observed two individuals of *Difflugia glòbulosa* with but a single nucleus in conjugation, and similar observations by Pénard ('90) upon a number of different species indicate that the phenomenon is widespread (Fig. 118, *A*). In *D. lobostoma*, Rhumbler occasionally found two mouth openings in one shell, and interpreted it as a case of fusion of shells as well as of protoplasm (*B*, *C*). Blochmann ('88) observed the fusion of two *Euglyphas* and the formation of a large double shell. In both cases there was but one functional nucleus observed, although in the latter form the peculiar behavior of the nucleus in one animal was very suggestive of maturation (*vide infra*).

The union of nuclei in temporary conjugants among the Flagellidia has been observed in at least one case, *Noctiluca miliaris* (Cienkowsky, '73, and Ishikawa, '91). Two individuals fuse, their nuclei come together, but do not fuse, and then separately undergo mitosis, which results in four daughter-nuclei. These separate and then fuse in the daughter-cells, two by two; thus nuclear fusion takes place some time after conjugation. In the simpler flagellates the nuclei fuse before spore-formation (Dallinger and Drysdale, '73).

Karyogamy is widely spread throughout the Infusoria, where conjugation in the different species is characterized by very similar features. Two individuals unite, usually by the anterior ends, with the mouth openings apposed. A protoplasmic bridge is formed between the two, through which there is an interchange of micronuclei. This interchange is followed by final separation of the conjugants, each of which regenerates the parts lost during the period of conjugation (oral cilia, macronucleus, etc.). After careful observations upon many different species of Infusoria, Maupas ('88, '89) found that certain conditions are apparently necessary in order that conjugation between two individuals can take place and lead to fertile results. These conditions are: (1) Sexual maturity, that is, the individuals must be removed by some generations from the last conjugating pair. Maupas established the fact that, in *Leucophrys patula*, only individuals of the three hundredth to the four hundred and fiftieth generation could be reinvigorated by conjugation; in *Onychodromus* only individuals between the one hundred and fortieth and the two hundred and thirtieth generation; and in *Stylonychia pustulata* only individuals between the one hundred and thirtieth and the one hundred and eightieth generations. If these individuals, when thus mature, are restrained from pairing, they become over-mature, after which, if they conjugate, the union is without result, and the individuals finally succumb to what Maupas calls "senile degeneration."

(2) A second condition is scarcity of food. Maupas also shows that an over-abundance of food causes the individuals to die from senile degeneration without developing the "sexual hunger."

(3) A third condition is diverse ancestry. Maupas arrived at the conclusion that two individuals from the same ancestor would not conjugate. "In many pure cultures of nearly related individuals," he says, "the fast to which I subjected them resulted either in their becoming encysted, or in their dying of hunger." "It was not until after senile degeneration had already begun to make inroads in the culture that I noticed that the conjugation of nearly related individuals occurred in the experimental cultivations. However, all such conjugations ended with the death of the Infusoria which had paired, but which were unable to develop further, or to reorganize themselves after they had fused. Such pairings are, therefore, pathological phenomena due to senile degeneration." [1]

The latter conclusion, drawn by Maupas, has not been entirely sustained. Bütschli was one of the first to question it,[2] and recently Joukowsky ('98) observed fertile conjugations among descendants of

[1] *Loc. cit.*, p. 411. [2] ('88), p. 1638.

the same individual. Maupas's conclusion, therefore, that cross-fertilization is necessary for Infusoria, as for Metazoa, appears to have been somewhat premature, although in view of the extreme care with which his observations and experiments were made, the objections which have been brought against it are not entirely conclusive.

It is evident from the foregoing review that, with the exception of sex-differentiation, all of the essential features which characterize fertilization are present in those forms of Protozoa where conjugation takes place between similar adult individuals. Here, also, a hint as to the significance of fertilization is seen in the fact that the form of the conjugating individuals is altered, thus indicating some change in the density of the protoplasm. Thus some Mastigophora and Sarcodina become viscous, and some Infusoria show unmistakable signs of exhaustion. Under these changed conditions the fusion of the cell-body is possible (plastogamy). This fusion may be partial (Cystoflagellidia, Gregarinida, Infusoria), or total (Monadida, Heliozoa, Rhizopoda), and it may or may not be accompanied by nuclear fusion (karyogamy). The same stages may be conceived for the union of the nuclei as for the union of the cell-bodies, the evidence appearing to show that as plastogamy is the outcome of cytotropy, or positive chemotaxis, so karyogamy is the outcome of karyotropy or nuclear attraction, and is made possible by plastogamy.

2. *The union of similar but different-sized individuals.*

Sexual differentiation is established when the conjugating organisms are of different size. No sharp line, however, can be drawn between conjugation in isogamous and anisogamous forms, but a number of instances might be cited in which the union of different-sized individuals is purely facultative, and the same result is accomplished either by the union of similar or of dissimilar forms. Thus, of two conjugating *Bodos* (*Heteromita*), one, which is formed by transverse division, is motile and becomes attached to a stationary form resulting from a longitudinal division, and anchored by one of its flagella. With the exception of these differences, which certainly indicate some internal difference in the gametes, the conjugants are identical. Here, for the first time, a distinction can be made between the more quiescent and the more motile conjugant, although it is not marked by difference in size. The latter condition, however, exists in *Polytoma*, where, according to Krassilstschik ('82) and Dallinger and Drysdale, normal free-swimming forms unite with smaller ones. Here, however, the process is purely facultative, for conjugation between similar-sized individuals also takes place. Bütschli does not consider this an indication of sex-difference, but merely the chance fusion of two *Polytomas* of different age. Nevertheless, the phenomenon is significant, and indicates that the smaller or less-grown individual has the capacity, whatever that may be, of conjugation, and may be regarded as an intermediate stage, at least, in the development of sexually differentiated forms (Fig. 120). A somewhat analogous process takes place in *Codosiga botrytis*, one of the Choanoflagellida, where, as first observed by Stein, an attached form conjugates with a free-swimming and somewhat smaller animal.

A similar but more definite sex-difference is seen in the peritrichous Ciliata, where the individuals are of dissimilar size. In all of these, with the exception of the genus *Zoothamnium*, a normal-sized individual fuses with a smaller one. Engelmann ('76) made the

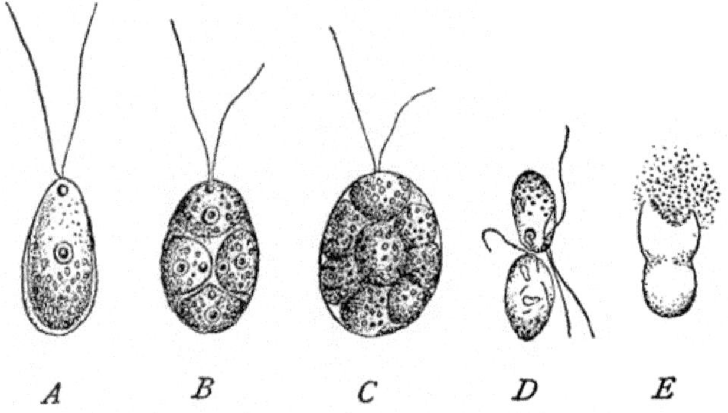

Fig. 120. — Conjugation of *Polytoma uvella* Ehr. [DALLINGER and DRYSDALE.]

interesting discovery that the larger form, or macrogamete, is always one whose sister-buds have given rise by division to smaller forms, or microgametes, which would certainly suggest that a particular condition of the plasm accompanies conjugation. In the genus *Zoothamnium*, alone, the macrogametes are considerably larger than the normal individuals (Trembley 1747, Ehrenberg, Greeff, Engelmann, and others).

The microgametes, which may arise by budding, as in *Lagenophrys ampulla* (Fig. 121), or by repeated divisions, as in *Epistylis* (Fig. 122), swim about freely until they come in contact with macrogametes, to which they finally adhere. Upon fusing, the microgametes gradually lose their definite structure, until finally they are absorbed.

Engelmann ('76), watching the process of sexual union in the ciliate, *Vorticella*, records the following interesting observations: "The buds, at the beginning, swarmed about with constant and considerable rapidity, rotating the while on their axes, but moving more or less in a straight line through the drop. This lasted from five to ten minutes or even longer without any special occurrence. Then the scene suddenly changed. Happening all at once in the vicinity of an attached *Vorticella*, a bud quickly changed its direction with a jerk, and approached the larger form, fluttering about it like a butterfly over a flower, and gliding over its surface here and there as though tasting. After this play, repeated upon several individuals, had gone on for some minutes, the bud finally became firmly attached." Again: "I observed another performance still more remarkable from its physiological and particularly from its psycho-physiological significance. A free-swimming bud crossed the path of a large *Vorticella* which had become free from its stalk in the usual manner and which was roaming about with great activity. At the instant of the meeting — there was no trace of a pause — the bud suddenly changed its direction and followed the *Vorticella* with great rapidity. It developed into a regular chase which lasted about five seconds, during which time

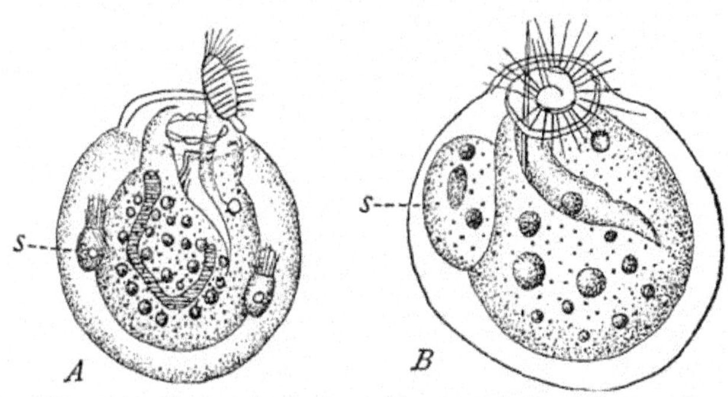

Fig. 121. — Conjugation in *Lagenophrys ampulla* St. [BÜTSCHLI.]
s. Microgamete attached to normal cell (*A*). *B.* Fusion of macro- and microgametes.

the bud remained about one-fifteenth of a millimeter behind the *Vorticella*, although it did not become attached, for it was lost by a sudden side movement of the larger form. The bud then continued its way as before. These processes are remarkable, since they demonstrate a fine and rapid perception, a rapid and safe will determination, and finely divided motor innervation. They show to what astonishing height and multiplicity physiological differentiation in animals can go, even within a single cell." [1]

The phenomena which Engelmann observed and regarded as evidence of psychic activity, have been shown to owe their origin for the most part to chemical and physical stimuli. But the sex-differentiation indicated by the diversity in size and activity of the gametes, and the fusion of the two cells which he described, are typical of fer-

[1] *Loc. cit.*, p. 583.

tilization, or of so-called sexual reproduction, throughout both animal and vegetable kingdoms.

3. *The union of swarm-spores (Isogamy and Anisogamy).*

It is but a short step from the primitive condition described above, where an ordinary individual unites with a smaller one, to the condition where both conjugants are reduced — indeed, both conditions may be present in the same organism. Thus the genus *Polytoma*, as described above, shows a facultative union between two normal-sized individuals, or between one normal-sized individual and a microgamete, or between similar microgametes. Other members of the group to which *Polytoma* belongs — the Chlamydomonadina — show similar indefiniteness, and in no case can it be positively stated that the union between a larger (ovoid) and a smaller (spermatoid) microgamete is obligatory. Goroschankin ('75) maintained that in *Chlorogonium pulvisculus* the spermatoid microgametes arise by an eight division and conjugate with the ovoid macrogametes which arise by a two or four division (Fig. 123), but Reinhardt ('76), on the other hand, observed conjugation between microgametes of the smaller size, although he noted a frequent difference in size. In *Polytoma*, while there is a facultative conjugation between individuals of diverse size, there is also, apparently, a tendency toward an obligatory union of microgametes. The differentiation goes a step further in *Phacotus lenticularis*, where, according to Carter ('58), an ovoid cell arising by two or four division in the normal manner, unites with a minute form which is the product of a sixty-four division.

Fig. 122. — *Epistylis umbellaria* Leeuw. [GREEFF.] Macrogametes (*M*) and microgametes (*m*).

The obligatory conjugation of microgametes is, however, safely established in a great many Protozoa, especially among the colonial forms of Mastigophora, and, to a less extent, in Sarcodina and in some Sporozoa. In the Sarcodina, macro- and microgametes are formed by many marine types, including Reticulariida and Radiolaria, and Brady, Brandt, and Haeckel do not hesitate to say that sexual reproduction is brought about by their union. In only one case, however (*Hyalopus*), has conjugation been actually observed. Here the cell-body spontaneously fragments into isogamous microgametes which swim away from the shell and conjugate (Schaudinn). In the Sporozoa, also, the recent results obtained by Siedlecki ('99) show an

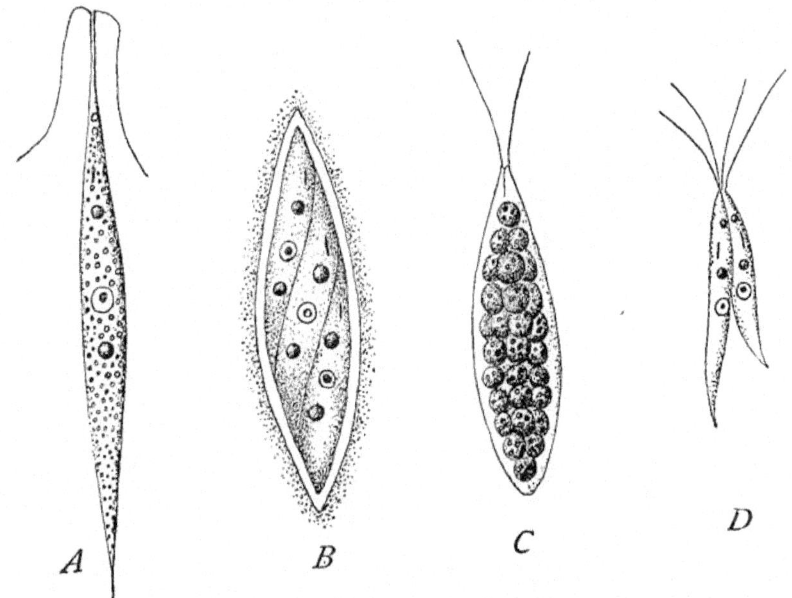

Fig. 123. — Conjugation in *Chlorogonium euchlorum* Ehr. [STEIN.]
A. Adult individual. *B.* Macro- and *C.* microgamete formation. *D.* Conjugation.

analogous phenomenon, and at the same time they throw considerable doubt upon Wolters's ('91) conclusions that the nuclei of two conjugating Gregarines unite.[1] According to Siedlecki, two similar individuals of *Monocystis ascidiæ* come together and secrete a common cyst within which they sporulate, each individual by itself. There is no union of nuclei as in *Actinophrys*, nor interchange of parts of nuclei as in the Infusoria, but the nuclei rapidly divide, and the subdivisions ultimately become the nuclei of minute cells resembling spores. These have been repeatedly observed in Gregarinida,

[1] Cf. p. 157.

and by most authors are called *sporoblasts*. The so-called sporoblasts become motile and move about with considerable freedom, a hitherto unrecognized phenomenon. But more remarkable still, the sporoblasts finally unite two by two, and after complete fusion of nuclei and cell-plasm, each double cell or copula divides into eight parts, the sporozoites (Fig. 124). These observations, which are the most conclusive that have yet appeared, place the Gregarinida in line with the Reticulariida and Radiolaria, and Siedlecki with Mesnil ('00) sees in this isogamous union a feature which distinguishes the Gregarinida from the Coccidiida.

The obligatory fusion of microgonidia is widely distributed in the colonial forms of Mastigophora, especially in the Phytoflagellida. It is to be regretted that we do not know the full life-history of the

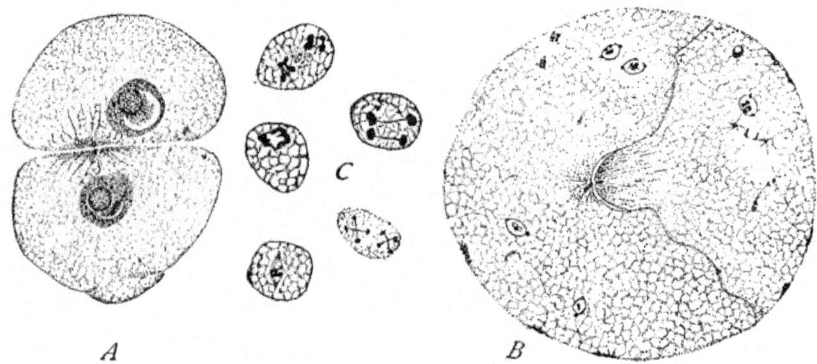

Fig. 124. — Conjugation of *Monocystis acidiæ* Lank. [SIEDLECKI.]
The two gregarines unite (*A*). The nuclei divide repeatedly, and many gametes are formed (*B*). These unite two by two, forming spores. Each spore divides to form eight sporozoites (*c*).

simpler and more indefinite colonies such as *Dinobryon*, *Anthophysa*, or *Synura*, where the aggregate arises through continued binary division, for in the higher types, the fertilized egg, as in Metazoa, passes by a regular cleavage into the adult form, the cells becoming separated only in the later stages. Nevertheless, in these more differentiated types some stages are unquestionably more primitive than others. In *Gonium pectorale*, a colony consisting of sixteen individuals, the colonies reproduce asexually by simultaneous division of all of the cells, four successive longitudinal divisions in each cell resulting in sixteen groups of sixteen cells each, and these groups form independent colonies (O. F. Müller, Cohn, Stein — Fig. 125).[1]

[1] The camera drawing from a permanent preparation shown in Fig. 125 throws considerable doubt upon this interpretation of the first three division planes as described by Stein. According to this one preparation the third division is horizontal, giving four cells above and four below. The plate form is assumed in the early sixteen-cell stage.

Under certain conditions some of the adult cells become separated from the colony and pass into a resting state, during which they divide into eight biflagellated microgametes, and these, as soon as liberated, conjugate in pairs (Hieronymous, Rostafinski, '75). There is no size differentiation between the conjugating microgametes. Nor is there

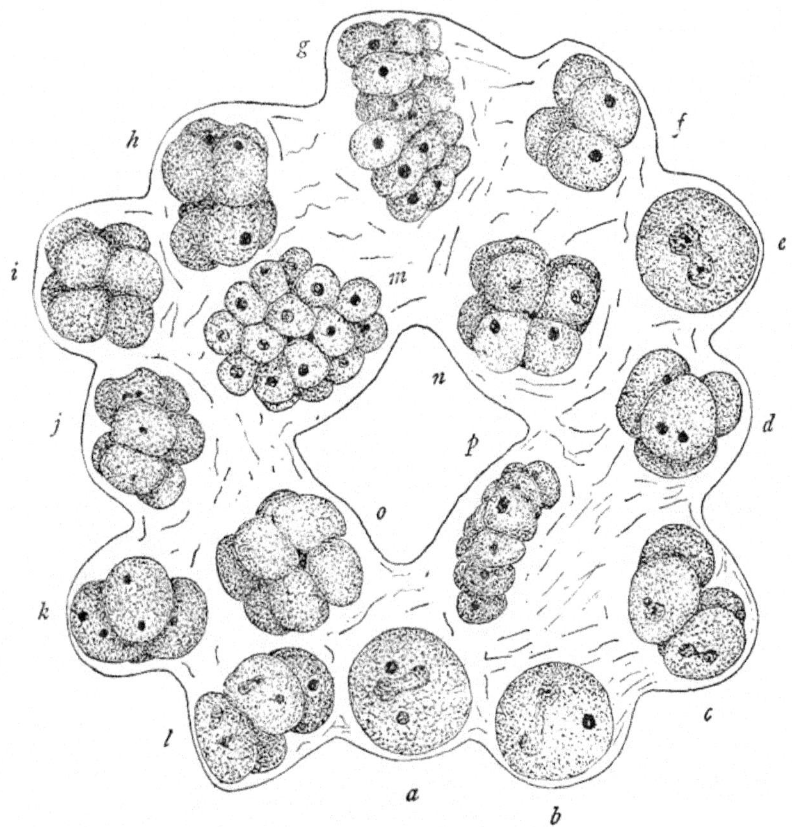

Fig. 125. — *Gonium pectorale* O. F. M., in division.

The third cleavage results in two layers of four cells (*i*, *n*, *o*). The flattening occurs in the 12–16-cell stage (*g*, *m*, *p*).

in the somewhat better-known form, *Pandorina morum*. In the latter, which is also a sixteen-cell colony, after a certain number of asexual generations, a generation appears in which each of the sixteen cells divides, not into sixteen parts for a new colony, but as in *Gonium* into eight gametes, which ultimately become free, conjugate, and pass into a resting zygote condition. After a longer or shorter period, either the cyst bursts, and a naked individual emerges, which by division forms the complete colony; or the zygote may first divide into two or three individuals, each of which forms a sixteen-cell colony. Pringsheim ('69) stated that a dimorphism exists in the gametes formed by large colonies and by small ones, and maintained that the larger gametes never conjugate amongst themselves, while the smaller ones can unite with each other or with the larger ones (Fig. 126).

An incipient sexual difference in the colony is thus indicated, although, as in chlamydomonads, the size difference in gametes is apparently facultative. The sexual difference is somewhat better marked in the genus *Eudorina*, although the most reliable authorities differ as to the details. According to Goroschankin ('76) and Dangeard ('89), the sexually mature colonies are easily distinguished as male and female, the latter resembling the ordinary colonies save for a slightly larger size. The male colonies are at first quite similar to the ordinary colony, but each of the sixteen cells divides to form a sixteen or thirty-two celled plate, and each of these cells gradually becomes long and spindle-formed and develops flagella at the pointed end. They ultimately become free and unite with larger ovoid cells. Carter's ('58) description differs so much from this that Bütschli doubts if he had the same species. He found that the colonies are hermaphrodite and divided into male and female portions. Four cells at one pole of the oval colony develop into spermatozoids, while the remaining twenty-eight cells become enlarged, and as ovoid cells, are probably fertilized by the spermatozoids. These observations, although conflicting, at least show the increasing complexity of the colony, and a differentiation, according to Carter, of male and female cells in the same multicellular individual, or according to Goroschankin, the colonies of *Eudorina*, become multicellular individuals in which the sex is definitely established. In these cases there is no distinction between germ and somatic cells, all being capable of assuming the plasmatic condition necessary for conjugation.

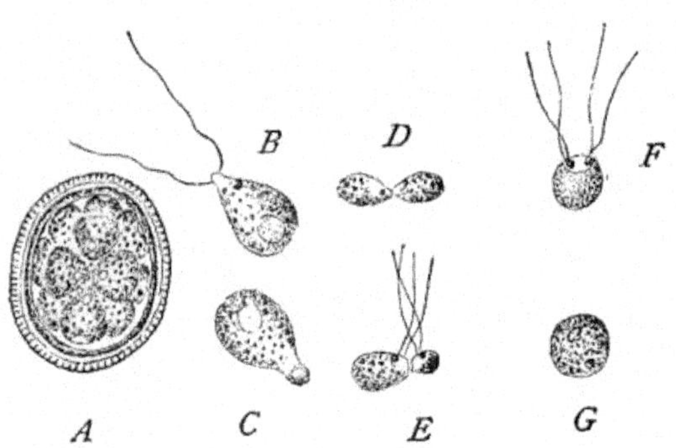

Fig. 126.—Conjugation of *Pandorina morum* Ehr. [PRINGSHEIM.]

Large gametes conjugate with small ones (*B*, *C*), or small ones conjugate with each other (*D*, *E*, *F*). The result is always a zygote (*G*).

4. *The union of eggs and spermatozoa.*

The union of large with small cells, which is only facultative in the majority of single Mastigophora, becomes obligatory in the Coccidiida. As in the colonial forms just described, *Coccidium* (schizont) increases usually by asexual reproduction (schizogony). A large number of *merozoites* are produced, each of which may repeat the cycle (Schaudinn). The asexual increase continues for five days (Schaudinn, '00), after which the merozoites give rise to sexually differentiated individuals, some of which are large (macrogametes), others small like spermatozoa (microgametes). The phenomena of fertilization have been carefully and independently worked out by Schaudinn ('96) and Siedlecki ('97) in the forms *Adelea ovata* and *Eimeria Schneideri*, and by Siedlecki ('98) in *Klossia octopiana* (Eberthi) (Fig. 127), and Schaudinn ('00) in *Coccidium*. The formation of the spermatozoids has already been described;[1] attention may be called, however, to the peculiar central, residual mass which remains after the microgametes are formed, suggesting the residual blastophore in the spermatogenesis of annelids. The microgametes or spermatozoids, when mature, are mere filaments of chromatin with a minimum of cytoplasm; they are pointed at each end and may or may not have flagella. They move very rapidly by serpentine undulations or by means of their flagella, until they come in contact with a female cell which is fertilized in the same manner as in the Metazoa. The development of the egg is similar to that of the microgamete, so far as the nucleus is concerned, but up to the present no maturation process has been recorded. Several microgametes cluster around the macrogamete, as spermatozoa group themselves around the egg (Fig. 128). The nucleus of the female cell moves toward the periphery, and one of the microgametes penetrates the nuclear vesicle, and its chromatin fuses completely with that of the female pronucleus, while the entire nuclear mass now moves back toward the centre of the cell. At the same time a peripheral portion of the cytoplasm becomes more dense, refractile, and finally thick and resisting, to form the membrane of the fertilized macrogamete, thus corresponding exactly with the vitelline membrane of a fertilized egg. After a thorough mixture of the chromatin in the cleavage nucleus, the latter divides by a peculiar method of mitosis, and a great number of spore-nuclei are formed.

In *Adelea ovata* fertilization is effected by the entrance of a male cell through a special opening, the *micropyle*, while the process is further complicated by the preparatory divisions which the nucleus of the microgamete undergoes before entrance. According to Schaudinn ('97) it divides twice while in contact with the macrogamete mem-

[1] See p. 159.

brane, and three of the resulting nuclei are eliminated, while one fuses with the female pronucleus, thus giving a striking analogue of the

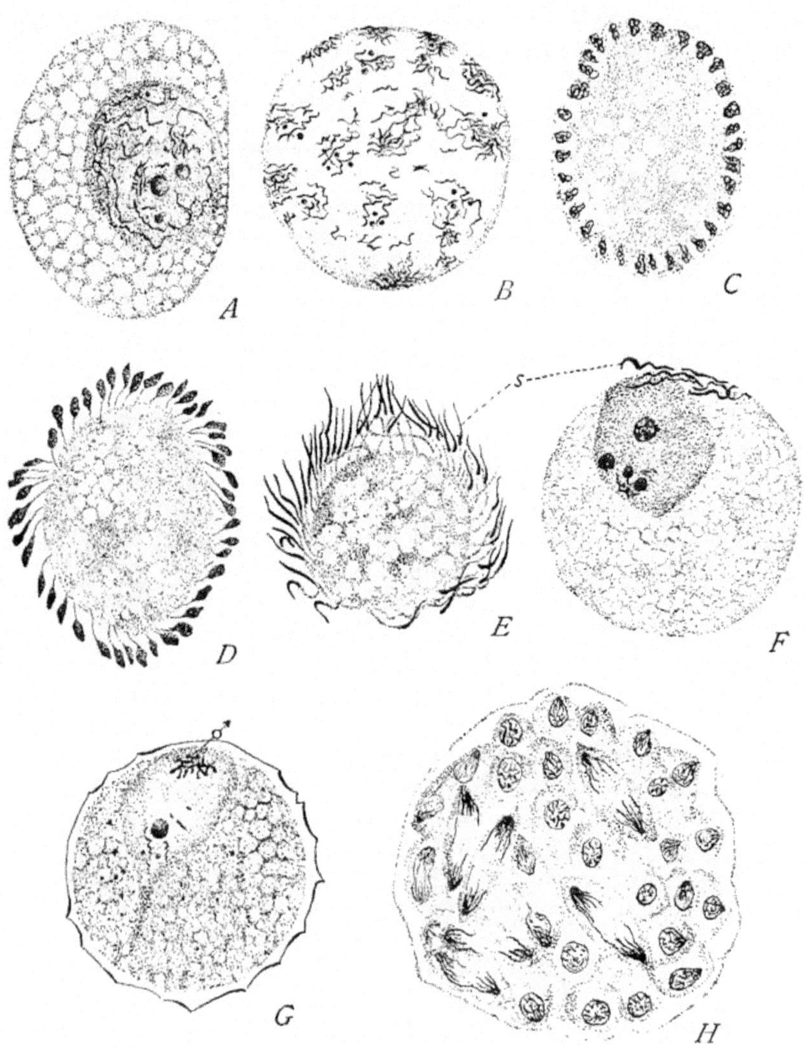

Fig. 127. — Formation of microgametes and fertilization of *Klossia octopiana* Schneid. [SIEDLECKI.]

The chromatin of the nucleus is distributed throughout the cell (*A*, *B*), finally forming nuclei of the future gametes (*C*, *D*, *E*). The mature microgametes (*s*) swim about, and join a macrogamete (*F*). The nuclei mix (*G*), and then the cleavage nucleus divides repeatedly by mitosis to form the spores (*H*).

polar-body formation in Metazoa, or of the degenerate final divisions of pro-conjugants in the other Protozoa. The formation of both macrogametes and microgametes occurs during the same day, and after a similar period of asexual increase, showing that the same ultimate causes probably operate in each. The macrogamete is distinguished from the schizonts by the possession of a reserve store of nutriment,

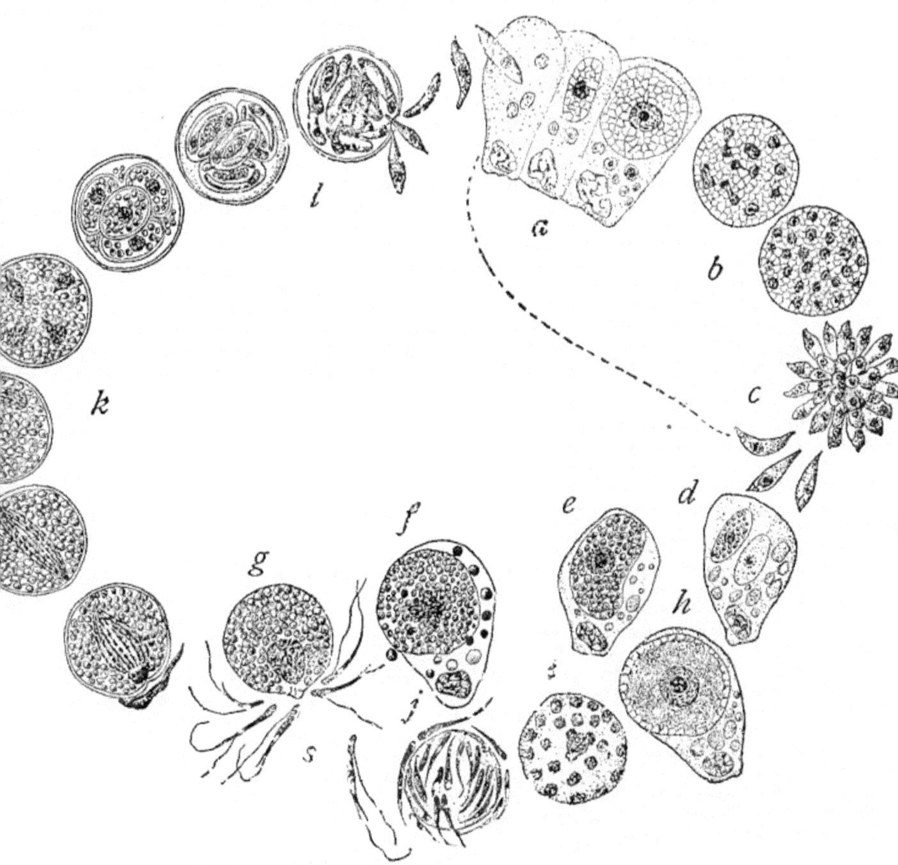

Fig. 128. — Life-history of a *Coccidium*. [SCHAUDINN.]
a, b, c, schizonts and asexual reproduction (schizogony). The merozoites at *c* repeat the cycle or pass on to the following stages: *d, e, f*, development of the female or macrogamete; *h, i, j*, development of the male gametes; *g*, copulation; *k* and *l*, stages in the formation of the spores and sporozoites.

in the form of granules, which are used in the later development. The storage of these granules begins in the young merozoite (Fig. 128, *c–g*) and in the close vicinity of the nucleus.[1] The nucleus is not essentially different from that of the schizont, but at one stage, according to Schaudinn, a large amount of chromatin is thrown out of

[1] Cf. yolk-formation in eggs of Metazoa; see Wilson, *The Cell*, p. 155.

it, a process regarded by him as a kind of reduction and found in some form or other in other Coccidiida (*Adelea*, Seidlecki, '99; *Coccidium proprium*, '98, *C. lacazei*, Schaudinn, etc.). The macrogamete is thus essentially a normal individual, in which a reserve store of food is deposited, and in which the quantity of chromatin is reduced. The microgamete, on the other hand, differs widely from the macrogamete, the schizont, and the merozoite.

The life-history of *Coccidium* is thus similar to that of *Gonium* or *Pandorina* and consists of periods of asexual reproduction, which alternate with periods of so-called sexual reproduction (Fig. 128). This conception of alternation of generations may be extended to all Protozoa, but in only one case (*Volvox*) is there a differentiation into germ and somatic cells in the same individual. *Volvox* affords an interesting intermediate stage between the generalized Protozoa and the specialized forms among the Metazoa and Metaphyta. Here the aggregate is differentiated into somatic and germ cells, although in the early stages there is no difference between them. Some of the latter (*parthenogonidia*) are sufficiently generalized to form new colonies by asexual reproduction, while others, apparently like the somatic cells, form the sexual reproductive elements; the remaining peripheral cells are specialized for feeding and motion.

It has been determined by Kirschner ('79), Carter ('58), and Stein ('78) that some colonies always produce male and others female. In the formation of male elements one of the ordinary cells of *Volvox minor* divides into 16 spermatozoids (Kirschner), or from 32 to 128 or more (Goroschankin, '75; Cohn, '75; and Stein, '68). The egg-cell, too, appears at first to differ but slightly from an ordinary cell, but it rapidly grows in size and becomes entirely different. Fertilization has been observed in both *Volvox minor* and *V. globator*, taking place in the inner spaces of the colonies, where the spermatozoa work their way to meet the eggs. In no case, however, have the inner processes of fertilization been observed.

Thus, in all Protozoa, with a single exception, apparently, of *Volvox*, where the beginnings of specialization are seen, all of the cells of a cycle, or of a colony, are equally capable of conjugation and of rejuvenescence, and the possibility of indefinitely continued life is open to them all. This power, as Weismann first pointed out, distinguishes the Protozoa from all higher animals and plants, where division of labor has resulted in specialization. The majority of the vegetative or somatic cells which form the organs and tissues of higher animals and plants, have, with their specialization, lost the power of rejuvenescence, and when the potential of vitality with which they start is exhausted they become degenerate through old age, and finally die. The germ cells, on the other hand, like the

Protozoa, retain the power to renew their vital activities, and with it the possibility of continued existence. Thus, the penalty of specialization appears to be death.

B. THE SO-CALLED MATURATION-PHENOMENA IN PROTOZOA

It is now a well-established fact that in the higher plants and animals the nuclei of the germ-cells, when ready for fertilization, contain only one-half as many chromosomes as the nuclei of the body-cells. This discovery, made by Van Beneden ('83), has been so widely extended in the animal and vegetable kingdoms that it is now generally regarded as a necessary condition of the union of sex-cells; for by this union, it is argued, the number of chromosomes would be doubled were it not for the preliminary reduction to one-half. The phenomena leading to reduction of the chromosomes to one-half the number characteristic of the species are known as the *maturation* or *reducing* phenomena. In the egg the nucleus divides twice, while near the periphery of the cell, and two minute cells, *abortive eggs* or *polar bodies*, are formed, which degenerate and disappear without further action. In the spermatozoön all of the four cells, which are formed by a similar double division, are functional. In both egg and spermatozoön one division of the nucleus, at least, results as in ordinary mitosis, in the equal partition of each chromosome. The chromosomes at this period frequently differ in appearance from the ordinary chromosomes, and from their peculiar shape are known as *tetrads* (*Vierergruppen*). At this point, however, the various observations differ and an animated controversy, which began with Weismann, is still vigorously maintained in regard to the exact way in which these maturation-chromosomes are formed and divided. The controversy, which in large part involves Weismann's speculations upon heredity, need not detain us here, however, for with the Protozoa, only the general aspects of reduction are in question.[1]

In the Protozoa, with the somewhat doubtful exception of *Paramæcium caudatum* (Hertwig, '89), there is no case on record of reduction in the number of chromosomes, and even in the few cases where actual chromosomes occur, their number is so great that counting is impossible. There are, nevertheless, certain phenomena antecedent to conjugation in Protozoa which warrant the belief that processes take place, even in these primitive animals, which are analogous to the formation of polar bodies in the Metazoa. In general, these processes consist in the elimination of one or more daughter-nuclei prior to conjugation, and are usually the same for both conjugants.

[1] For a discussion of the problems of reduction, see Wilson, *The Cell*, 2d ed., 1900, p. 233.

It is a significant fact, first pointed out by Maupas ('89), that micronuclei in Infusoria which have about exhausted their potential of vitality, disappear by atrophy and absorption in the cytoplasm, while the latter continues to live and even to divide for a limited time (*e.g. Onychodromus*). Thus, the loss of vitality operates upon the nucleus as upon the remaining protoplasm, and the final divisions of the nucleus, like the final divisions of the cells, result in daughter-nuclei, which have not the power to live, and which disappear in the cytoplasm. Furthermore, if two individuals can fuse only when the body protoplasm is in the proper condition (*Difflugia, Amœba*, etc.), the same may be true of nuclei. It is certainly true that nuclei of the plasmodia of Mycetozoa, or of *Actinophrys* in plastogamy, or of adult multinuclear *Actinosphæria*, do not fuse. But when two *Difflugias* are in the proper condition for conjugation, the nuclei do fuse. When, however, two individuals are in the proper condition, so far as the cytoplasm is concerned, it does not follow that the nuclei are ready, and each nucleus may undergo division prior to fusion, one of the daughter-nuclei disintegrating and disappearing (*Monocystis, Actinophrys*), or several of them disappearing (Infusoria, *Actinosphærium*). In *Coccidiida* a certain amount of the nuclear material is budded off and disintegrates while in the plasm, and the conjugating nuclei, without membranes, fuse to form the cleavage nucleus. Such cases may be interpreted as evidence of loss of vitality to such an extent that the nucleus has the power to divide, but not to stimulate division of the cell-body.

The loss of vitality is, I believe, the principle which lies at the bottom of the so-called maturation-phenomena among the Protozoa. Briefly reviewing some of these processes in Protozoa, it will be seen that in some cases, as in senescent Infusoria, the nucleus atrophies, while in other cases the nucleus divides once or twice previous to fusion, thus simulating the maturation of Metazoa. The former has been observed in *Euglypha* and *Actinosphærium*, the latter in *Actinophrys, Actinosphærium*, some Coccidiida, and in Infusoria.

In *Euglypha*, Blochmann ('88) observed the conjugation of two individuals, which had become united by their mouth parts. The cytoplasm of one conjugant passed into the shell of the other and fused with the cytoplasm there, but the nucleus was left behind. A pseudopodial process from the fused cytoplasm finally picked up the rejected nucleus, and its position in the cytoplasm was restored. It did not live, however, to fuse with the other nucleus, but disintegrated in the plasm (Fig. 129). These observations have been recently confirmed by Prowazek ('00) and are regarded by Bütschli and Blochmann as an indication of reduction previous to conjugation; but it may be interpreted as an instance of degeneration of the nucleus, while the failure to unite with the other nucleus leads to disintegration, and may be regarded as *facultative reduction.*

The loss of nuclei through disintegration and absorption is better established in the case of *Actinosphærium.* Gruber ('83) and Brauer ('94) maintained that the number of nuclei in old *Actinosphæria* is reduced by fusion until only a few remain, but Hertwig ('98) shows that with the exception of about 5 per cent, all of the nuclei atrophy and are absorbed in the cytoplasm. The remaining nuclei, which he calls the "sexual nuclei," then undergo so-called maturation divisions, and fuse to form the cleavage nuclei of new cycles.

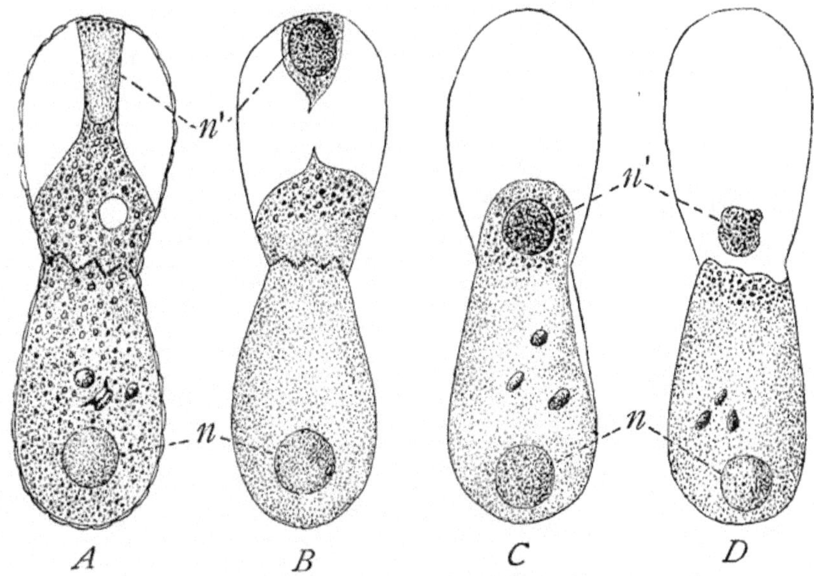

Fig. 129. — Conjugation of *Euglypha alveolata* Duj. [BLOCHMANN.] *n*, functional nucleus. *n'*, degenerating nucleus.

The simplest case of maturation comparable to that in Metazoa is shown in two rather widely separated forms, *Actinophrys*, a heliozoön, and *Monocystis*, a gregarine. In the former, Schaudinn ('96) observed that, although the cytoplasms of two individuals fuse, the nuclei remain apart. They finally divide by mitosis, and two of the daughter-nuclei fuse, while the other two (polar bodies) disintegrate and are absorbed in the cytoplasm (Fig. 130). Similarly in *Monocystis*, Wolters ('91) has shown that two similar individuals join in a common cyst and that the nuclei divide, one of the daughter-nuclei in each individual disappearing in the plasm.

In both of these cases, only one daughter-nucleus is eliminated in each conjugant. In the other forms mentioned, there are at least two, and the process approaches still more closely to maturation in Metazoa.

In *Actinosphærium*, Hertwig describes a process which may be summarized as follows: after most of the nuclei are absorbed in the cyto-

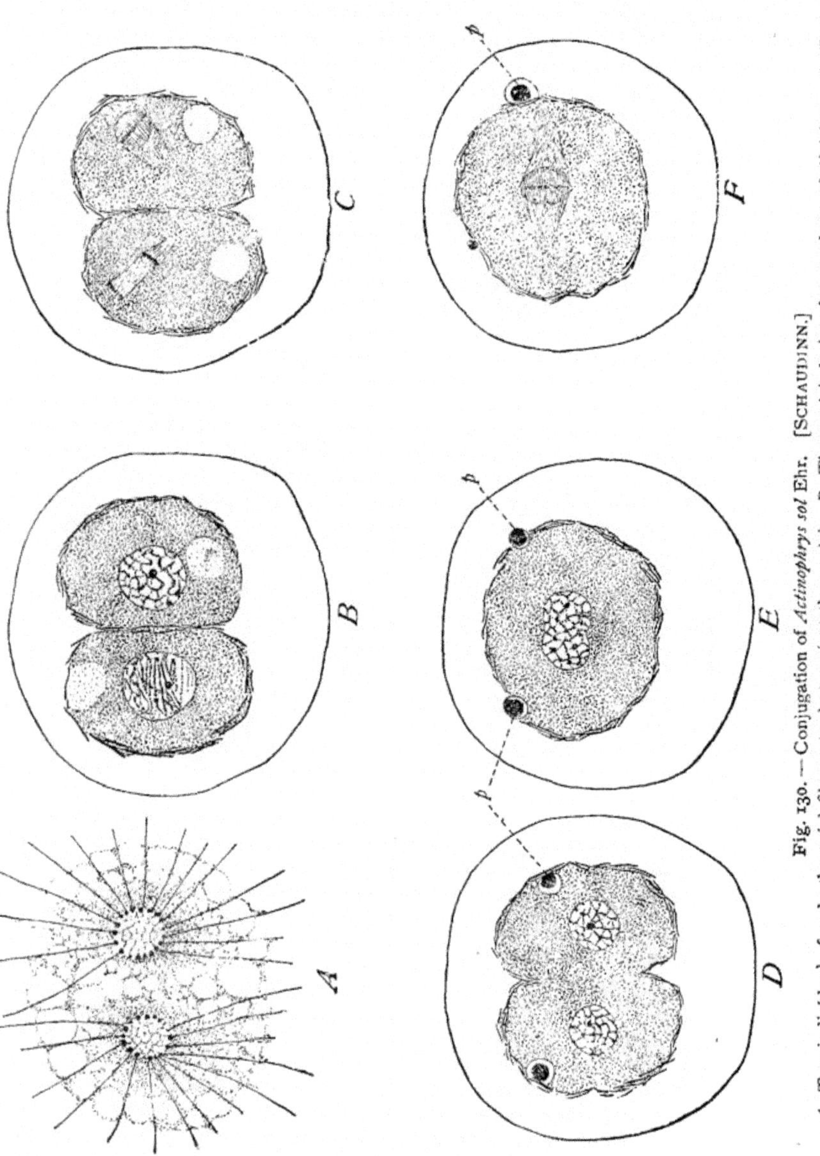

Fig. 130. — Conjugation of *Actinophrys sol* Ehr. [SCHAUDINN.]

A. Two individuals fused; the axial filaments abut against the nuclei. *B.* The nuclei during the prophase of division. *C.* For-

plasm, the "mother-cyst" divides into as many parts (primary cysts) as there are nuclei remaining. In small animals the entire mass may form one primary cyst with one nucleus, but large specimens may have twenty or more primary cysts. Each primary cyst is surrounded by a special gelatinous mantle; the nucleus of each divides once (primary mitosis) into two nuclei characterized by radiations at one pole, where a centrosome is developed from nuclear threads which grow out into the surrounding cytoplasm. During the formation of the centrosome, the body of the primary cyst divides, giving rise to the secondary and mononucleate cysts.

Each secondary cyst forms two "polar bodies" in the following manner. In each the nucleus divides by mitosis into two nuclei, one of which degenerates while the other prepares for a second division by mitosis. After this division one of the resulting nuclei also degenerates ("2d polar body"), leaving one ripe nucleus in each secondary cyst. Both polar bodies consist exclusively of nuclear matter, and they rapidly degenerate. Both polar or maturation mitoses are characterized by the presence of centrosomes derived from the original centre. After the two secondary cysts formed from the two primary cysts are ripe, they fuse together, protoplasm with protoplasm, and nucleus with nucleus, thus re-forming the primary cyst. The fusion product (germ-sphere) is distinguished from the older primary cyst, not only in the changes described in ripening and fertilization, which, after the formation of the polar bodies, would hardly be noticed, but in the following additional points: (1) The body concentrates into a denser structure and smaller circumference. (2) The last remnants of vacuoles disappear. (3) The silicious particles at first diffuse, collect at the periphery and finally form a compact outer coating. (4) A protective membrane (yolk membrane), extremely impervious to killing agents, is secreted inside of the protective coating. After a week's rest the germ spheres creep out, each animal holding six or eight nuclei, formed by mitosis. Before beginning to grow, the spheres apparently divide into mononucleate forms.

Again in *Adelea ovata*, one of the Coccidiida, a somewhat similar elimination takes place, but in only one of the conjugants. Prior to conjugation, the nucleus of the male cell divides twice, and only one of the resulting nuclei penetrates and fertilizes the egg, while the other three disintegrate and disappear (Schaudinn and Siedlecki, '97). This case is very similar to the well-known maturation-divisions among the Infusoria, although in the latter both conjugants undergo the same processes. In the different species of Infusoria the maturation-phenomena, while differing as to details, agree in their essential features, and for purposes of illustration *Paramœcium caudatum* may be selected, for with its single macronucleus and single micronucleus it is simpler than the majority of other forms. The details of the process which have been made out by Maupas ('88, '89) and Hertwig ('89) are as follows. The micronucleus of each conjugant divides twice to form four daughter-nuclei, of which one only remains active, while the other three degenerate and disappear. The active daughter-nucleus divides again and one of the resultant nuclei migrates to the other organism, while the other resultant nucleus remains quiescent. The migrating nucleus in each conjugant unites with the quiescent nucleus of the other individual, and thus effects fertilization, after which the organisms separate (Fig. 131).

The three daughter-nuclei (*corpuscules de rébut*) which are eliminated in each individual before the union of the "germ nuclei," are undoubtedly analogous to the polar bodies of the Metazoa. The agreement in number with the polar bodies of the Metazoa is interesting, but it may be doubted whether this agreement has any real significance, for the number of nuclei eliminated in other genera of Ciliata varies considerably, *e.g.* *Vorticella*.[1]

Mark ('82), followed by Bütschli ('85), explained the formation of polar bodies in the Metazoa as aborted eggs resulting from the attempt to form many individuals as the early sperm cells do, and the latter regarded it as a reminiscence of the colony-formation of the Protozoa, basing his view upon the sexual relations of *Volvox* and *Pandorina* (see above), while Hartog ('91) expressed a somewhat similar view in the statement "the page of morphological history, revealing that the oogamete was primitively one of a brood of at least four, has not been obliterated from the ontogenetic records of the Metazoa."[2]

Richard Hertwig ('98), after summing up the evidence in Protozoa of the homologue of polar bodies, came to the conclusion that in all cases there is probably a double division before, after, or during fertilization, which represents a physiological condition of the cell and without phylogenetic significance. Excepting for the Infusoria the double division is, however, established in only one form (*Actinosphærium*), and it is rather difficult to believe that observations have been incomplete in all other reported cases where only one such division is found. Hertwig's view, however, that the preliminary division of the nuclei prior to fertilization represents a physiological condition of the protoplasm, is in no way opposed to the facts in Protozoa.

Up to the present time no such structures as tetrads have been found in any Protozoa; indeed, in only a few cases have chromosomes been observed, and then they are so numerous as to render counting impossible. Hertwig ('89) somewhat doubtfully asserted that the number of chromosomes in *Paramæcium caudatum* is reduced from 8 or 9, to 4 or 6, and, again ('98), he maintained that in *Actinosphærium* the

[1] Cf. Maupas. [2] *l.c.*, page 66.

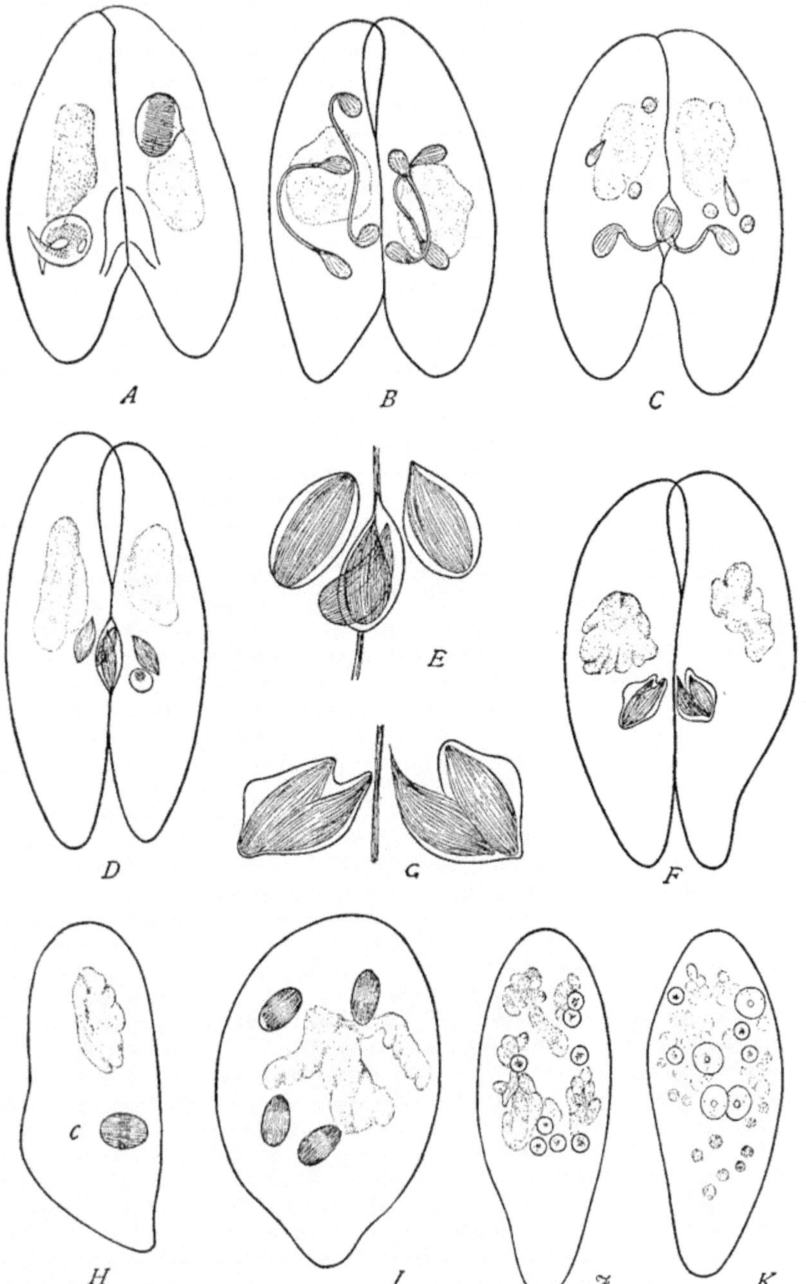

Fig. 131.—Conjugation of *Paramœcium caudatum.* [*A–C*, after R. HERTWIG; *D–K*, after MAUPAS.] (The macronuclei dotted in all the figures.)

A. Micronuclei preparing for their first division. *B.* Second division. *C.* Third division; three polar bodies or "corpuscules de rébut," and one dividing germ-nucleus in each animal. *D.* Interchange of the germ-nuclei. *E.* The same, enlarged. *F.* Fusion of the germ-nuclei. *G.* The same enlarged. *H.* Cleavage-nucleus (*c*) preparing for the first division. *I.* The cleavage-nucleus has divided twice. *J.* After three divisions of the cleavage-nucleus; the macronucleus is breaking up. *K.* Four of the nuclei enlarging to form the new macronuclei.

number of chromosomes is not reduced in the maturation-divisions although the amount of chromatin is reduced by half.

It must be pointed out also that in many cases of conjugation among Protozoa no maturation processes are known. Thus all of the Flagellidia, with their complicated division of labor and high grade of sex-differentiation, offer not a single instance of nuclear reduction, and the conclusion is suggested that the maturation of forms in other divisions of the Protozoa show no genetic relation to analogous processes in Metazoa, but are independent expressions of the same unknown vital forces which cause the formation of polar bodies, or the double division of tetrads.

C. General Considerations

In the foregoing review of the phenomena of conjugation and maturation, it is only too apparent that many gaps in the series, and the incompleteness of the observations in various instances cited, prevent any far-reaching generalizations. It may be pointed out, however, that more or less similar conditions characterize all of the diverse phenomena, and afford a basis for future explanations.

The various conjugation-phenomena seen in the Protozoa seem to show that each cycle starts with a certain potential of vitality which is gradually exhausted in the vegetative activities of the long line of individuals formed by simple division, or by spore-formation. As Bütschli long since suggested, such a cycle can be compared to the ontogeny of a metazoön where the somatic cells, starting with an initial vitality, finally die from senile degeneration. An important difference, however, lies in the fact that each protozoön has the inherent power of conjugation, and when senile degeneration has progressed to a certain extent, the vitality can be restored by this process. The most careful observations on the Protozoa, as upon all other animals and plants, have failed to demonstrate the nature of this renewal of vitality, or the reason why the temporary or permanent union of two exhausted cells should result in one or two rejuvenated ones.

An explanation of the many diverse modifications in the form of the conjugants, and of the various maturation-phenomena, does not, however, appear at the present time quite so far out of reach, for these phenomena can be observed, and the similar characteristics point toward a common interpretation. The evidence in Protozoa, while not conclusive, certainly points, I believe, as Minot long since suggested, toward the phenomenon of decreasing vitality as the underlying condition which indirectly brings about conjugation, sex-differentiation, and maturation.

Attention may again be called to the fact, repeatedly observed by Bütschli, Engelmann, Gruber, Maupas, and others, upon the Infusoria, that if conjugation is prevented, continued cell-division leads to degeneration, and that this process is cumulative, resulting eventually in the atrophy of external and internal organs (including vibratile membranes, cilia, micronuclei) and, finally, in death. "There is no question," says Bütschli, "that conjugation is a process which, if prevented, leads to death of the ciliates, just as a race of Metazoa dies out if restrained from sexual reproduction.[1]" The cell-body becomes

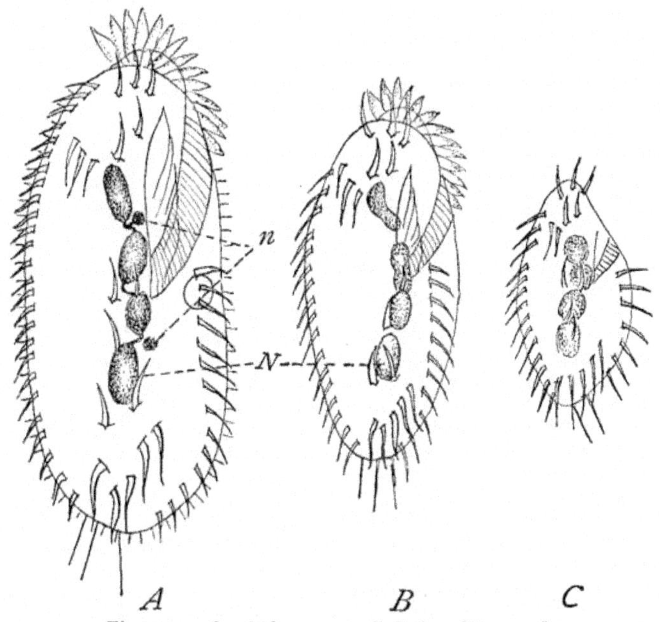

Fig. 132. — *Onychodromus grandis* Stein. [MAUPAS.]

A. Normal individual. *B*. Smaller form without micronuclei; degenerate. *C*. A still more reduced and degenerate form. *N*, macronucleus; *n*, micronucleus.

smaller, more plastic, and less resistant, and assumes various irregular shapes (Fig. 132). If degeneration has not gone too far, the normal size of the individual, the organs, and the general vitality are restored by conjugation, which is effected through the union of the soft mouth parts, while the internal phenomena consist of the interchange of micronuclei and possibly of some cytoplasm.[2] The phenomena of degeneration may therefore be considered in two categories: (1) the effect upon the plasm and the form of the cell-body; (2) the effect upon the nuclei. The former, I believe, leads to the reduced size of the conjugants, the latter to the so-called maturation in Protozoa.

[1] ('88), p. 1638. [2] Cf. Maupas ('88), Bütschli ('88).

In forms of Protozoa other than the Infusoria, there have been no careful observations and experiments to determine the limits of the potential of vitality. Nevertheless, certain facts have been recorded which appear to be of the same order of phenomena as those seen in the Infusoria. Thus, in some Monadida (*Cercomonas*, *Tetramitus*, etc.) the loss of the customary body-form and the assumption of a plastic condition prior to conjugation may be interpreted as an indication of degeneration, and comparable to the loss of membranelles, cilia, etc., in the Infusoria. A similar interpretation may account for the fusion of two cells of *Difflugia lobostoma* (Fig. 118), which under normal conditions will not unite, but which fuse readily under "certain conditions of the plasm" (Rhumbler). Also in *Polytoma* and *Chlorogonium*, where facultative conjugation occurs between two full-sized cells, between a full-sized and a smaller cell, or between two smaller cells, the difference in size may be interpreted as an expression of degeneration, comparable to that which is indicated by the senescent, small-sized individuals of *Onychodromus* or *Paramæcium*, where fusion may also occur between two small conjugants or between a normal-sized cell and a small one (Figs. 120 and 132). In all cases the protoplasm has reached that state which, for the want of a better term, may be called mature, when conjugation is possible. The sex-difference, which is facultative in *Polytoma* and *Chlorogonium*, becomes obligatory in the allied form, *Phacotus*, a condition which may be compared with that of the peritrichous Ciliata, where, in the Vorticellidæ, the free-swimming microgametes are dwarfed forms of the normal individuals with which they fuse (Fig. 122).

In the more complex colonies of Flagellidia, and in the Coccidiida, the microgametes, although they arise by spore-formation, are to be compared with spermatozoa of Metazoa, and the macrogametes are equivalent to eggs. The obligatory sex-difference is here indicated by the storage of a reserve food supply with concomitant quiescence in the larger forms (females), and by loss of cytoplasm and concomitant increase of motion in the smaller ones (males). The researches upon Coccidiida have established the fact that, as in Infusoria, there is a period of conjugation which alternates with a period of spore-formation, and the changed form of the conjugants into spermatozoa and eggs may be taken as evidence of exhaustion as in the less specialized cases among the Protozoa; although with our present knowledge an interpretation of these conjugants as degenerate cells can only be based upon analogy with other forms, where a similar cycle brings about more or less similar conjugating individuals.

It is apparent from the facts cited that the study of Protozoa can throw but little light upon the question of sex-difference. It may be due either to differences in nutriment or to some more deeply lying cause in the protoplasm, but in any case, it cannot be of primary importance, for the one essential requisite demanded by the lowest flagellate and the highest animal or plant is fusion with another cell, and by such fusion the restoration of an exhausted vitality.[1]

We must finally, with Hertwig, distinguish clearly between fertilization and reproduction. Fertilization of the ovum in Metazoa is so closely followed by development of the embryo, that it has grown to be considered a direct act of reproduction as well as its cause, and the same view would be equally applicable to the Coccidiida or Infusoria. But, when applied to conjugation in other forms of Protozoa, *i.e.* to fertilization in its more general aspects, it is evident that the phenomenon has some deeper significance. Bütschli ('76) long since pointed out that, during the period of conjugation, the two individuals of conjugating *Paramœcia* might have formed many others by simple division, and he, with Engelmann, showed that the phenomenon, in all probability, could not be a reproductive act. So, too, *Onychodromus* might divide thirteen times during the period of conjugation (Maupas). The facts of facultative conjugation, *e.g.* in *Polytoma* or in the Rhizopoda, are arguments in the same line, for, in these cases, an individual may or may not fuse before forming spores. Maupas ('89) maintained that division is the only method of reproduction, and pointed out that division is more rapid before conjugation than after, and Hertwig ('98) showed that, in *Actinosphærium*, reproduction actually precedes fertilization, and he says: "A reproductive process is bound up with the encystment of *Actinosphærium*, whereby a mother-cyst gives rise to several primary cysts, each primary cyst to a germ-sphere, each germ-sphere to new individuals. Reproduction here precedes fertilization, and the latter has no effect upon the former." He further states that fertilization, which has no direct connection with reproduction, occurs in both animals and plants, and he adds: "Since we know reproduction without fertilization and fertilization without reproduction, there must be processes combined in 'sexual reproduction' which in their essence do not belong together. To distinguish methods of increase as sexual and asexual, awakens false impressions. There is, in fact, only one kind of reproduction, *i.e.* division in the widest sense of the word. What is known as sexual reproduction among Metazoa, is division which is combined with

[1] Watase's ('92) interesting suggestion as to sex-differentiation is significant in the light of the foregoing facts. Regarding sex-differentiation as a manifestation of irritability (*l.c.*, p. 485), he says: "The organism is either a male or a female, not by the difference of 'primary sexual characters' alone, but by the difference which saturates the whole of its entire structure. Such a difference is, however, neither absolute nor permanent. It is a temporary differentiation of protoplasm into one of two different directions, and sooner or later comes back to the original neutral or non-sexual state from which it started, thus manifesting the phenomenon characteristic of all protoplasmic irritability" (p. 493).

sexual activity. The egg is a potential individual as shown by the facts of parthenogenesis, where, without fertilization, the egg develops to form a new individual. The first appearance of this individual goes back to the germ-glands, where it results from the division of an oögonium. Here is the reproductive act proper; fertilization is only the liberation of an inhibited development through which the potential individual becomes actual. It leads to no increase, but rather to decrease of individuals, since two potential individuals — egg-cell and sperm-cell — join to form one body."[1]

The phrase "liberation of an inhibited development," if applied to the effect of conjugation among the Protozoa, might be translated into Bütschli's "Verjüngung" ('76), or Engelmann's "Reorganization" ('76), or Maupas's "Réjeunissement." In no case does word or phrase explain the phenomenon. From all the facts known at the present time, the only conclusion that can be drawn is that conjugation, apparently, is not the cause of reproduction, but as Bütschli, Engelmann, and Minot long since pointed out, in some unknown way provides the energy for continuing the functions of the individual, including the power of reproducing.

SPECIAL BIBLIOGRAPHY VII

Bütschli, O. — Studien über die ersten Entwicklungsvorgänge der Eizelle, der Zelltheilung und die Conjugation der Infusorien. *Abh. d. Senckenb. naturf. Gesell. Frankfurt a. Main*, X., pp. 213–452, 1876.

Engelmann, Th. W. — Ueber Entwickelung und Fortpflanzung von Infusorien. *Morphol. Jahrb.*, pp. 573–635, 1876.

Geddes and Thompson. — The Evolution of Sex. *London*, 1889.

Hartog, M. M. — Some Problems of Reproduction, etc. *Q. J. M. S.*, XXXIII., pp. 1–81, 1891.

Hertwig, R. — Ueber Kerntheilung, Richtungskörperbildung und Befruchtung von Actinosphærium Eich. *Abh. d. K. bay. Akad. d. Wiss. München*, II. Kl. XIX., 1898.

Hertwig, R. — Mit welchem Recht unterscheidet man geschlechtliche und ungeschlechtliche Fortpflanzung? *Sitz. Ber. d. Ges. f. Morph. u. Physiol. München*, II., 1899.

Maupas, M. — Le réjeunissement karyogamique chez les cilies. *Arch. d. zool. expér. et génér.*, (2) VII., pp. 149–517, 1889.

Minot, C. S. — Growth as a Function of Cells. *Proc. Bost. Soc. Nat. Hist.*, XX., 1879.

Schaudinn, F. — Ueber den Zeugungkreis von Paramœba Eilhardi. *Sitz. Ber. Akad. Wiss. Berlin*, 1896.

Siedlecki, M. — Ueber die geschlechtliche Vermehrung der Monocystis ascidiæ R. Lan. *Bull. d. l'Acad. d. Sci. d. Cracovie*, December, 1899.

Wilson, E. B. — The Cell in Development and Inheritance. 2d ed. *New York*, 1900.

[1] *loc. cit.*, p. 97.

CHAPTER VIII

SPECIAL MORPHOLOGY OF THE PROTOZOAN NUCLEUS

"By *cell-division*, accordingly, the hereditary substance is split off from the parent-body; and by cell-division, again, this substance is handed on by the fertilized egg-cell or oösperm to every part of the body arising from it. Cell-division is, therefore, one of the central facts of development and inheritance." — WILSON.[1]

It has been shown in the foregoing chapters that local modifications of the general protoplasmic basis of protozoan cells give rise to cell-structures of considerable complexity, and, when compared with cells in the tissues of Metazoa, these complex forms often appear to be the more highly differentiated. These differentiations have mainly arisen by modifications of the cortical plasm, in response, probably, to the action of the environment and the mode of life. The differentiations of the inner cell-plasm of the Protozoa, on the other hand, are apparently much less complex than in Metazoa, and the nucleus, especially, appears to be structurally much more simple in the Protozoa. In the division of protozoan cells, structures occur which are undoubtedly similar to those of tissue-cells, but are, on the whole, of a simpler and more generalized character. For this reason the careful study of these more primitive organs cannot fail to throw light upon morphological problems which, in Metazoa and the higher plants, are more difficult of solution.

The changes which the nuclei undergo during division of the cell have received little attention from the phylogenetic standpoint, although the similarity of these changes, even in the most diverse tissues of unrelated animals and plants, bespeaks a common origin. If the structures involved in these changes can be traced back to more generalized organs in the Protozoa, a key may perhaps be found, — not, indeed, to ancestry of the Metazoa, as some writers have maintained, — but to some of the problems of higher cell-differentiations. It is my purpose in the present chapter, therefore, to give somewhat in detail a comparison of the various structures found in protozoan cells with those of the cells in differentiated tissues of Metazoa and Metaphyta, and so far as our present knowledge permits, to point out the possible origin of different cell-organs in higher types, from the more primitive structures as they appear in Protozoa.

[1] The Cell, p. 63.

An ordinary cell in Metazoa and the higher plants consists of protoplasm, which is typically differentiated into *cell-body* or *cytoplasm*, and *nucleus*. These areas of protoplasm differ considerably in their chemical composition,—the former being rich in proteid, of which albumins play the most important part, and poor in phosphorus; the latter, on the other hand, being rich in phosphorus, which is bound up in a substance called *nuclein*, and poor in albumins.

Both nucleus and cytoplasm, as seen under the high power of a microscope, have a complicated structure. In both there appears to be a general ground substance, the *cytolymph* in the cell-body, and the *karyolymph* in the nucleus. Throughout this ground substance, in both the cell-body and the nucleus, extends an alveolar meshwork, the intra-nuclear portion being known as the *linin reticulum* (Fig. 133). The cytoplasm frequently contains other structures, such as plastids, vacuoles, metaplasmic bodies, crystals of various kinds, etc., which may or may not be permanent in the cell. A more important structure is the *centrosome*, which is usually present in the Metazoa and which has been generally considered as playing a leading rôle in cell-division. This body, which is extremely minute, is, as a rule, surrounded by a granular area of cytoplasm known as the *attraction-sphere*, *centrosphere*, or simply as the *sphere*. The nucleus, in addition to the linin reticulum and the karyolymph, contains *chromatin*, a substance which, more than anything else, distinguishes the nucleus from the cytoplasm. As a rule, chromatin appears in the form of small granules distributed through the linin network. These granules in the resting nucleus may be close enough together to give the appearance of a chromatin reticulum, while during division of the nucleus they become fused, forming structures known as chromosomes. The latter, save in certain stages in the formation of the sex-cells, are of definite number and appearance for the same species. The chromatin network is frequently thickened to form relatively large masses of chromatin known as *net-knots*, or, together with the plasmosomes, as *nucleoli*. The *plasmosomes*, or true nucleoli, are comparatively large, spherical masses suspended in the karyolymph and are of unknown function. The nucleus, finally, is inclosed within a definite membrane which usually disappears during mitosis.

Although the main features of indirect nuclear division, or *mitosis*, in animals and plants are so similar that a common type of mitotic or division-figure can be described, there are, nevertheless, confusing variations and differences, especially in the so-called "achromatic" portions of the mitotic figure.[1] Broadly speaking, these variations in mitotic figures may be reduced to three main types: (1) Forms with

[1] See Wilson, The Cell, Chap. II.

centrosome and spindle-figure in which central spindle, mantle-fibres, and astral rays can be distinguished. (2) Forms with centrosome and spindle-figure in which astral fibres are present, but no distinction between central spindle and mantle-fibres can be seen. (3) Forms without centrosome, but with a spindle-figure which cannot be resolved into mantle-fibres and central spindle.

In all of these cases the granular chromatin of the resting nucleus is arranged in the form of a reticulum upon a linin network, and, with few exceptions, the prophases of division are similar, in that the granules are brought together and concentrated in the form of deeply

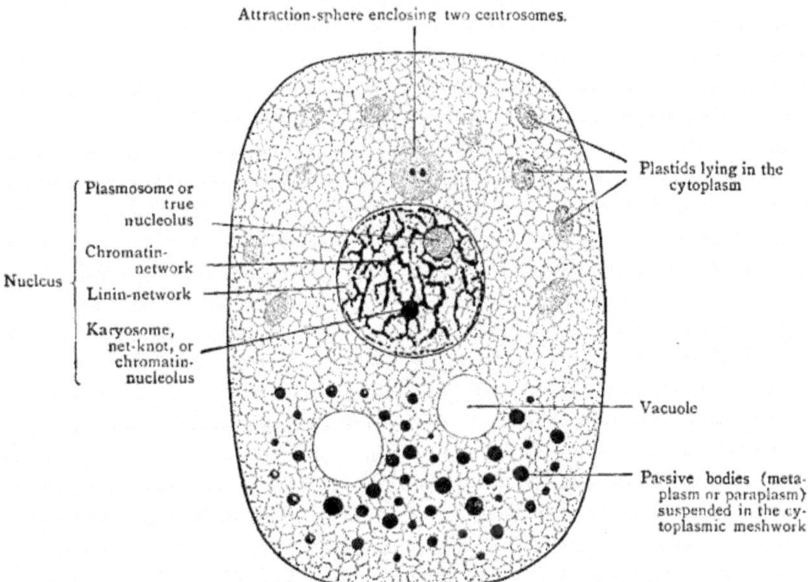

Fig. 133. — Diagram of a cell. Its basis consists of a meshwork containing numerous minute granules (*microsomes*) and traversing a transparent ground-substance. [WILSON.]

staining rods, the *chromosomes*, which are of definite number, shape, and size for each species. Finally, the nuclear membrane disappears and the chromosomes are left naked in the cytoplasm and connected by spindle-fibres with the two poles of the mitotic figure.

As the variations in the three types have mainly to do with the achromatic structures, further mention of the chromatin changes may be omitted. In the first type considered, the approach of division is signalized by the division of the centrosome into two daughter-centrosomes connected by fibres which form a small spindle called by Hermann ('91) the *central spindle*. This continually enlarges as the centrosomes diverge, other fibres (mantle-fibres) meantime growing from the centrosome and pushing in the nuclear membrane, which finally disappears, leaving the chromosomes in the cytoplasm. The mantle-fibres then become attached to the chromosomes, and the latter, finally, surround the central spindle like a ring. The origin of these fibres is variously interpreted. The mantle-fibres according to some observers arise from cytoplasmic material, according to others, from linin substance in the nucleus. The origin of the central spindle-fibres is also in dispute. Some observers believe that two opposing groups of fibres starting from the centrosomes meet to form the spindle (Drüner, '95; MacFarland, '97; etc.); others believe that the diverging centrosomes are connected from the first by the central spindle-fibres (Hermann, '91; Flemming, '91; Heidenhain, '94; Kostanecki, '97; etc.), and Boveri ('88) and Heidenhain maintain that the fibres are formed from the substance surrounding the centrosome (archoplasm of Boveri; substance of the "centrodesmus" of Heidenhain).

Several transitional stages between the first and the second types have been described. In some of these (many egg-cells), the central spindle appears first as very delicate fibres between the dividing centrosomes, but these break later, and no connection remains. The complete spindle in such a case consists of fibres which apparently pass from pole to pole, with chromosomes strung upon them, while central spindle-fibres, if present, must be newly formed and intermingled with the other fibres, and both sets must have the same origin.[1]

Finally, the third type of mitosis differs from the first in the absence of central spindle and centrosomes, and from the second by the apparent absence of centrosomes. While great difference of opinion exists as to the presence or absence of centrosomes in plants higher than the Fungi, some observers denying, others affirming, its presence in the same species, the balance of opinion at present appears to be toward the negative side, and evidence is certainly accumulating in support of this view. According to Strasburger, Mottier, Osterhout, etc., the spindle in plant mitoses arises by the gradual convergence of rays which make their appearance tangential to the nuclear membrane. Arising, as it were, from the substance of the cytoplasm, and converging to form a bi-polar mitotic figure, the spindle-fibres are supposed by Strasburger and his followers to be formed from a definite and distinct substance to which he gave the name *Kinoplasma*. The nuclear membrane, as in the other types, always disappears before the nuclear plate is formed, and nuclear division proceeds in the usual way.

Turning now to the Protozoa, the heterogeneity in form and structure of the nuclei is particularly suggestive. All of the several parts

[1] Cf. Wilson ('96), *Toxopneustes*.

which characterize the typical nucleus of Metazoa are rarely present here in one and the same nucleus. The nuclear membrane in some cases is absent, in other cases well defined and persistent throughout active as well as resting stages. The linin reticulum, with its inclosed chromatin granules, is frequently absent, while the ground substance of the nuclei, or the karyolymph, although it has not been critically examined, is undoubtedly present in the majority of cases. Nucleoli or plasmosomes have apparently been found in but one instance (*Actinosphærium*, Hertwig, '98). The nuclei of Protozoa are, as a rule, further distinguished from those of the Metazoa by the presence of another intra-nuclear body, which apparently corresponds to the sphere and centrosome of the Metazoa, and, as in the latter, plays a prominent part in division. Boveri ('01) has recently proposed the name *Centronucleus* for nuclei with these division centres. We can distinguish, then, in the various nuclei of Protozoa, (1) a nuclear membrane; (2) a nuclear reticulum or linin network; (3) nucleoli, or plasmosomes (in rare cases); (4) chromatin, occasionally in the form of granules strung upon a linin reticulum; (5) spheres or division-centres analogous to the extra-nuclear kinetic centres in Metazoa, but for the most part intra-nuclear in the Protozoa.

A. The Nuclear Membrane

The usual definition of a nucleus includes the membrane as an integral part, sharply separating cytoplasmic and nuclear structures. In a few Protozoa there is no indication of a membrane, and the usual constituents of the nucleus are distributed throughout the cell. In others, notably in the majority of the Sporozoa, the nuclear membrane, as in most Metazoa, disappears during mitosis. In all other cases, however, the nuclear membrane persists, in part at least, throughout active as well as resting phases. It frequently appears as a very faint structure scarcely to be distinguished from the cytoplasmic reticulum (many Flagellidia), although increasing grades of density may be obtained, which culminate in thick resisting membranes like those of *Noctiluca*, *Amœba proteus*, and Ciliata. The membrane of an isolated macronucleus of the Ciliata becomes separated from the nuclear contents and ultimately dissolves in the water (Bütschli). The great rigidity of this membrane and its persistence during division of the nucleus led to the view that it is not identical with a nuclear membrane of a higher cell (Bütschli, '76, '83; Hertwig, '76), and some observers even regarded it as highly differentiated protoplasm in the form of chitin (Stein, '54, *Opercularia*), or cellulose (Brandt, '82, *Amœba*).

In the Cystoflagellate *Noctiluca miliaris*, one part of the membrane which is ordinarily thick and resisting disappears during division, and the edges which are left gradually fade away into the cytoplasmic reticulum, where they cannot be distinguished from the outer network. In this case apparently, and possibly in other Protozoa, the membranes arise from the substance of the cytoplasmic reticulum. Another interpretation, however, is possible. Thus in some Coccidiida (*Klossia*), Labbé ('96) and Siedlecki ('98) described the membrane as very delicate and staining in the same manner as the chromatin network, and Siedlecki adds: "In fact, it represents only a more condensed part of this network."[1] Hertwig ('98) held a similar view of the membrane in the case of *Actinosphærium*, regarding which he says: "In well-preserved specimens the nuclear reticulum and membrane stand in very close connection, and frequently the two cannot be distinguished."[2] The best evidence, however, of the nuclear origin of the membrane is given by Schaudinn ('94) in the developing nuclei of *Calcituba polymorpha*, a rhizopod belonging to the family Miliolidæ (Fig. 134, *A*). Here the young nucleus appears as a solid homogeneous sphere of chromatin which gathers fluid from the surrounding plasm and forms peripheral vacuoles. These vacuoles then pass to the inside, and soon the entire nucleus appears vacuolated, a mere network of chromatin. The chromatin then segregates into a central mass connected by fibres with a peripheral layer of chromatin which forms the nuclear membrane. The further history consists of the separation of bits of the central mass, which pass along the radial lines to the membrane, where they ultimately form a layer of chromatin similar to the original. By the bursting of the membrane these are liberated and recommence the cycle.

In other forms, notably in Sporozoa, the nuclear membrane is apparently of minor importance in the cell. Wolters ('91) described a delicate membrane in some forms of *Monocystis* (*e.g. M. agilis*); other forms, as *Monocystis ascidiæ*, have a thick membrane composed of deeply staining fibrils,[3] while others of the same genus have none at all. Various other forms of Protozoa also have no nuclear membranes, the chromatin in such cases being distributed throughout the cell. Examples of this type of nucleus are found in all classes; among the Sarcodina, Gruber ('84) and Frenzel ('91), among the Mastigophora, Bütschli ('96) and Calkins ('98), among the Ciliata, Balbiani ('60), Grüber ('84), and Bergh ('89) have described them. In these various descriptions it is not always made clear whether the distributed chromatin is merely a diffused nucleus or whether each of the parts is not a single small nucleus which divides by itself, as in the dividing granules described by Schewiakoff

[1] Siedlecki ('98), p. 806. [2] *Loc. cit.*, p. 635. [3] Cf. Siedlecki, '99.

('93) in *Achromatium.* A careful distinction should be made between these types. In *Loxophyllum* (Balbiani, '60), *Urostyla* (Bergh, '89), and *Tetramitus* (Calkins, '98), the chromatin granules come together and form a single nucleus prior to division.

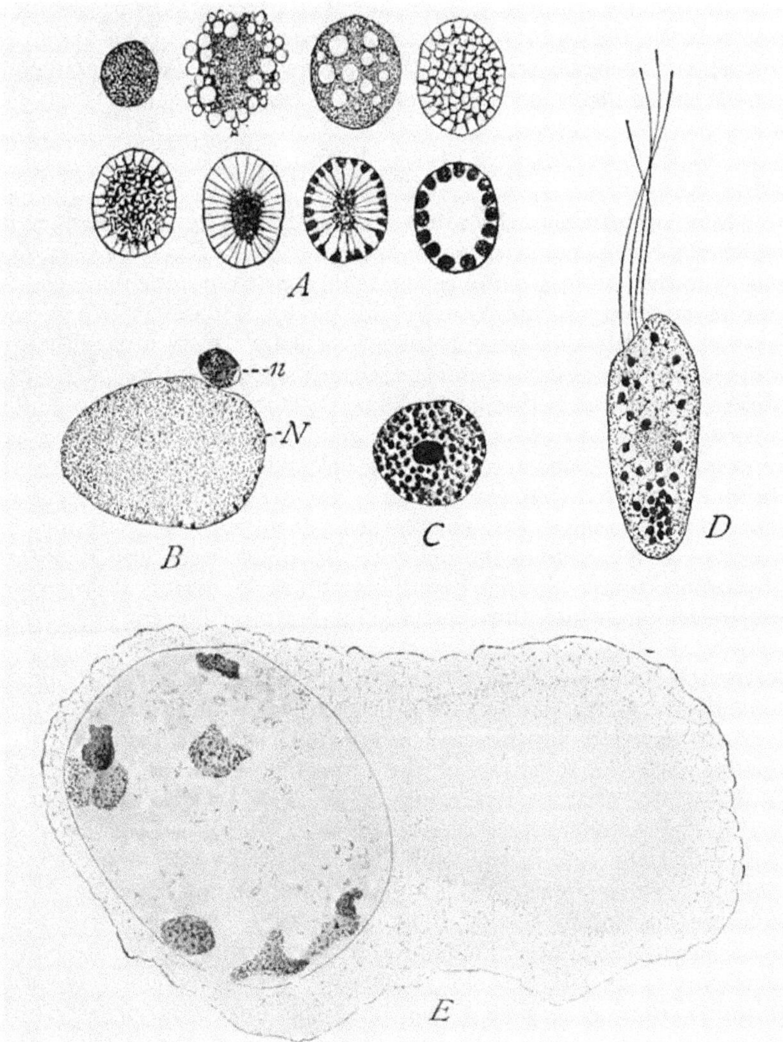

Fig. 134.—Types of nuclei. [*A. Calcituba polymorpha* Roboz, from SCHAUDINN. *B. Colpidium colpoda*, from a preparation. *C. Euglena viridis* Ehr. from a preparation. *D. Tetramitus chilomonas*, n. sp. *E. Noctiluca miliaris* Sur., from a preparation.]

A single karyosome (*A*) becomes vesicular, and ultimately gives rise to several daughter-karyosomes (so-called "fragmentation" Schaudinn). Several karyosomes in *Noctiluca* (*E*) hold the chromatin, the rest of the nucleus is filled with "achromatic" granules. In *Tetramitus chilomonas* (*D*) the chromatin is scattered throughout the cell; the lighter-colored body in the centre of the cell is the homologue of the deeply stained central body in *Euglena* (*C*).

It appears, therefore, that the nuclear membrane in Protozoa can have no great significance, since it may be absent altogether (distributed chromatin), present only during resting phases (Sporozoa), or persistent throughout all changes of the cell. When present, it may be formed, apparently, from the cytoplasmic reticulum (*Noctiluca*), or from the nuclear reticulum (*Actinosphærium*, Reticulariida, Coccidiida), and these observations are in line with the results obtained by numerous workers upon the nuclei in Metazoa.

B. The Linin Network

Many protozoan nuclei are permanently in the condition represented by Schaudinn's first stage of the nucleus of *Calcituba* (Fig. 134), and like this possess neither membrane nor linin reticulum, consisting throughout of chromatin (many phytoflagellates). When, however, a membrane is present, there is usually a clear area with little or no trace of structure (some Heliozoa, Pénard, '90), or else a well-defined reticulum between the membrane and the chromatin. The reticulum has been observed in nuclei of every class of Protozoa, as a fine meshwork similar to the cytoplasmic network, and to the linin reticulum of tissue nuclei. In many cases, however, the chromatin granules are not embedded in such a reticulum, but form a central mass in the nucleus. In other cases, and frequently during preparation for division, the granules are dispersed throughout the reticulum, which then appears like a typical nucleus (*e.g. Actinosphærium*, Fig. 140). In some forms the linin-threads are so extremely minute, and the chromatin granules so large, that it appears incredible that the latter should be inclosed in the former, as is generally believed to be the case in the nuclei of Metazoa. In many macronuclei, the structure of the linin and chromatin, on the other hand, is like that of tissue-cells. In nuclei with a firm and resisting membrane, as in *Amœba proteus*, there is little room for belief that the cytoplasmic and nuclear contents are connected, and the two parts appear to be quite different in structure. This nucleus and a few others show no trace of the linin reticulum, but contain at least two kinds of granules which make up the bulk of the chromatin and ground substance or karyolymph. *Noctiluca miliaris* (Fig. 141, *E*) has a nucleus of this type with large granules which stain intensely with acid dyes (Ishikawa, '94; Calkins, '98). The granules may possibly represent the œdematin-granules which Reinke ('94) distinguished in the nuclei of Metazoa, or possibly they represent a diffuse or distributed nucleolus, as Balbiani ('90) assumed in regard to the nucleus of *Loxophyllum meleagris.* Doflein ('00) maintained that this nucleus possesses a distinct linin network which stains with Berlin blue; in poorly preserved material the outlines of the granules appear to form a reticulum.

C. The Nucleolus

A distinct plasmosome or true nucleolus comparable to the analogous structure in Metazoa apparently exists in no case save possibly in *Actinosphærium*, and even here it is limited to a passing phase during mitosis (Hertwig, '98). It is probable that the structures which have been almost universally but erroneously called nucleoli, do not belong at all to this category of nuclear elements, but represent either the functional chromatin which is aggregated into a central mass (*karyosome*) during the quiescent or vegetative period of cell-life, or the intranuclear division centre.

D. The Chromatin

The form which the chromatin assumes in Protozoa gives rise to a great variety of nuclei which, at first sight, appear to have little in common with each other or with nuclei of tissue-cells. There is, however, a certain relationship between the different types, from extremely simple conditions, to structures as complex as in Metazoa, and through them there is a possibility of ultimately explaining the chromatin-changes in Metazoa.

Five types of nuclei, based upon the disposition of the chromatin, can be distinguished. Of these the most primitive is, (1) the solid sphere or karyosome (*Binnenkörper* Rhumbler), which has neither linin reticulum nor membrane (*e.g. Calcituba*). An advance is shown in (2) nuclei having one such karyosome surrounded by karyolymph, the whole inclosed within a membrane (*vesicular nuclei*, Gruber, '84), while still higher types are: (3) nuclei with several karyosomes (two to thirteen or fourteen), with membrane, karyolymph, and with or without a nuclear reticulum (*e.g. Noctiluca*); (4) nuclei with a large number of smaller masses of chromatin inclosed in a definite membrane with or without a linin reticulum (*e.g. Amœba proteus*); (5) nuclei consisting of granules of chromatin unconfined by a nuclear membrane and spread over the entire cell (distributed nucleus), or aggregated about a central body ("intermediate" type; Calkins, '98; *e.g. Tetramitus*).

Many Phytoflagellida, Choanoflagellida, some Sarcodina, and sporozoites, among the Sporozoa, have chromatin in the form of a single homogeneous karyosome. They are usually very minute, and little is known about them save that they are undifferentiated, and that they divide by a simple constriction into two equal parts.

A transition to the second type of nucleus is shown in *Calcituba* and other marine Rhizopoda (Fig. 134). Here the result of nuclear activity is the formation of a nuclear membrane and a nuclear reticulum, the meshes of which are filled with karyolymph apparently derived from the cytoplasm. Many of the stages which Schaudinn described had already been observed in various Sarcodina by Hertwig, F. E. Schultze, Bütschli, Gruber, Rhumbler, Hofer, and others, but the consecutive stages in the nuclear disruption had not been seen.

The second type of nucleus is particularly well marked in the group of Sporozoa. Schneider as early as 1881 described the nucleus of *Klossia eberthi*, one of the Coccidiida, as spherical, filled with "nuclear sap" and inclosing a great "nucleolus," formed of a dense cortical layer and a central alveolar portion. He observed no reticulum, but recent observers, Mingazzini ('92), Labbé ('96), Siedlecki ('98), described the nucleus of Coccidiida as containing a distinct reticulum, in which granules of chromatin and achromatin (oxychromatin in various stages of development) can be made out. Similar nuclei are found among the Gregarinida (Wolters, '91; Marshall, '93; Siedlecki, '99). The chief interest of these nuclei, however, centres in the chromatin mass, the "karyosome" of Labbé ('96), which, as in similar nuclei among the Sarcodina, has been described under several names.[1] It colors so strongly with nuclear dyes that it often appears dense and homogeneous (Fig. 135), but Schneider ('83), Labbé ('96), and Siedlecki ('98) agree that it is composed of at least two parts, an outer cortical portion, thick and resisting in texture, and an inner granular part. The cortical portion consists of chromatin with an exceedingly fine alveolar structure, the thick walls of the alveoli being frequently pressed so closely together that the striated appearance, early described by Schneider, results. The internal granules, on the other hand, take an intense acid stain, and are identified by Labbé as oxychromatin granules.

The history of this karyosome is strikingly similar to that of the isolated chromatin mass in *Calcituba*. The observations of Siedlecki and Labbé agree on this point, and a basis is thus secured for the comparison of these primitive nuclear changes. As in *Calcituba*, the young sporozoite (*A*) has an homogeneous nucleus, *i.e.* a naked karyosome, but, in the epithelial cell of its host, it very quickly assumes the adult structure (Siedlecki, '98). The intermediate stage between the homogeneous nucleus and the formation of the intra-nuclear karyosome has not been observed, but, as in *Calcituba*, the chromatin reticulum presses more and more toward the periphery, where it is finally condensed into the nuclear membrane. As the karyosome

[1] *Chromatosphérite* of Schneider, *Binnenkörper* of Rhumbler and Schaudinn, *Morulit* of Frenzel, *Chromatin reservoir* of Calkins, *Karyosome* of Siedlecki, *Nucleolus* of many authors.

condenses, there is left a small bud, usually attached by a short peduncle, and this bud, which Siedlecki calls the secondary karyosome, seems to be the seat of the subsequent changes in the disruption of the nucleus. The secondary karyosome enlarges as though a certain amount of chromatin had reached it through the peduncle. When it has reached the volume of the first karyosome, it is also similar in structure. A third karyosome is then formed in the same way, and so on until in some cases more than twenty have been developed. The chromatin of the original body breaks into granules, and in this state penetrates the connecting thread into the daughter-karyosomes, formed by budding. It is rather difficult to determine the function of this peculiar process, unless with Schneider ('83) and Mingazzini ('92) we assume that it is an antecedent or preliminary phase of spore-formation. Labbé, on the other hand, regarded the karyosome as a reserve of chromatin which at first contains all the chromatin of the nucleus, and which increases constantly by the addition of other nuclear particles (oxychromatin, Labbé, '96), the increase going into the formation of secondary karyosomes. He considered these phenomena only stages in the "purification" (*l'épuration*) of the nucleus, but Siedlecki returns to the view of Schneider and Mingazzini, and shows that they are but preparatory phases of reproduction, in fact the beginning of division of the nucleus, leading to the formation of microgametes or male reproductive cells.[1]

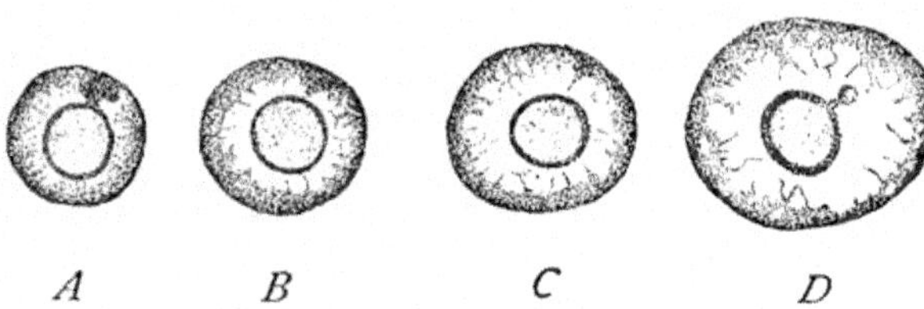

Fig. 135. — The nucleus and karyosome in *Klossia eberthi*. [SIEDLECKI.]

In other groups of the Protozoa, nuclei with one karyosome are somewhat more complicated and obviously of a higher type. The Heliozoa, for example, furnish nuclear types which, at first sight, apparently agree with the above forms. The agreement is, however, only superficial; for although similar in their general features to the nuclei of Sporozoa, they contain achromatic structures lacking in the latter, while their development is altogether different. A distinct

[1] The stages which Siedlecki has described may be briefly outlined as follows: The chromatin network reappears in the vicinity of the membrane, and constantly increases by the addition of the smaller karyosomes. It is then drawn out into long threads and at the same time the nuclear membrane disappears. The karyosomes are almost entirely used up in the formation of these elements, although a small portion remains as the characteristic residual mass. The chromatin then collects at the periphery of the cell, where little nuclei are formed, and each nucleus becomes the chief part of a filamentous microgamete (Fig. 127).

linin reticulum occurs around the central karyosome, the history of which has been recorded by Schaudinn ('96) in *Actinophrys* and *Acanthocystis*, and by Brauer, Gruber, and Hertwig in *Actinosphærium*. In the latter form, which may well serve as a type, the karyosome in nuclei of ordinary vegetative forms is distinctly granular, the chromatin granules being grouped in a variety of ways. There is a constant tendency, however, for the granules to extend outward along the course of the linin reticulum, and toward the nuclear membrane, while during mitosis the granules are grouped together in the centre of the nucleus and along parallel lines.

In nuclei with from ten to twelve karyosomes, the history is practically the same as in the simpler cases with one. Here also each karyosome breaks down into numerous granules, but there is no linin reticulum, and the granules unite in lines to form the chromosomes (*Noctiluca*, Figs. 138, 141). The formation of the numerous granules takes place by division of the karyosomes instead of by budding as in the Sporozoa.

Another type of nucleus is characterized by small-sized chromatin granules, which may or may not be strung upon linin threads. Familiar examples are seen in *Euglena viridis* and the majority of the flagellates. Here the chromatin is in the form of small rod-like granules connected by linin threads and surrounding a central division centre, all being inclosed within a nuclear membrane (Fig. 136). Upon division, the rods of chromatin separate into two groups. Neither the observations of Blochmann nor of Keuten show any indication of splitting of the granules. In allied forms the nuclear membrane may be either present or absent; when absent, the granules of chromatin may permanently surround a central body (*Chilomonas*, some trachelomonads), or may be scattered throughout the cell, collecting only for division (*Tetramitus*, *Urostyla*). In *Tetramitus* the granules collect, without fusing, about such a centre, but in *Urostyla* the numerous portions of the macronucleus fuse into a single nucleus for division (Balbiani, '60; R. S. Bergh, '89).

Among the Protozoa as a whole, there is no general agreement in the chromatin changes preliminary to division. The object of such changes, as Roux first pointed out, is apparently to get the chromatin in the best position for equal division,[1] and the formation of granules from the larger masses is a widely spread if not universal phenomenon leading to this end. Three well-marked types of granule formation have been described (1) by Schewiakoff (*Euglypha*); (2) by Gruber, Brauer, and Hertwig (*Actinosphærium*); and (3) by Calkins (*Noctiluca*). In the first type the process is described by Schewiakoff ('88) as conforming to the metazoan type; the granules form a distinct spireme

[1] Cf. Gruber ('84).

which, as in some Metazoa, splits lengthwise before it is broken into chromosomes (Fig. 23, p. 55). A distinct spireme has also been described in micronuclei of the Infusoria by Bütschli, Pfitzner, Maupas, Hoyer, and Bergh, and in Heliozoa by Schaudinn, while some macronuclei, as in *Stylonychia* and *Loxophyllum*, are apparently in a permanent spireme stage (Fig. 104). The latter case is particularly interesting, for, as in the nuclei from the salivary glands of the *Chironomus* larva (Balbiani), the spireme consists of alternating disks of chromatin and a non-staining substance which Balbiani ('90) considered achromatin with the characteristics of plasmosomes.

In *Actinosphærium* and in *Spirochona*, *Kentrochona*, etc., the mass of chromatin breaks down into granules which collect in lines across the nucleus to form primitive chromosomes (Fig. 139).

Noctiluca is apparently about midway in complexity between *Actinosphærium* and *Euglypha*. The ten or eleven karyosomes, as in Sporozoa and Rhizopoda, break down into an immense number of minute chromatin granules, which collect, at first, in groups in the region of the reservoirs from which they were derived (Fig. 141); but they are later distributed about the nucleus. The granules are then collected in lines which radiate inward from one side of the nucleus. By the constant addition of granules, thick chromosomes are formed, which split down the centre. Again there is no evidence to indicate division of the granules or the presence of a definite spireme stage.[1]

From the foregoing review, the facts at present appear to indicate that the most primitive nucleus is the single mass of chromatin without membrane or reticulum (Coccidiida, Reticulariida). The simplest nuclear membranes are formed directly from this chromatin (Schaudinn in *Calcituba*), as is one type at least of the nuclear reticulum (Reticulariida and Sporozoa). The primitive nuclear mass breaks down into granules in preparation for reproduction, a phenomenon which is almost universal in Metazoa and Protozoa. In the latter group, the essential feature, apparently, is the division of the chromatin mass, rather than of the chromatin granules, as seen by the collection of chromatin granules of distributed types (*e.g. Tetramitus*) into one group which is halved, or in the formation of primitive chromosomes as in *Actinosphærium* or *Noctiluca*. The division of the chromatin granule is apparently not necessary, as shown by the thick chromatin aggregates in *Noctiluca* and *Actinosphærium*, and in the peculiar relations in some Ciliata (*Urostyla*), where the long lines of chromatin are divided into separate segments, each of which forms a nucleus.

[1] Doflein ('00) has recently given a very different account of chromosome-formation in *Noctiluca*, based upon incomplete observations.

E. KINETIC STRUCTURES

The confusion which has arisen concerning the relation of the centrosome to the surrounding structures in Metazoa, and the uncertainty which arises from an ambiguous terminology, are magnified many times when we come to consider analogous structures in Protozoa. It is well to say at the outset that in the Protozoa there are but few structures that can be compared with the complex astral systems such as those in the leucocyte, or at the spindle-poles of dividing animal cells. If, however, we regard the centrosome (including the centriole) and attraction sphere as a unit, and as representing a structure concerned in cell-division, we have a basis for comparison with structures in Protozoa which play a corresponding rôle. In the present chapter, I shall, for the sake of brevity, speak of these structures in Protozoa as the "division-centres." They are not always of definite form and size, although consisting probably of an analogous substance or substances. In some cases the material or substance which corresponds to that in a more definitely formed division-centre, is apparently spread throughout the nucleus, and in other cases even into the cytoplasm. It will be convenient, therefore, to speak of the "division-centre" not only as a body, but also as a substance which is intimately connected with mitosis.

In some cases the division-centre is intra-nuclear, in others extra-nuclear, and in still others it may be sometimes one, sometimes the other. In some of the more primitive Protozoa, notably in some of the marine Rhizopoda, and in a few Sporozoa, there appears to be no such element present during the resting periods of the cell. During division, however, there appears to be a substance which is derived from the chromatin and which functions as a division-centre. This is particularly interesting in the case of *Monocystis ascidiæ*, recently described by Siedlecki ('99). Here, during the formation of the conjugating gametes or sexual reproductive bodies, there appears in the nucleus a distinct, deeply staining granule which divides while against the inner nuclear membrane, to form two division-centres. These become the poles of the division-figure, and during division, there is a connecting strand of deeply staining material. There are other cases, also, especially among the Sporozoa, in which a division-centre appears during nuclear division, although it cannot be made out in the cells when at rest. Thus, I have seen a faintly striated spindle in the dividing nucleus of a polycystid Gregarine of the leech *Clepsine* (Fig. 138), while the resting nucleus shows no trace of a substance similar to that forming the spindle, unless, indeed, it be a portion of the karyosome, which, in several Sporozoa, has been shown to contain two differently staining substances.[1]

In still other Sporozoa there are distinct spindle-fibres, which in some cases (*e.g. Monocystis agilis*, Wolters, '91) consist of two sets, and their presence in resting nuclei cannot be made out. In all Sporozoa, however, the origin of the spindle-substance and the division-centre is difficult to make out because of the disappearance of the nuclear membrane during division. In the case cited by Siedlecki, however, there appears to be no doubt that the division-centres, with their spindle-fibres, arise within the nucleus.

In other forms of Protozoa there is usually some evidence of the division-centre in the resting cell as well as in the mitotic figure. The substance of such centres differs from the chromatin of the nucleus in its different staining reactions, and from the substances in the cytoplasm by its more intense coloration. In some cases, such centres are found within the nuclear membrane during all phases of cell-life (the majority of Mastigophora and Infusoria and many Rhizopoda). In other cases they are permanently outside of the nuclear membrane (*Noctiluca, Paramœba, Heterophrys, Sphærastrum*). Again, they may be intra-nuclear during some phases and extra-nuclear during others (*Tetramitus, Actinosphærium, Acanthocystis*).

1. *Intra-nuclear Division-centres.*

The least-differentiated division-centres are those which are permanently within the nucleus. They are found in all classes of the Protozoa, and have been variously interpreted. For a long period they were erroneously regarded as nucleoli, but since their true function was first suggested by Keuten ('95) they have been variously regarded as "nucleolus-centrosomes" (Keuten), centrosomes (Hertwig, '96), and spheres, equivalent to centrosome plus attraction-sphere (Calkins, '98).

In the majority of the Flagellidia the division-centre is clearly defined and distinct from all other parts of the cell. Its changes in form, which were first made out in *Euglena* by Blochmann ('94) and by Keuten ('95), can be easily followed in almost any species of *Euglena*. It lies in the centre of a spherical group of chromatin granules which are connected with one another by a linin reticulum, the whole being inclosed within a firm nuclear membrane (Fig. 136). During resting stages of the cell it is globular or ellipsoidal in form, but during

[1] The "centrosomes" which Labbé ('96) described in *Klossia eberthi, Bananella lacazei*, and *Pfeifferia gigantea* must be regarded with doubt until their relation to the nucleus in division is made out.

division, it elongates to form a dumb-bell-shaped body, the two ends remaining connected until the end of division, by a strand. After division the daughter-centre rounds out and resumes its customary form and position within the group of chromatin granules.

An essentially similar process was described by Schaudinn ('94) in the division of the rhizopod *Amœba crystalligera*, and although he recognized that the intra-nuclear body plays the chief rôle in division, he regarded it as the nucleolus (Fig. 137). It has since been found in the majority of Mastigophora, in many Rhizopoda, Heliozoa, and in

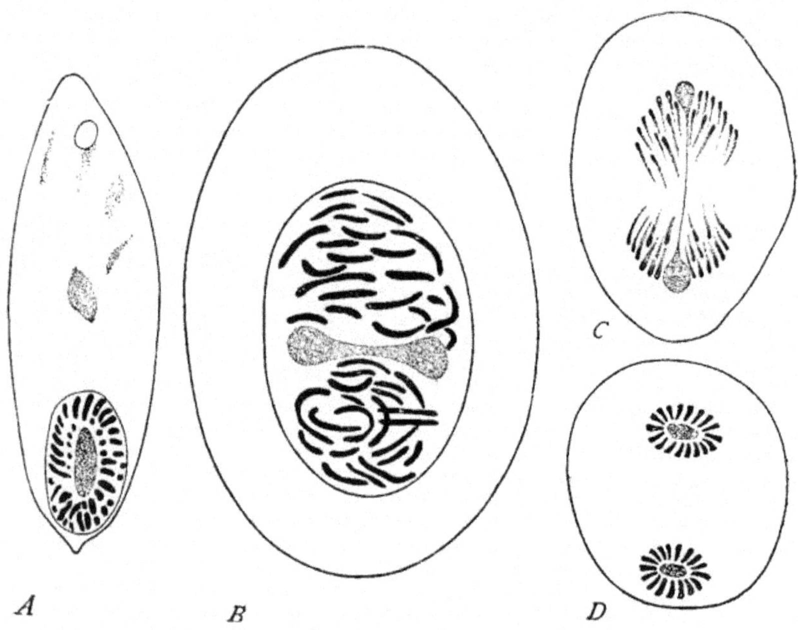

Fig. 136. — Mitosis in *Euglena*. [WILSON after KEUTEN.]

A. Preparing for division; the nucleus contains a "sphere" or division-centre, surrounded by a group of chromosomes. *B*. Division of the sphere to form an intra-nuclear spindle. *C*. Later stage. *D*. The division completed.

Infusoria. In many cases it loses its distinct outline and becomes more or less indistinct, although reappearing during cell-division as the division-centre. In the dinoflagellate *Ceratium hirundinella*, Lauterborn ('95) described it as indistinct and apparently without the usual function.

The intra-nuclear division-centre becomes difficult to see, at least in the resting phases, in micronuclei of the Infusoria, in the rhizopod *Euglypha*, in the heliozoön *Actinosphærium*, and in the ciliate *Spirochona*, forms which may be selected as showing the variations in the intra-nuclear division-centre.

The micronuclei of Infusoria were first observed in division by Balbiani ('58, '59), but were incorrectly interpreted as bundles of spermatozoa, an error which Bütschli ('76) was the first to point out. Bütschli showed that these supposed bundles are nothing more than the micronucleus in division, and he correctly interpreted the micronucleus as analogous to the nuclei of tissue-cells. In the figures which were made at this time, there is a distinction between chromatin and achromatin (*e.g. Paramæcium*), and a mass of "achromatic" substance is pictured at each pole (cf. Fig. 138, *b*). Subsequent

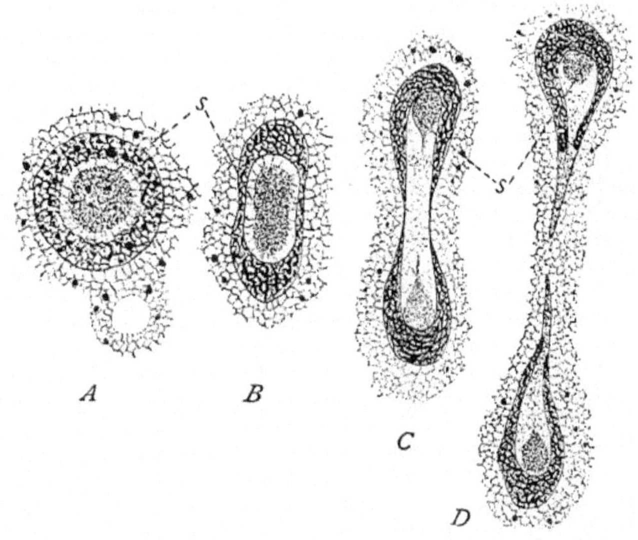

Fig. 137. — Division in *Amœba crystalligera* Schaud. [SCHAUDINN.] *s*, division-centre.

observers have obtained similar results. Maupas ('89) showed that the spindle figure is made up of two distinct parts, and Hertwig ('89, '95) made out the granular character of the chromatin, and in his later work ('95) showed that the resting nucleus as well as the dividing nucleus has an achromatic body. The relation of the intranuclear body in the resting nucleus and in dividing forms was not made out, although Hertwig assumed that the spindle in the latter is derived from the division-centre. The later stages of the dividing nucleus show the chromatin massed at the two ends in front of a cap (pole-plate) of achromatic material, while a strand similar to that in *Euglena*, connects the two ends (Fig. 139, *D–H*).

It was Hertwig ('77), also, who gave the first account of the origin of the achromatic spindle-figure in the division of the macronucleus of the peritrichous ciliate *Spirochona gemmipara*. In this nucleus he described two distinct parts, one homogeneous, and with the exception of a central granule, which he called the "nucleolus," without structure; the other, chromatin in the form of granules, which surround the homogeneous portion. In preparation for budding, the granule within the homogeneous portion swells, sends out pseudopodial processes, and at the same time becomes more indistinct, until, finally, the substance of the so-called nucleolus is lost in the surrounding granules of chromatin. Shortly afterward, two heaps of homogeneous substance (*pole-plates*) appear at the ends or poles of the division-figure, and the chromatin granules are arranged in lines between these masses. Even at this early date Hertwig compared the pole-plates or "end-plates," as he called them, with the *Polkörperchen* (centrosomes) in the mitotic figures of Metazoa (Fig. 139, *A–C*), and concluded that they are derived from the original "nucleolus."

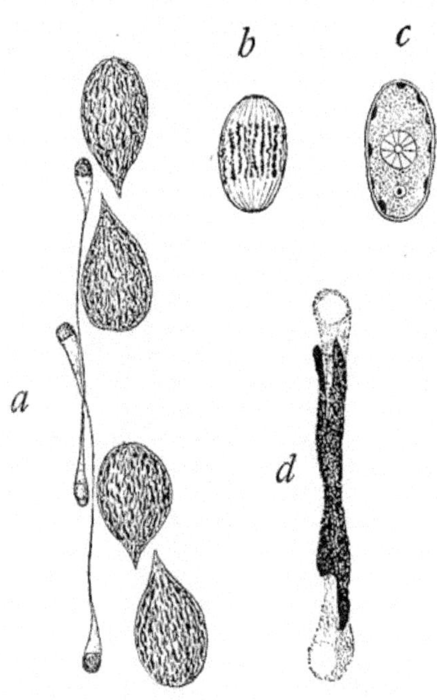

Fig. 138. — *a.* Macronucleus and micronucleus of *Stylonychia* in division. *b.* Micronucleus of *Paramæcium aurelia* in division. *c.* The nucleus of *Chilodon cucullulus.* [BÜTSCHLI.] *d.* Dividing nucleus of *Clepsidrina sp.*

Spirochona has been repeatedly examined since 1877, and Hertwig's main conclusions are confirmed. There is a difference of opinion, however, in regard to the origin of the "nucleolus" or division-centre, as we may be justified in calling it. Plate ('86) thought that the corpuscle is formed anew after division, by the accumulation of chromatin which penetrates the homogeneous portion in a state of solution. He also distinguished an inner structure in the supposed homogeneous portion, and maintained that the pole-plates are formed from its substance. Balbiani ('95) more recently came to a somewhat similar conclusion. He found that the so-called homogeneous part is made up of short and fine fibrillæ which do not stain with the chromatin dyes, and are, therefore, to be classed as "achromatic" structures. He maintained that the "nucleolus" is formed by the aggregation of several granules of chromatin during the later stages of division.

Two substances, therefore, are present in the nucleus of *Spirochona*, which play some rôle in the process of division. The apparent relation of these substances to one another is strikingly similar to the relations between the centrosome and the attraction sphere in Metozoa, but so many points remain obscure that no safe conclusions can be drawn. In one other form, *Chilodon cucullulus*, there is an intranuclear structure which recalls that of *Spirochona*, but nothing is known about its division-phases (Fig. 138, *C*).

In all other forms with well-developed "pole-plates" there is a similar mystery regarding the material which enters into their formation. Despite the numerous observations, there is no case on record

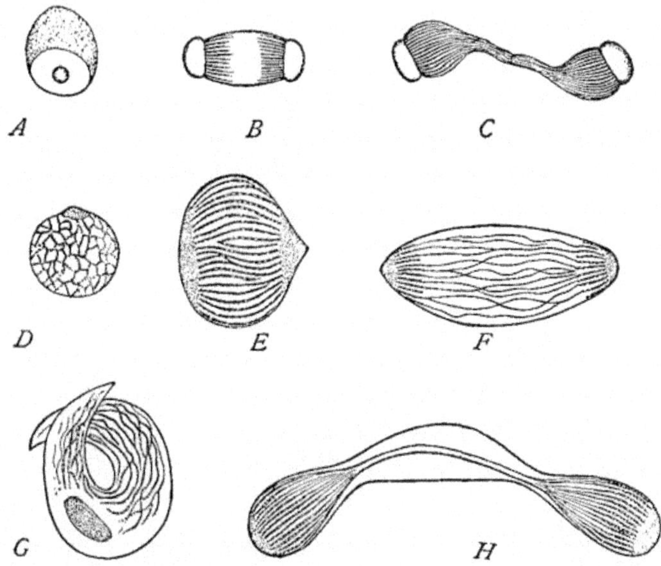

Fig. 139. — Mitotic division in the Infusoria. [From WILSON after R. HERTWIG.]

A–C. Macronucleus of *Spirochona*, showing pole-plates. *D–H.* Successive stages in the division of the micronucleus of *Paramœcium*. *D.* The earliest stage, showing reticulum. *G.* Following stage ("sickle-form") with nucleus. *E.* Chromosomes and pole-plates. *F.* Late anaphase. *H.* Final phase.

in which the origin of the pole-plates has been definitely made out. In all cases, however, it is assumed that the so-called nucleolus or "achromatic body," which is generally present in the resting cell-nucleus, becomes modified in some manner (Hertwig, '95, assumed by the addition of water or other fluid substance) until it is much enlarged, when it forms the pole-plates. This general view is based upon the supposition that, like a centrosome, the achromatic body divides, half going to one pole and half to the other.

In Sarcodina, as in Infusoria, there is no direct evidence to determine the history of the "achromatic" body within the nucleus, although two classical objects, *Actinosphærium* and *Euglypha*, have been repeatedly examined. Nuclear division in the former was first described by Gruber ('83), then by Hertwig ('84), by Brauer ('94), and, finally, reëxamined in great detail by Hertwig ('98). All agree as to the general features of division, but disagree widely in details. In some stages (before the "primary mitosis," Hertwig) the chromatin is in a single large karyosome which incloses a faintly staining achromatic mass (Gruber, Hertwig). In addition to these there is

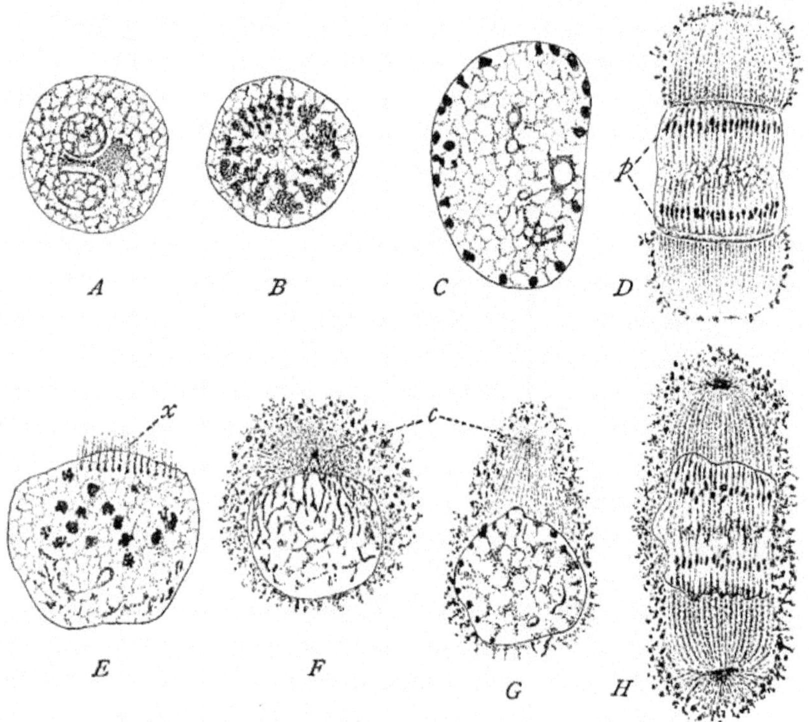

Fig. 140. — Nuclear division in *Actinosphærium*. [HERTWIG.]

A, *B*. Ordinary vegetative nuclei. *C*. Prophase of division. *D*. Ordinary mitosis. *E*. Transfusion of intra-nuclear substance (*x*) to the outside. *F*, *G*, *H*. "Maturation-mitoses." *p*, pole plates; *c*, centrosome.

a conspicuous linin network (Brauer, Hertwig), which Brauer and Hertwig regard as forming the pole-plates during division. During mitosis peculiar and as yet unexplained masses of protoplasm (achromatic) are formed on the outside of the nucleus (*Protoplasmakegel*). Brauer and Hertwig agree that these masses are nuclear in origin, but neither gives a satisfactory explanation of their function in division (Fig. 140). The pole-plates are connected by fibres of "achromatic" substance, which are formed from the linin reticulum (Hertwig) and upon which the chromatin granules are strung. At a "certain period in mitosis" (Brauer), a period which Hertwig identified as "maturation mitosis," minute bodies, analogous to centrosomes, emerge from the nucleus and take a position in this extra-nuclear mass of achromatic material (Fig. 140, *F*). At this stage of division, therefore, there are structures which, more than anything else in the Protozoa, resemble the astral system in the cytoplasm of many metazoan cells. They occur, however, only at certain periods and are not characteristic of ordinary vegetative mitosis.

Nuclear division in *Euglypha* as described by Schewiakoff ('88) is characterized by stages which are remarkably similar to those in Metazoa. The chromatin passes through a spireme stage and breaks into chromosomes by transverse division of this spireme; a spindle is formed from "achromatic material," and *Polkörperchen*, which are strikingly suggestive of centrosomes, form the poles of the division-figure (Fig. 23, p. 55). The origin, again, of the spindle-fibres and of the pole-bodies was not determined. Schewiakoff assumed that the latter are derived from the cytoplasm.

In all of these cases, with the exception of certain maturation phases of *Actinosphærium*, the entire division-figure remains inside of the nuclear membrane. The substance of the spindle, therefore, must arise from within the nucleus, and although it has not been definitely proved, there is good reason to believe that this substance is contained, during the resting phases, in the so-called "nucleolus" or intra-nuclear sphere (division-centre). There is no doubt upon this point in regard to the division-figure of the simple flagellates (*Euglena*, etc.), for the clearly defined achromatic body can be traced throughout all stages. The connecting strand in *Euglena* is not fibrillated, and therefore a true spindle is lacking, but there seems little room to question the analogy between such a strand and the fibres in forms like *Actinosphærium*, *Spirochona*, etc. Indeed, the relation of the single strand to the fibrillated spindle is apparently well marked by an intermediate form in *Paramœcium* (Hertwig, '96), where the central portion of the division-figure is a single strand which widens and becomes fibrillated at the ends (cf. Fig. 139, *H*).

2. *Extra-nuclear Division-centres.*

We now pass to a consideration of more complicated types of protozoan nuclei, in which, during the resting phases, the kinetic structures are outside of the nuclear membrane.

The *Centralkorn* in Heliozoa, with its radiating fibres which form the axial filaments of the pseudopodia (cf. p. 82), have often, and probably justly, been compared with the centrosome and astral rays of Metazoa. This central granule, first observed by Grenacher ('69), was shown to be connected with the pseudopodia by Greeff in the same year. Subsequent observations by F. E. Schultze ('74), Hertwig ('77), and many others, have confirmed these results, and demonstrated the widespread occurrence of the central granule among the Heliozoa. Bütschli ('92) was the first to suggest the similarity to the centrosome with its radiating fibres, a view which O. Hertwig ('93) and Schaudinn ('96), and with them the majority of cytologists, have accepted. The most complete observations have been made upon species of *Acanthocystis* and *Sphærastrum* by Schaudinn ('96). In the latter form (Fig. 144, *A*) the corpuscle is distinctly granular, and in fixed preparations has a definite alveolar structure. The beginning of division is signalized by the withdrawal of the pseudopodia; the central granule then divides, and with it the entire astral system.[1] The daughter-centres are joined together by a connecting strand which Schaudinn regarded as a possible central-spindle (Fig. 144, *C*).

The extra-nuclear central granule in these Heliozoa thus acts like the intra-nuclear division-centre of *Euglena* and other flagellates. The extra-nuclear division-centres in the rhizopod *Paramœba* and in the cystoflagellate *Noctiluca,* on the other hand, are much more like the centrosphere in Metazoa; and in *Noctiluca* especially, the mitotic figure, while of the protozoan type, is more like that of the Metazoa than of any other known single-celled form.

On the outside of the nucleus in *Noctiluca,* in the cytoplasm, and close against the nuclear membrane, is a large, faintly staining spherical mass, which acts as a division-centre. During the early stages of nuclear activity, the sphere divides into two similar halves, connected by a strand composed of fibres which are formed from the substance of the sphere. These fibres compose the central spindle, and are homologous in every way with the central-spindle fibres of the usual type of mitosis in Metazoa (Fig. 141, *C–E*). The nucleus then elongates in a direction at right angles to the central spindle, and at the same time it bends in the centre in such a way that the central spindle sinks into a depression in the nucleus, which surrounds it upon three sides. In this way the nuclear plate is finally wrapped about the central spindle in the form of an incomplete ring, a condition which obtains in all higher mitotic figures where the central spindle is present. The nuclear membrane then disappears[2] in that part of the nucleus which is turned toward the central spindle, while it is retained unbroken in all other parts of the nucleus

[1] The division of the central granule was first observed by Sassaki ('93) in the marine form, *Gymnosphæra.*

[2] Cf. Calkins ('98), p. 15; Ishikawa ('99), p. 244.

Fig. 141, *E, F*). Thus the chromosomes, as in the higher types, are brought in contact with the central-spindle fibres. They then split longitudinally, and through the agency of mantle-fibres are separated into two equal groups, each group drawn toward its own daughter-sphere. Within the sphere the fibres are focussed in a centrosome,

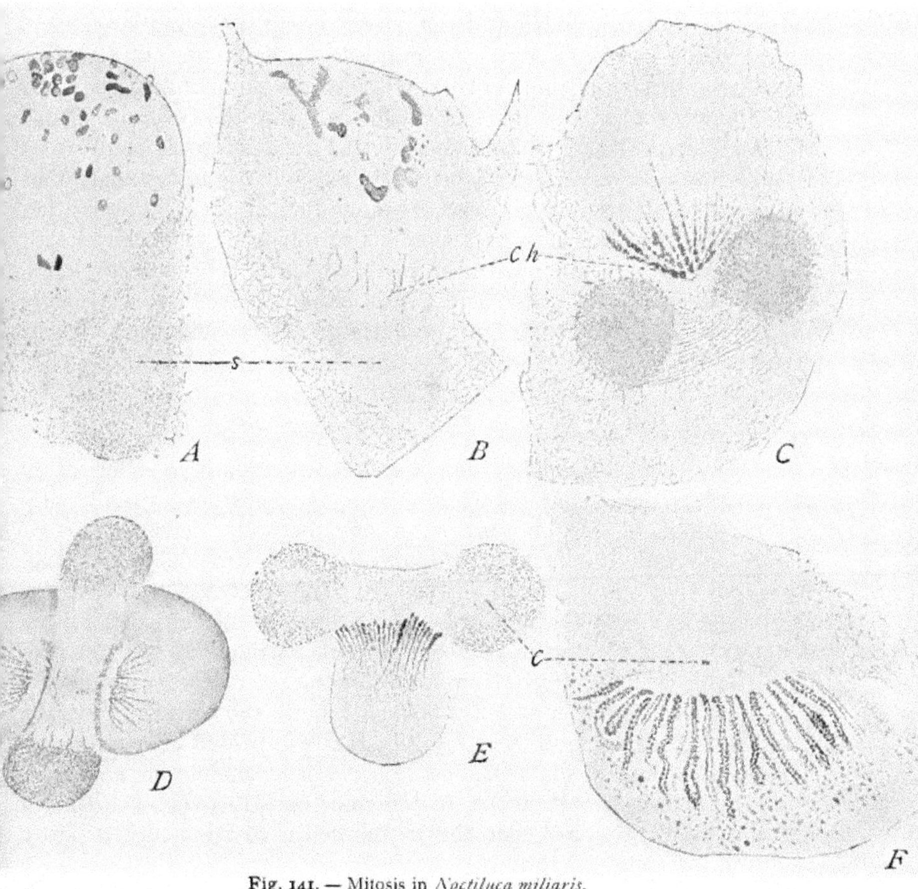

Fig. 141. — Mitosis in *Noctiluca miliaris*.
s, sphere; *c*, centrosome; *ch*, chromosomes.

which, at this period, can be demonstrated with the greatest ease. The division is finally completed by the separation of the remainder of the nucleus and the re-formation of the daughter-nuclei, while the chromosomes disintegrate into granules, which again form the large chromatin reservoirs characteristic of *Noctiluca* (Fig. 134, *E*).

As Ishikawa ('94) first pointed out, the centrosome within the sphere, the mantle-fibres and their insertion in the chromosomes, the origin of the central spindle from the substance of the sphere, are all features obviously common to the analogous structures in Metazoa.

An extra-nuclear body, somewhat similar to the sphere of *Noctiluca*, has been described by Schaudinn ('96) in the rhizopod *Paramœba*. *Paramœba* reproduces by swarm-spores (cf. p. 93), which are formed by the spontaneous division of the parent organism into a large number of small parts. Before this multiple division, the extra-nuclear achromatic mass (division-centre) divides into a number of parts equal to the number of spores to be formed, and after the several portions are distributed about the cell, the nucleus divides into as many parts as there are portions of the achromatic mass. It is unfortunate that the minutiæ of division are not given, and until future observation confirms Schaudinn's interpretation, his results must be received with some scepticism. The swarm-spores themselves reproduce by longitudinal division, and the nuclear processes involved are extremely suggestive of the relations of the extra-nuclear to the intra-nuclear division-centres.

3. *The Relation of Extra-nuclear to Intra-nuclear Division-centres.*

In view of the fact that the division-centres are in some cases extra-nuclear and in others intra-nuclear, the question naturally suggests itself as to the connection, if any, between them. No one conversant with the facts will doubt that they are analogous structures, but that one has been derived from the other is not so obvious. Hertwig ('95) held that "these centrosomes of the egg-cell (Echinoderms) are not specific cell-organs, but portions of the nucleus which have become freed from the chromatic nuclear substance."[1] This view of the origin of extra-nuclear kinetic substance in Metazoa is difficult to accept, and in forming it, Hertwig passed over too many questionable intermediate stages. There is considerable evidence, however, among the Protozoa, to indicate that Hertwig's conception has a basis of fact, and that the extra-nuclear division-centres arose from the intra-nuclear forms.

As indirect evidence of such an origin of the extra-nuclear division-centres, it might be pointed out that in all mitotic figures in which there is a central spindle, a portion, at least, of the spindle substance is surrounded by chromatin, and may be said to be intra-nuclear in position. This is certainly the case in all Protozoa, and the relations of the extra-nuclear centres to the chromatin is particularly suggestive in the dividing swarm-spores of *Paramœba*, in *Noctiluca*, in *Tetramitus*, and in *Actinosphærium*, while, according to Schaudinn's

[1] *Loc. cit.*, p. 53.

observations, in two cases at least, there is positive evidence that the extra-nuclear centres originate in the nucleus (*Acanthocystis*, *Oxyrrhis*).

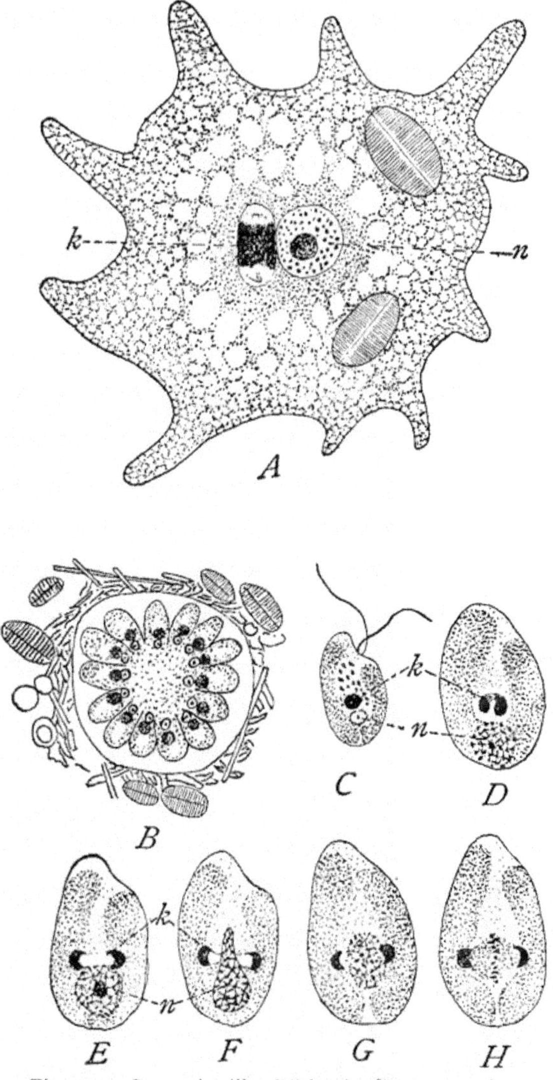

Fig. 142.—*Paramœba eilhardi* Schaud. [SCHAUDINN.]
A. Section. *B.* Sporulation. *C–H.* The flagellated swarm-spores in process of division. *k*, the *Nebenkörper*; *n*, the nucleus which wraps itself around the division-centre (*Nebenkörper*).

In *Paramœba*, although the division-centre (*Nebenkörper*) apparently plays no part in the nuclear division of the mother-animal, its daughter-parts play the same rôle in division of the swarm-spores as that of the intra-nuclear division-centre in *Euglena* (Fig. 142). In *Noctiluca* the central spindle, which is derived from the extra-nuclear division-centre, takes a similar intra-nuclear position during division of the nucleus, and when the nuclear membrane disappears in the adjacent portions of the nucleus, the kinetic substance is again intranuclear, although the centres of attraction are outside (Fig. 141, *D*, *E*). In *Actinosphærium*, both Brauer ('94) and Hertwig ('98) explained the cytoplasmic accumulations (*Protoplasmakegel*) as coming from the nucleus, and Hertwig pictured the outer mass and the inner achromatic substance as connected through openings in the membrane (Fig. 140). Confirmatory evidence is also shown in other Heliozoa. In *Actinophrys* there is no extra-nuclear division-centre, and Schau-

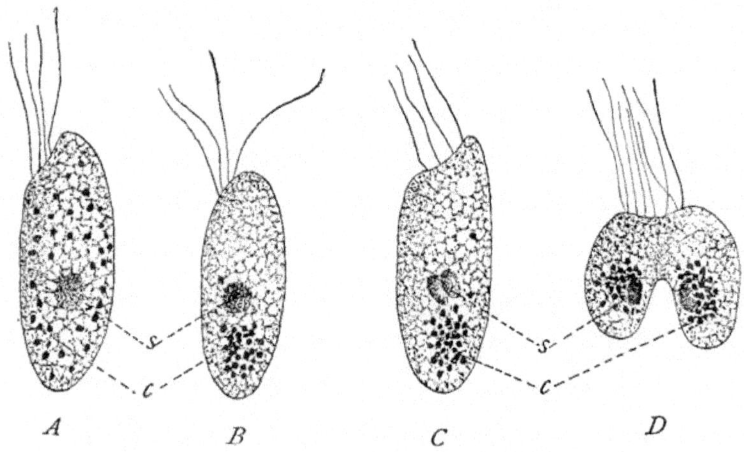

Fig. 143. — Mitosis in *Tetramitus chilomonas*.

A. Ordinary form with distributed chromatin (*c*) and division-centre (*s*). *B*. The chromatin granules are collected prior to division. *C*. The division-centre has divided. *D*. Later stage in division; each daughter-nucleus is surrounded by a group of chromatin granules.

dinn ('96) interprets the achromatic spindle figure as nuclear in origin (Fig. 130, p. 236).[1]

The flagellate *Tetramitus* shows an apparently similar division-centre. During the resting phases, the chromatin is distributed throughout the cell, while an indefinite "achromatic mass" appears to be in direct connection with the cytoplasmic reticulum. Immediately before division, however, the chromatin granules collect about this body, and then, save for the absence of a membrane, the aggregate resembles the nucleus of *Euglena*. Division takes place as

[1] A significant fact is that in *Actinophrys* the radiating axial filaments centre in the nucleus, while in other Heliozoa with a "*Centralkorn*" the radiating axial filaments centre in this extra-nuclear body. This indicates that in *Actinophrys* the attraction-centre is within the nuclear substance and presumably in the "achromatic substance."

in *Euglena*, the intra-nuclear division-centre dividing first. After division the chromatin granules again disperse and the division-centre becomes again cytoplasmic (Fig. 143).[1]

Still more convincing evidence is shown by the history of the division-centres of *Acanthocystis* and the flagellate *Oxyrrhis marina* (Schaudinn, '96). *Acanthocystis* has a permanent extra-nuclear division-centre which divides and forms a complete spindle (Fig. 144, *A–D*).

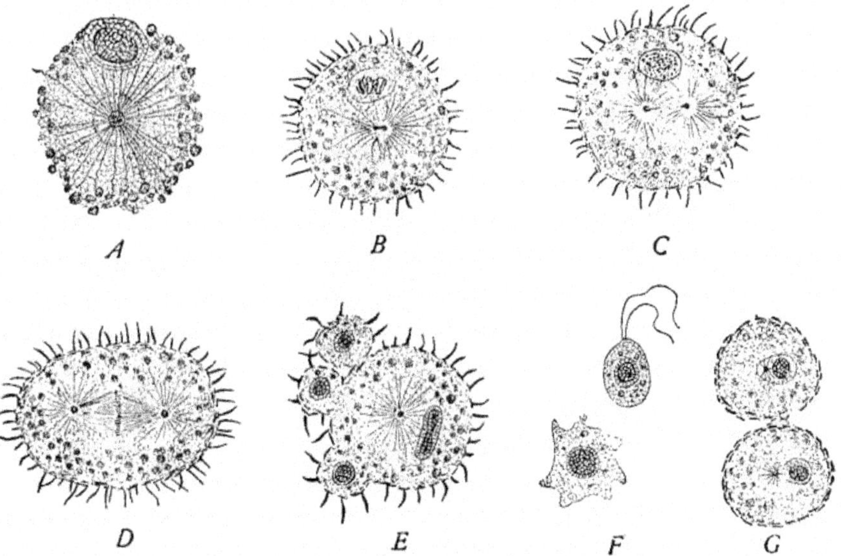

Fig. 144.—Nuclear division and spore-formation in Heliozoa. [SCHAUDINN.]

A. A vegetative cell of *Sphærastrum* with the axial filaments focussed in a central-granule (division-centre). *B–D.* Division of the nucleus in *Acanthocystis*. *E, F.* Flagellated and amœboid swarm-spores formed by budding. *G.* Exit of the division-centre from the nucleus.

Like *Paramœba* this heliozoön reproduces by swarm-spores; the division-centre, however, takes no part in their formation, but remains intact while the nucleus divides without mitosis. The buds, therefore, contain no portion of the original material of the division-centre, nor is there any evidence in them of such a centre until about the fifth day. At this time Schaudinn found a "*Centralkorn*" within the nucleus of each swarm-spore, which passed later through the nuclear membrane and into the cytoplasm, where it developed into the well-known division-centre of the adult *Acanthocystis* (*G*).

The extrusion of the kinetic substance of the nucleus which thus

[1] An intermediate stage between this condition and the condition in *Euglena* is shown by some species of *Chilomonas* and *Trachelomonas*, in which there is no nuclear membrane, but in which the chromatin remains permanently aggregated about the division-centre (Calkins, '98).

takes place under normal conditions in *Acanthocystis* can be brought about in *Oxyrrhis marina*, as Schaudinn has demonstrated, by placing the flagellate in an abnormal medium. Ordinarily this form has an intra-nuclear division-centre like that of *Euglena*, but if it is transferred to a more dilute salt solution, the substance of the division-centre is forced out of the nucleus and into the cytoplasm, where it swells to many times its usual size, and may even divide outside of the nucleus, forming a large spindle in the cytoplasm (Schaudinn, '96).

4. *The so-called "Centrosomes" in Protozoa.*

In addition to the division-centres or spheres described above, there are occasional centres in Protozoa resembling centrosomes of the Metazoa. So little positive knowledge is at hand, however, that the centrosome question in Protozoa must rest in abeyance. A number of division-centres have been described as centrosomes, but in almost all cases these are more like the spheres described above, than the centrosomes of Metazoa. In a number of instances, however, the primitive division-centres contain central, deeply staining granules which form the foci of the spindle fibres, and these would seem to be analogous to the centrosomes of higher forms. Such granules are found in *Spirochona*, *Kentrochona*, *Actinosphærium*, and *Noctiluca*, and probably they exist in other forms as well; these, however, are the only forms that have been critically examined.

In *Spirochona* and *Kentrochona*, the granules in question appear only at a late stage in division (Balbiani, '95; Doflein, '96), and they remain in evidence until the next division period. In *Actinosphærium* similar granules appear during the maturation mitosis, but are not found at other times (Hertwig). In *Noctiluca* they appear at each mitosis and form the insertion points for the mantle-fibres (Fig. 141, *F*).

The origin of these so-called centrosomes still remains obscure, although in each case it has been maintained that they arise from the chromatin. Plate ('86) and Bütschli ('88) assumed that in the case of *Spirochona gemmipara* they are formed from chromatin in solution, while Balbiani ('95) maintained that they are actually granules of chromatin broken off from the polar ends of the chromosomes. Hertwig ('98) described a similar origin in the case of *Actinosphærium*, where the granules are supposed to arise by budding of the chromatin (Fig. 140). Brauer ('94), on the other hand, held that the centrosomes come from the pole-plates. In *Noctiluca*, finally, Ishikawa ('94, '99) and Calkins ('98) concluded that the centrosomes were nuclear in origin, while the spheres are extra-nuclear at all times. The conclusion, however, could not be supported by positive evidence and, at best, has only the value of an assumption.

F. General Considerations

The mitotic figure in plants, higher animals, and in Protozoa is composed of two parts, of which one, the chromatin, is widely held to be the primary agent in heredity, while the other, forming the "achromatic structures," is generally regarded as the agent by means of which the chromatin is divided equally between the two daughter-cells. The two parts have more or less independent antecedent phases, and it is possible, therefore, to conceive the two portions unequally developed. In Protozoa both portions are relatively simple in structure, although complex chromosomes may accompany relatively simple achromatic specializations, and *vice versa*.

The stages in chromosome formation in different types of Protozoa may be briefly summarized as follows: (1) The most primitive nucleus is, apparently, in the form of a compact sphere of chromatin, the multiple division of which is the prelude to reproduction of the cell. (2) A higher type comprises nuclei with membranes, and with chromatin in one (Sporozoa) or in many (Noctiluca, etc.) karyosomes, which break up by multiple division into granules. The granules thus formed unite secondarily into lines forming primitive chromosomes. (3) In still higher forms the granules do not return to the karyosome stage, but are widely distributed over the linin reticulum (Flagellidia, *Actinophrys*, some stages of *Actinosphærium*, Metazoa). (4) The highest type contains chromatin granules embedded in a linin reticulum, or aggregated to form net knots or karyosomes analogous to the more primitive chromatin spheres. Like the primitive karyosomes, these net knots break up into granules which come together in lines for division (spireme stage of some Protozoa, of plants, and Metazoa), and these lines segment into chromosomes of definite number and size (Metazoa and Metaphyta).

In some Protozoa the nuclei remain permanently in one or the other stage described. Thus in many of the Phytoflagellida they are permanently in stage 1; in the simple Monadida they are typically in stage 3, in the Rhizopoda in stage 2, while the higher types of nuclei pass through nearly all of these stages during preparation for division. The net knots thus show a return to the primitive condition, the chromatin granules to the permanent granular state in Flagellidia, or to the disruption of the karyosomes of type 2; the fusion of the granules into spiremes, to the primitive chromosome formation in *Noctiluca*. The definite chromosomes, finally, represent the highest grade of chromatin specialization.

While it is quite probable that the chromatin of plant nuclei, of metazoan nuclei, and of protozoan nuclei is everywhere essentially the same substance, the similarity is not so obvious in case of the "achromatic material," regarding which wide difference of opinion prevails. If, with van Beneden ('83), Bütschli ('92), and many others, we assume that the spindle in Protozoa arises by the local modification of the protoplasmic network, we leave unexplained all of those division-centres in Protozoa which are unmistakably permanent throughout resting and active phases of the cell, and from which spindles are formed (flagellates, *Noctiluca*, etc.). Nor can the interesting hypothesis of Rabl ('89), even when supported by the evidence which Heidenhain ('94), Bühler ('95), and Kostanecki ('97) have added, explain the facts in Protozoa, unless, indeed, it be assumed with Rabl that the spindle fibres, never losing their identity, but merely modified to form portions of the protoplasmic network, remain in Protozoa, in the form of a compact and definite division-centre. Could Rabl's suggestion be thus adapted, the result would be an hypothesis which agrees essentially with the *archoplasm* theory of Boveri ('88). Boveri maintained that the kinetic structures of the cell are derived from a specific substance which he called *archoplasm*. At first ('88) he held that the archoplasm, in the form of granules, is a permanent substance, but in a subsequent paper ('95) he modified the theory by the suggestion that the archoplasm might be distributed about the cell in the form of a homogeneous material which cannot be readily demonstrated, and which under the influence of the centrosome may be crystallized out from the protoplasm. An essentially similar hypothesis was formulated by Strasburger ('92) in connection with plant-cells. He held that in these cells the protoplasm is composed of two essential substances, one of which, the *trophoplasm*, is alveolar in structure and is especially concerned with the processes of nutrition; the other, *kinoplasm*, is fibrillar in structure and is devoted to the formation of the active portions of the cell, spindle fibres, cilia, flagella, and outer cell-covering (*Hautschicht*).

The facts in Protozoa fit in best with the archoplasm hypothesis. Here, in a great many forms, is a definite structure composed of a specific substance which, during cell-division, forms the spindle-figure. Hertwig ('96) favored the view that the cytoplasmic and nuclear reticula in the various phases which this portion of the protoplasm can assume, are sufficient to explain the several division-centres, without calling upon a special kinoplasm or archoplasm. He did not, however, in my opinion, give sufficient weight to the simpler division-centres in Protozoa, but based his views upon the relatively complex phenomena of division in *Actinosphærium*. He believed that in this form the spindle fibres and the pole-plates (which he homologized with centrosomes of the Metazoa) are derived from the reticulum of the nucleus.[1] No satisfactory explanation was given of the "*Plastinge-*

[1] *Loc. cit.*, p. 59.

rüst," which he described in the vegetative nucleus, although he recognized a distinction between it and both the linin reticulum and the chromatin. The achromatic figure in *Actinosphærium* is so large as compared with the nucleus of resting cells, that it is difficult to accept Hertwig's view, that it all comes from the linin of the resting nucleus (cf. Fig. 140, *A*, *B*, *D*). The writer, in a recent publication ('99), proposed the application of Boveri's theory of archoplasm to the division-centres of Protozoa; and, applicable, apparently, to all types of the division-figure in Protozoa, it may serve for the time as a working hypothesis. Briefly stating this hypothesis, it was held that the division-centre, consisting of archoplasm, retains its definite form and size in the nucleus of many of the primitive forms (flagellates), but under certain conditions may become enlarged or diffuse. Thus, in *Tetramitus*, the division-centre is much more diffuse during the resting phases of the cell than during division (cf. Figs. 134, 143); and in *Oxyrrhis marina* Schaudinn ('96) found that the division-centre becomes much enlarged when the cells are transferred to a medium of less density, and conversely, when placed in a denser medium, the structures become reduced in size and more definite. It was held that the intra-nuclear division-centre becomes similarly diffused. The nucleus of *Amœba proteus*, for example, contains chromatin in the form of minute granules, which are arranged about the periphery of the nucleus, while the central portion is occupied by a large homogeneous mass, which can be explained as an enlarged or diffuse intranuclear division-centre. The pole-plates, also, which are widely distributed throughout the Protozoa, may be explained as a temporary accumulation of this ordinarily diffuse archoplasmic substance, and thus homologous with the "nucleolus-centrosome" or division-centre of *Euglena*, and with the centrosphere of Metazoa.[1] The substance which may thus become diffused through the nucleus may also penetrate the nuclear membrane, until accumulations on the outside of the nucleus result. Hertwig described such transfusion of achromatic material from the nucleus of *Actinosphærium* to the aggregates (*Protoplasmakegel*) on the outside (Fig. 140, *E*). Finally, just as it becomes diffused throughout the nucleus in Protozoa and Metazoa, so it may become diffused throughout the cytoplasm in Metazoa, as postulated by Boveri.

A number of ingenious theories have been made to account for the origin of the division-centres of the Metazoa. Bütschli ('91) was the first to suggest that the micronucleus of Infusoria might be the protozoan analogue of the metazoan centrosome. Hertwig ('92) and Heidenhain ('94) accepted the suggestion, and the latter, in particular, worked out a complicated theory of phylogeny upon it. The

[1] *Loc. cit.*, p. 224.

improbability of this theory, from the phylogenetic standpoint, was convincingly shown by Boveri ('95); and Lauterborn ('96), admitting this difficulty, proposed the theory in a slightly modified form. He assumed that micronucleus and centrosome may have had a common ancestor in some primitive bi-nucleated protozoön, such as *Amœba binucleata* (Schaudinn), and that intermediate stages may be seen in certain existing Protozoa, such as *Paramœba eilhardi* (Schaudinn) and *Noctiluca miliaris.* He considered the micronucleus to have arisen from one of these primitive nuclei, but by a differentiation quite different from that which gave rise to a centrosome. This theory, also, has been shown to be untenable by Boveri ('01). A different theory elaborately worked out has appeared in several of the recent publications of R. Hertwig ('95, '96, '98). In its latest form, Hertwig's theory may be summarized as follows: (1) The achromatic substance is at first uniformly distributed in the resting nucleus, but appears during division as pole-plates, the analogue of centrosomes; (2) the achromatic substance becomes permanently an intra-nuclear centrosome; (3) it is extruded from the nucleus to form an extra-nuclear centrosome.

Still another point of view was suggested by Schaudinn ('96) after his very remarkable observations and experiments on *Acanthocystis* and *Oxyrrhis marina.* He suggested the following two possibilities: Either the division-centre (his centrosome) is a structure formed within the nucleus, becoming subsequently cytoplasmic in position, as it does in the buds of *Acanthocystis,* or as the nucleolus-centrosome does in the nucleus of *Oxyrrhis* when immersed in dilute sea-water; or else it is normally cytoplasmic in position, and becomes secondarily intranuclear.

Not only in Protozoa, but also in Metazoa, the sphere and centrosome have been described, in some cases, as coming from the nucleus. The most noteworthy of these is Brauer's description of *Ascaris megalocephala univalens* ('93) and Rückert's ('94) description of its nuclear origin from the germinal vesicle of a copepod.

In these two cases, however, it seems obvious that such an origin among higher metazoan cells can have little value in problems relating to the historic origin of the centrosomes; for here differentiation is fully as complete as in any other cells, and the origin of such centrosomes can give no light on the problem. To a certain extent, Hertwig's view, also, is open to the same criticism.

Schaudinn's other alternative has opened the way for another view,[1] which, based upon the lower Flagellidia and Rhizopoda, explains more primitive conditions in more primitive organisms. It is not

[1] Cf. Calkins ('98, '99).

improbable that the condition of the distributed nucleus, as found in *Tetramitus*, may have been widely spread among the lower forms of life. In *Tetramitus* the division-centre becomes more compact and distinct during the preparatory division stages, while the chromatin granules collect in a small aggregate in its immediate vicinity. The sphere then divides and the chromatin aggregate separates into two portions. This stage of the division probably corresponds with the stage described by Schaudinn in *Paramœba*, where the chromatin granules form a ring about the divided sphere. The swarm-spores increase by simple division, as in *Tetramitus*, and a dumb-bell structure results. The nucleus, which is a well-defined body, then moves around the connecting strand of the daughter-centres until it surrounds it as the nucleus of *Noctiluca* surrounds the central spindle (Fig. 142, *F*, *G*). The connecting strand of the daughter-centres of *Paramœba* is analogous, therefore, to the central spindle of *Noctiluca* and of the Metazoa.

It is a temporary stage like this in *Tetramitus* or *Paramœba*, that lends plausibility to Schaudinn's other alternative, although the reverse view may also be conceived, the diffusion of the chromatin granules taking place by reason of the disappearance through gradual degeneration of the nuclear membrane. If the latter view of the origin of diffused chromatin be accepted, then the theory of the original intra-nuclear position of the sphere, for which there are certainly a great many supporting facts, is warranted. There is considerable evidence, on the other hand, to support the argument that the division-centre was originally a distinct cytoplasmic structure, which only secondarily became connected with the nucleus (*e.g.* *Tetramitus* and *Paramœba*). Intermediate stages between the temporary stage in *Tetramitus* and *Paramœba*, and the permanent intra-nuclear division-centre, may be seen in numerous Flagellidia, such as *Chilomonas* and some species of *Trachelomonas*, where no nuclear membrane surrounds the chromatin granules. Even in these low forms, the sphere apparently exerts some force of attraction, perhaps chemotactic, upon the chromatin, and this may or may not be strong enough to keep the granules permanently aggregated. If not, the distributed nucleus may result; if so, the intra-nuclear condition of the sphere is the outcome. In *Paramœba*, *Noctiluca*, in diatoms, and in the majority of Metazoa and plants, a nuclear membrane is formed and the sphere remains in its cytoplasmic or extra-nuclear condition, and, as a cytoplasmic body, or as a kinetic substance (archoplasm or kinoplasm), may undergo further differentiations leading to the complicated mitotic figures of higher animals and plants. In the majority of cases the nuclear membrane disappears during mitosis, and the primitive conditions are thus repeated. By rupture of the membrane, the substance of the division-centre comes into direct contact with the chromatin, and is, in part at least, surrounded by it.

The facts point toward the conclusion that the centre of activity in the division of the protozoan cell, as in Metazoa, resides in a special structure, which, to avoid confusion in terminology, has been called the division-centre. In some cases this structure resembles the astral system of Metazoa, in consisting of an outer spherical mass with radiating processes (astrosphere), and an inner focal granule or granules (centrosome). The evidence further tends to show that the division-centre in Protozoa consists of a specific substance different from the chromatin and from the cytoplasm, and possessing above all other portions of the cell an active rôle in division. No conclusive evidence is forthcoming to show whether this substance is permanent in all cells, or whether it was originally nuclear or cytoplasmic in origin, although the widespread intra-nuclear condition favors the view that it originated there.

The origin of the central granule (centrosome or centriole) within the division-centre is even more uncertain; the few data at hand lead us provisionally to believe with Balbiani and Hertwig that it is derived from the chromatin. The observations, however, upon which this conclusion is based are scanty, and further research must be undertaken before we can hope for a sufficient basis of facts upon which to generalize.

SPECIAL BIBLIOGRAPHY VIII

Boveri, T. — Zellenstudien. *Jenaische Zeitschrift*, XXII. and XXIV., 1888 and 1890, and Part IV., 1901.

Brauer, Aug. — Ueber die Encystierung von Actinosphærium Eich. *Z. w. Z.*, LVIII., 1894.

Bütschli, O. — Ueber die so-genannten Central-körper der Zelle und ihre Bedeutung. *Vehr. d. Nat. Med. Ver. z. Heidelberg*, N. F., IV., pp. 535–538, 1892.

Calkins, G. N. — The Phylogenetic Significance of Certain Protozoan Nuclei. *Ann. N. Y. Acad. Sciences*, XI., 1898.

Hertwig, R. — Ueber Kerntheilung, Richtungskörperbildung und Befruchtung von Actinosphærium Eich. *Abh. d. K. bay. Akad. d. Wiss. München*, II. Kl. XIX., 1898.

Ishikawa, C. — Ueber die Kerntheilung bei Noctiluca miliaris. *Ber. Naturf. Ges. Freiburg*, VIII., 1893.

Schaudinn, F. — Ueber die Centralkorn der Heliozoën, eine Beitrag zur Centrosomenfrage. *Verh. d. deutsch. Zool. Gesell.*, 1896.

Strasburger, E. — Histologische Beiträge. IV., *Jena*, 1892.

Wilson, E. B. — The Cell in Inheritance and Development. 2d *Edition, New York*, 1900.

CHAPTER IX

SOME PROBLEMS IN THE PHYSIOLOGY OF THE PROTOZOA

"Die psychischen Vorgänge im Protistenreich sind daher die Brücke, welche die chemischen Processe in der unorganischen Natur mit dem Seelenleben der höchsten Thiere verbindet." — VERWORN.[1]

ALL animals are subject to disintegration and waste of substance through oxidation, and to reintegration and renewal of substance through the addition of new materials. Waste and renewal are usually spoken of together under the head of metabolism, and together they constitute one of the essential properties by which living matter is distinguished from non-living. When renewal of substance exceeds its waste, the phenomenon of growth results, and growth leads to reproduction. In the higher animals these various functions are distributed among many organs, but in the Protozoa they are all performed by the single cell. The simplest of living forms, these organisms naturally invite a comparison of their functions, on the one hand, with those of the higher animals, and, on the other hand, with the physical and chemical operations of inorganic nature. A modern attitude on the second of these comparisons is taken by Verworn, an ardent opponent of the old conception of a specific vital force, different from the forces of the inorganic world. "An explanatory principle," says Verworn, "can never hold good in physiology, in reference to the physical phenomena of life, that is not also applicable in chemistry and physics to lifeless nature. The assumption of a specific vital force in every form is not only wholly superfluous, but inadmissible."[2] This extreme reaction from the old vitalistic point of view appears to be somewhat premature; for, as Driesch, Whitman, Wilson, and many others have suggested, there exists in every organism a power of adaptation and certain coördinating factors by which the organism acts as an individual or a unit, notwithstanding the fact that its body is composed of a great number of chemically different substances. At the present time these coördinating factors and the power of adaptation transcend physical or chemical analysis, and raise the lowest protozoön immeasurably above inanimate objects, and perhaps justify, in a modern sense, the much-abused term "vitalism."

[1] *Psycho-Protisten Studien*, p. 211.

[2] Lee's *Verworn*, p. 46.

Among the most interesting problems suggested by the Protozoa are those relating to their apparently conscious activities. Consciousness has often been ascribed to the Protozoa in order to account for certain actions which appear to be voluntary. Thus they are often described as "selecting" their food, of "choosing" building material for their shells and tests, or of "voluntarily" moving around an object, etc. Dujardin was one of the first to discredit consciousness in the Protozoa, and, with a truly modern point of view, he wrote as follows: "If one invokes the faculty which the Protozoa have of directing themselves in the liquid, and of wilfully pursuing their prey, at least will it be necessary first to verify the reality of this faculty which I believe as fabulous as everything else reported as instinct on the part of these animalcula."[1] Nevertheless, it is generally recognized that consciousness, like life itself, could not have arisen at once in the higher animals, but must have developed by a slow process of evolution from some property of protoplasm, which, if the principle of genetic continuity involved in the doctrine of evolution holds good for the lowest forms, must be present in some form in all Protozoa. Of the many functions which make up the vital activities of Protozoa, that of irritability or response to external stimuli undoubtedly stands nearest to the basis of consciousness of the higher animals, and it may be expressed either by the general protoplasm or by specialized "sensory" ectoplasmic modifications.

The other functions of Protozoa, notably those of nutrition and excretion, can be treated with greater assurance, and can be more readily compared with similar functions in higher animals. In the present chapter we may first inquire how closely these more elementary functions agree with those of the higher forms, and then consider some of the evidence upon which consciousness has been attributed to these primitive forms.

A. Intra-cellular Digestion in Protozoa

A distinction must be made at the outset between the digestive processes of most Metazoa and of Protozoa. In the former, with some exceptions, the digestive fluids are poured out from the epithelial cells which line the digestive tract, into the lumen of that tract, and the food is digested in the stomach or intestine. In the Protozoa and in some Metazoa (*e.g.* the Cœlenterata), on the other hand, the food is taken directly into the cells and there digested. The former method of digestion is said to be inter-cellular, the latter intra-cellular.

[1] ('41), p. 111.

Only those Protozoa which take in solid food have been satisfactorily studied in this connection. The processes of digestion and assimilation in parasitic, holophytic, and saprophytic forms are in the main unknown. Among the carnivorous forms, however, some advance has been made, although it cannot be said that the digestive processes even here are fully understood. The classic objects for research in this direction have been the large predatory ciliates (*Stylonychia*, *Prorodon*, *Climacostomum*, *Cyrtostomum*, *Stentor*, etc.), the Heliozoa (*Actinophrys*, *Actinosphærium*), and the rhizopods (*Amœba*, *Polystomella*, and *Pelomyxa*). The first observations upon intra-cellular digestion in these forms were made in the last century (Corti, 1774; Goeze, 1777), when it was seen that living ciliates or flagellates, when taken into the cell-body of another protozoön, soon cease their struggling and die. The details of the process and the causes of death have been made out in the last quarter-century, and it is now safely determined that an acid secretion plays the most important part in the killing and subsequent digestion of the captives.

The length of time which an ingested protozoön can live in a gastric vacuole of another form varies, according to the organisms, from five or ten minutes to several hours. After the prey has become quiet, the digestive processes, indicated by the disruption of the body of the captive and gradual absorption of its digestible parts, go on more or less rapidly. The indigestible parts, in the form of a granular residue, are voided to the outside.

It has been determined upon pretty safe evidence that the chief and probably the main source of nutriment of the Protozoa consists of the proteid substances of the organisms taken in as food, while, with a few exceptions, carbohydrates and fats are not assimilated. Carbohydrates, in the form of ordinary starch, appear to be untouched by the body fluids of many Rhizopoda (Greenwood, '86; Meissner, '88; Fabre-Dumergue, '88); although the recent observations of Stolc ('00) indicate that in one form, at least (*Pelomyxa palustris* Greeff), starch grains are easily corroded and at least partially digested. Certain kinds of starch are more easily digested than others; rice starch is much more soluble than potato starch in the digestive fluids of *Pelomyxa*. The Infusoria seem to have a much more developed power of starch dissolution, for the grains of potato starch are all more or less disfigured (Fig. 145), resembling in this respect the partial digestion of the starch grains in the higher animals (Meissner, Fabre-Dumergue, Greenwood). The starch after solution forms a dextrin (Meissner) or an erythrodextrin (Fabre-Dumergue), but the transformation of these into glucose has not been made out.[1]

[1] "In man, according to recent investigations, starch is said to be broken up by diastase into five successive hydrolytic cleavage products, as follows: (1) Amylodextrin $(C_{22}H_{20}O_{10})_{54}$,

The emulsification of fats has never been observed, and opinions differ as to its possibility. Meissner ('88) and Greenwood ('87) asserted upon empirical grounds that it does not take place in Rhizopoda, Heliozoa, or Infusoria. Their evidence is mainly based upon the observations that the fat particles which had been taken in were thrown out unaltered after a number of days. Fabre-Dumergue ('88) and Bütschli ('83) take a less positive view, saying that the possibility of emulsification is certainly not excluded, and that probably a certain amount of the ingested fat is digested.

Chitin, cellulose, and the shells, tests, etc., of other Protozoa pass through the body plasm with little or no change. Chlorophyl also may be placed in the same category (Meissner, Le Dantec, '92), although a few well-authenticated observations show that in some

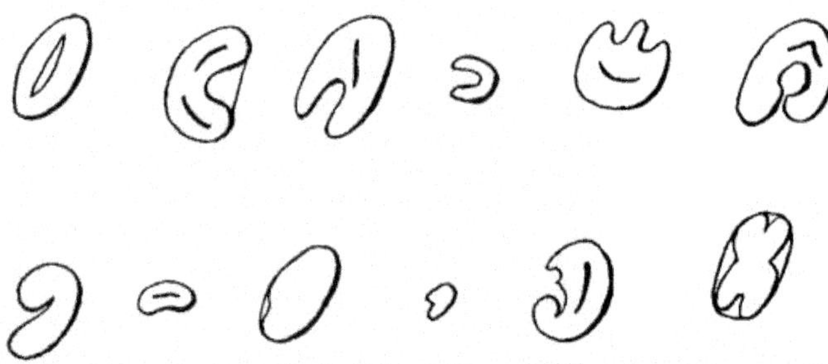

Fig. 145. — Starch grains in Ciliata, after partial digestion. [MEISSNER.]

cases, at least, chlorophyl becomes red, yellow, brown, and finally black (Perty, '52), thus indicating some change in its constitution. There is, however, no evidence to show that it becomes dissolved,[1] while the careful observations of Le Dantec ('92) show that in some cases, far from being injured by contact with the plasm, the plant-cells containing chlorophyl may actually thrive in it. Single cells of the alga which Beyerinck ('90) identified as *Chlorella vulgaris*, belonging to the order Protococcaceæ, when taken into the protoplasm of *Paramœcium bursaria*, are surrounded by a vacuole like any food particle. Soon, however, the vacuole disappears and the plant-cell is left in direct contact with the plasm, where it divides to form a layer of symbiotic algæ characteristic of this species of Paramœcium. Thus, in this case

a substance giving a deep blue color with iodine. This is next changed to (2) Erythrodextrin $(C_{12}H_{20}O_{10})_{18} + H_2O$, or $(C_{12}H_{20}O_{10})_{17} \cdot (C_{12}H_{22}O_{11})$, which is readily soluble in water, and gives with iodine a reddish brown color. Erythrodextrin is converted into (3) Achroödextrin $(C_{12}H_{20}O_{10})_6 + H_2O$, or $(C_{12}H_{20}O_{10})_5 \cdot (C_{12}H_{22}O_{11})$, which is likewise very soluble, tastes slightly sweet, but gives no coloration with iodine. Achroödextrin now breaks up into (4) Isomaltose, which through change in configuration is transformed to its isomere (5) maltose." From Howell ('97), p. 1007.

[1] Cf. Butschli, p. 1802.

at least, symbiotic forms arise through the inability of the animal protoplasm to digest the plant; either can live equally well without the other, although it is probable that there is some mutually beneficial action between the two when together. Beyond the fact of the indigestible cellulose coating, there is nothing to show why the plant-cell is not acted upon by the dissolving fluids of the body, as is the case when the same plant-cells are taken in by other Protozoa, and the subject approaches very near to the time-honored but yet unsolved problem why the stomach does not digest itself.

Considerable interest attaches to the nature of the fluid which causes the disruption and digestion of the proteids in living protoplasm of plants and animals serving as food. It has been asserted (Maupas, '83) that the protoplasm of the prey becomes a part of the protoplasm of the captor without further change, the implication being that there is no especial fluid created by the carnivorous organism, as in higher animals, to digest the food. A considerable body of evidence has grown up, however, showing that this is not true, and that a definite acid is formed, by means of which the solid food particles are disintegrated and dissolved. The method of procedure in determining this point is based upon the same principle of differential staining as that employed by cytologists in working out the chemical nature of the various parts of the cell in higher animals. The application, however, is very much more limited, for only those chemical substances can be used which by *intra-vitam* application have no deleterious effects upon the organisms. As far back as 1879, Engelmann, observing that blue litmus granules turn red and remain so in the protoplasm of some rhizopods and ciliates (*Amœba proteus*, *Paramœcium aurelia*, *Stylonychia mytilus*, and *S. pustulata*), credited the color-change to an acid in the cytoplasm. Subsequent work by numerous physiologists has fully substantiated the results obtained by Engelmann, and similar but more refined methods in the hands of Meissner, Fabre-Dumergue, Metschnikoff, Greenwood, and Le Dantec have given a firm basis for the belief that digestion of proteids in Protozoa is similar to that in the Metazoa, but beyond the fact that the dissolving fluid is a mineral acid, nothing further concerning it is known.

The bare statement of Engelmann's view of an acid cytoplasm is subject to misinterpretation. While it is undoubtedly true that an acid is present in the cytoplasm, it by no means follows that the cytoplasm itself is acid. On the contrary, it was soon demonstrated that in some forms at least (ciliates, Metschnikoff, '88, and Meissner, '88, as well as *Amœba* and *Actinosphærium*, Meissner), the cytoplasm has a decidedly alkaline reaction while the acid is confined to the fluid in the gastric vacuoles. Dujardin ('41) long before had noted that a considerable quantity of water accompanies the food particles, whatever they might be, into the protoplasm, and forms the fluid of the gastric vacuole, and later physiologists have found that this water becomes acid by gradual secretion from the surrounding protoplasm. (Fabre-Dumergue, Meissner, Metschnikoff, '89; Le Dantec, '90).[1] It thus appears, if these observations be complete, that in these cases at least the food particles never come in direct contact with the pro-

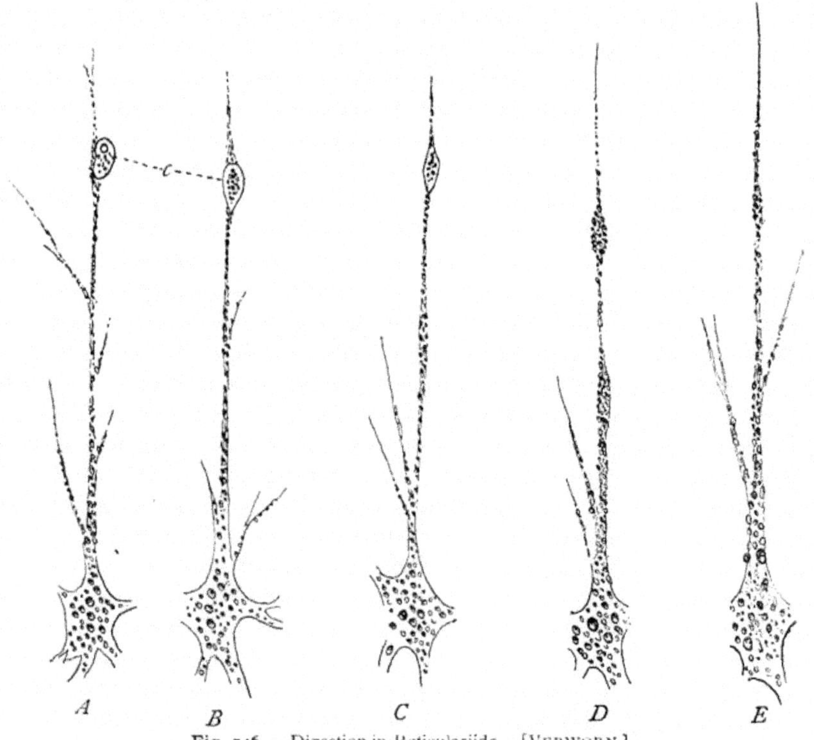

Fig. 146. — Digestion in Reticulariida. [VERWORN.]

A, B, C, D, E, successive stages in the disintegration of a ciliate (*Colpoda*) (*c*), in a pseudopodium of *Lieberkühnia*.

toplasm, but are always suspended in the liquid of the vacuole. There are, however, certain exceptions to this rule, and with these in mind, it is not yet possible to regard the subject as definitely established. Food-taking in certain marine Rhizopoda is apparently accomplished without the formation of a gastric vacuole, and the prey, possibly a small ciliate, disintegrates while in contact with the plasm, and the disintegrated parts move about in cyclosis with the endoplasmic granules (*Lieberkühnia, Gromia fluviatilis*, Fig. 146). In these cases the digestive fluids must be in any and all

[1] Cf., however, Greenwood ('94).

parts of the protoplasm, to be called out, perhaps, by the stimulus of the ingested substances, and the conclusion is obvious that the vacuole or improvised stomach in the various forms is not absolutely essential.

Greenwood ('94) has given a very careful description of the gastric vacuoles in the ciliate *Carchesium polypinum*, one of the Vorticellidæ. Here the gastric vacuoles, after leaving the mouth opening, pass downward into an area bounded by the horse-shoe-shaped macronucleus (Fig. 147), where they pass into a state of "storage" (characterized by the loss of the water taken in with the food), and the food particles are thus left, for the time being, in direct contact with the protoplasm. This stage, which may last from one to twenty hours, is eventually ended by the formation of a vacuole again, about the ingesta.[1] Preceding the period of vacuole formation there is a sudden concentration of the peripheral food particles of the ingesta into a central, solid ball. This condition—Greenwood calls it the "aggregation" stage,—is brought about, she believes, as in clotting, by the accumulation and concentration of some retractile substance. The fluid freshly secreted to form the reappearing vacuole has a decidedly acid reaction, and a powerful solvent action

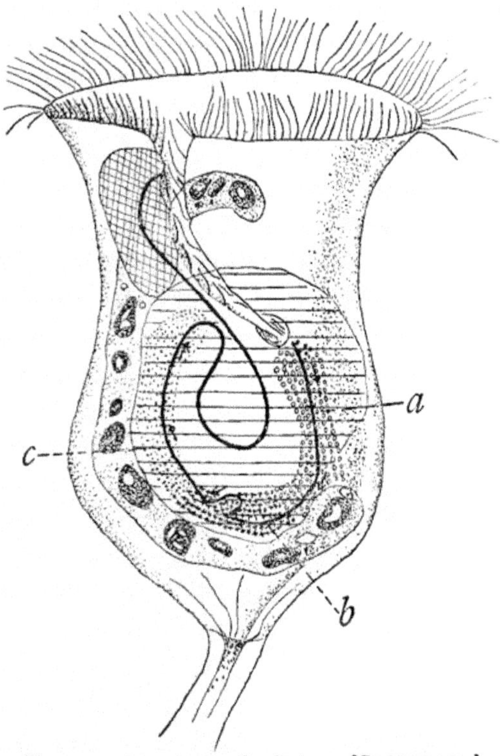

Fig. 147.—Digestion in *Carchesium*. [GREENWOOD.]
The path which the food takes is represented by dots; *a* (circular round marks) represents the position of storage; *b* (crosses) represents the position of rest; *c* (dots), the region of the later changes.

[1] Le Dantec ('95) does not believe that the vacuole disappears during the so-called quiescent phase, but is invisible merely because at this time it has the same index of refraction as the surrounding plasm, becoming visible again after diffusion of the digested parts. The strength of Le Dantec's criticism is taken away by his own observations upon *Gromia fluviatilis*, where the ingesta is digested without the formation of a vacuole.

upon proteids, so that in this secondary vacuole the real disintegration of the food substance takes place.

According to Greenwood's observations, therefore, it appears that the original vacuole does not become the digestive vacuole, but that the food particles first become free in the protoplasm, to be re-collected and digested in an acid-holding gastric vacuole.

There are no observations to indicate the changes which take place in the disintegrated food particles from the time they leave the gastric vacuole until the absorption, by intussusception, of the nutritive parts contained in them. Most observers, however, are agreed that there must be a sort of chyme formed which mixes with the protoplasmic fluids and is absorbed by them.

Closely connected with the metabolism of the protozoön cell, but as yet of unknown origin, are the so-called "excretory granules" (*Assimilationskörperchen* of Rhumbler, *Excretkörner* of Bütschli, Schewiakoff, *corpuscules biréfringents* of Maupas, etc.). The wide distribution of these granules, or crystals, their formation and disappearance under varying conditions of the organisms, make it probable that they play an important part in the physiology of the Protozoa. Whether, however, they represent a final stage in the processes of digestion, or represent the products of katabolic metabolism, has not been satisfactorily made out.

The granules in question, with or without a definite crystalline form, have been described in almost every class of Protozoa. In shelled and in naked fresh-water rhizopods (Auerbach, '55; Carter, '64; Lankester,'79; F. E. Schultze,'75; Maupas,'83; Schewiakoff,'88), in Heliozoa (Hertwig & Lesser, '79; Maupas,'83), in Flagellidia (Bütschli, '78, '83), and in Infusoria (Maupas, '83; Stein, '59; Wrzesniowski, '79; Rhumbler, '88; Schewiakoff '94), they occur in varying numbers and positions. In the Ciliata, where they have been most thoroughly studied, they may lie well distributed about the endoplasm (Fig. 148), or may be concentrated at the two ends of the animal in the vicinity of the contractile vacuoles (*Paramæcium;* cf. Schewiakoff, '94). When distributed about the body, they lie in vacuoles which move freely with the protoplasmic flow. They vary considerably in form (*C*), but, as a rule, have a crystalline appearance, while the larger granules are striated by radial or parallel lines, indicating, Schewiakoff believes, the coalescence of needle-form crystals. They vary in size according to the degree of aggregation, but measure, on the average, from 0.003 to 0.014 mm. in length (Schewiakoff).

Numerous suggestions as to the significance of these crystals, and a few valuable experiments to determine their chemical composition, have been made. From analogy with other animals it was early supposed that they represent concretions which correspond to uric acid crystals in Metazoa (Stein, '59; Wrzesniowski, '79; Entz, '79; Maupas, '83; Rhumbler, '88), while from the form of the crystals and chemical reactions, Bütschli[1] at first regarded them as crystals of oxalic acid, but later[2] was content to regard them as the main end product of the metabolism of proteids in the body.

While numerous observers have experimented in various ways to determine the chemical nature of these questionable bodies, the most convincing results have been obtained by Schewiakoff ('94). After repeated experiments, many of which only confirmed the earlier con-

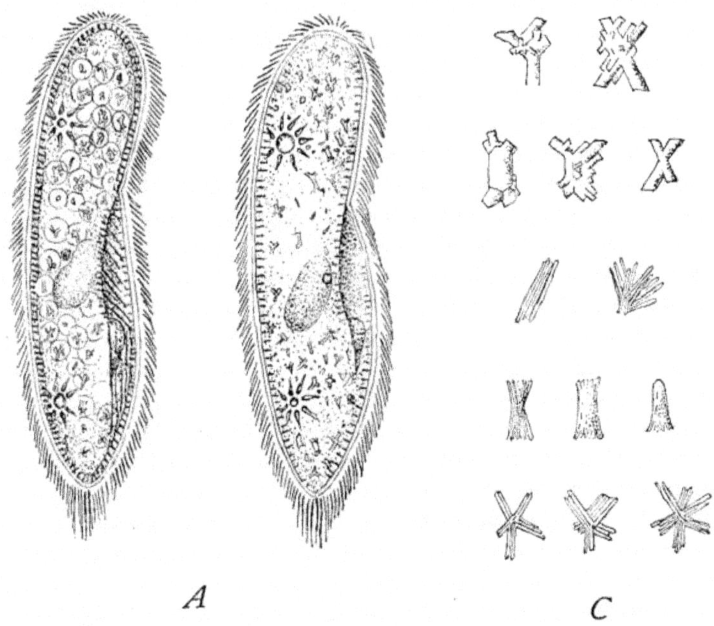

Fig. 148. — Excretory granules in *Paramæcium*. [SCHEWIAKOFF.]
C. The isolated crystals.

clusions of others, he obtained the following results: (1) the crystals are insoluble, in the ordinary sense, in water; (2) slightly soluble in concentrated acetic acid and in dilute ammonia; (3) more soluble in solutions of different salts, weak acetic acid, and ammonia; (4) easily soluble in mineral acids and alkalies; (5) insoluble in alcohol, ether, and carbon bisulphate; (6) negative results with staining agents showed that they can be neither albuminate nor carbohydrate in nature; (7) reactions to osmic acid, alcohol, and ether excluded the possibility of fats; (8) delicate tests showed that the crystals were composed of an unorganized substance; (9) final tests showed this

[1] Protozoa, p. 103.

[2] *Loc. cit.*, p. 1484 ('88).

to be a calcium salt, which was determined as calcium orthophosphate ($Ca_2H_2(PO_4)_2$).[1]

While there are many chances for error in Schewiakoff's work, it is probable that he has come very near to the correct interpretation of these bodies. Their origin, however, as well as their significance, remains in doubt. Their disposal also has not been satisfactorily accounted for. Stein reported their defecation with the undigested remains through the anus, but Entz, Maupas, and Schewiakoff believe, apparently on justifiable grounds, that they are dissolved and pass to the outside through the contractile vacuole.

B. Respiration

The assumption that Protozoa take in oxygen and liberate carbon dioxid rests almost entirely upon indirect evidence, which, however, is so strong as to leave little reason to doubt the validity of the assumption. An infusorian, for example, moving rapidly day and night during its entire life, and eating constantly throughout this period, must undergo continual waste in the liberation of energy by combustion. A constant supply of oxygen and a constant excretion of the waste products of combustion appear to be equally necessary for the continuance of this activity. With remarkable intuition, Spallanzani (1776) suggested the contractile vacuole as the organ of respiration by means of which the waste matters are thrown out, and later observers have offered no evidence of value to disprove the suggestion. In a ciliate, for example, the volume of a contractile vacuole at complete diastole is about one-tenth of the volume of the animal itself, and, contracting every two or three minutes, the vacuole must in half an hour expel to the outside a volume of water equal to that of the entire animal. The oxygen-laden water which is thus expelled must have entered the body through the mouth opening or by osmosis through the body walls. The important rôle which respiration plays in the physiology of the Protozoa, and the agency of the contractile vacuole in this process, was first clearly recognized by Schwalbe ('66) and Zenker ('66), then by Wrzesniowski ('69), and later by Rossbach ('72), Bütschli ('77), Limbach ('80), Maupas ('83), and Fiszer ('85). Before the period of Schwalbe and Zenker, the contractile vacuole had been interpreted as a heart and the centre of a circulatory system (Corti, 1774; Gleichen, 1778; Wiegmann, '35; Siebold, '48), etc., and Pouchet ('64) went so far as to describe colored blood pumped by the vacuole throughout the body. This view, which now

[1] "Calcium is by far the most abundant metallic element in the body. . . . It is found in all the cells and fluids of the body, probably loosely combined with proteid." — Howell, *loc. cit.* p. 967.

has but an historic interest, has recently been rather feebly advocated by Greeff ('91) and Pénard ('90), who adopted it without the supposed justification which the older naturalists had in comparing the contractile vacuole of Protozoa with the water vascular system of the flatworms. Until Leydig ('57) demonstrated the excretory function of the water vascular system, it was supposed that the flatworms obtained oxygen from a stream of water taken in by the vascular system from the outside, in a manner analogous to the air supply from the tracheæ of insects, and Schmidt ('67), following Dujardin, and followed by Balbiani ('60, '61) and Maupas ('79), attempted to explain respiration in Protozoa in the same manner. At the present time, while the probability is very strong that the contractile vacuole expels water in which the oxygen has been replaced by carbon dioxid, there have been singularly few actual observations to confirm it. Certes's ('85) experiments show that some alteration takes place in the water after its entrance into the protoplasm. This was demonstrated by placing Infusoria in water colored by dissolved aniline dyes; the water of the contractile vacuole remained clear and colorless, although the surrounding medium was intensely colored. The only direct observations of the presence of carbon dioxid was made by Brandt ('81) upon *Amœba*. Placing these organisms in a medium colored by dilute hæmatoxylin, he found that the water of the vacuole became yellow and then red, thus showing the characteristic reaction of hæmatoxylin in the presence of an acid. The observation is not conclusive, however, for the presence of uric acid might also give this reaction.

Numerous attempts have been made to describe the series of events which lead up to, and cause, the contraction of the vacuole. Being entirely hypothetical, they may be dismissed with a brief mention. Schwalbe, Rossbach, Engelmann, Maupas, and many others explained the bursting as due to the contractility of the protoplasm. Schwalbe attempted to trace the impulse or stimulus of contraction to the products of destructive metabolism, which become stored up in the vacuole so that the latter, when full, presses upon the protoplasm and causes it to contract. Rossbach ('72) more obscurely attempted to trace the stimulus to the chemical change which takes place at the moment of oxidation. Each oxidation forms an oxidation product which, as soon as formed, incites the stimulus. Zenker ('66) was more clear in describing the process as due to the attraction of protoplasm for oxygen-holding water, and repulsion for water without oxygen, the result being that when the oxygen is removed from the imbibed water, the latter is expelled from the body. Rhumbler ('98) gave a similar interpretation and adduced experiments with inorganic fluids simulating the contractile vacuole. Bütschli ('83) regarded the systole as due to a simple physical force,—surface tension. The liquid of the vacuole mixes quickly with the surrounding water, as a small drop fuses with a larger mass, as soon as the intervening layer of protoplasm gives way before the pressure of the growing vacuole. Delage and Hérouard ('96), on the other hand, held that contractility of the protoplasm brings about the contraction just as it causes the expulsion of undigested food matters.

When present, carbon dioxid is probably dissolved in the water of the contractile vacuole, although, in some cases, especially among the Sarcodina, gas vacuoles containing, probably, carbon dioxid have been repeatedly observed, first by Perty ('49) in *Arcella*, and subsequently by Bütschli ('74), Engelmann ('78), Entz ('78), and others, not only in *Arcella*, but in other rhizopods as well. According to Engelmann, the gas is secreted very rapidly, but at irregular intervals, and he, with other modern observers, accepted and confirmed the suggestion made by Perty, that the gas vacuoles serve as an hydrostatic apparatus, by means of which the organisms can raise and lower themselves in the water.

Although the contractile vacuole appears to play an important rôle in respiration, it is not absolutely necessary for the performance of this function, for a great many forms have no such organ. The marine rhizopods and Radiolaria, and some of the fresh-water forms of Rhizopoda (*e.g. Pelomyxa*), have no contractile vacuoles. Respiration in such cases must take place by osmosis. An interesting series of forms are the Opalinidæ, parasitic ciliates of which some genera have a contractile vacuole with numerous feeding canals (*Anoplophrya*, *Hoplitophrya*), while others have no contractile vacuole, although the canals are present (*Opalina*, according to Fabre-Dumergue). In none of these forms is there a mouth, and respiration must take place by osmosis.

There is, on the whole, very little direct evidence to support the conclusion which on *a priori* grounds appears indisputable, that, like other organisms, the Protozoa take in oxygen and give off carbon dioxid. The fragmentary evidence which we have tends to the conclusion that, when present, the contractile vacuole, probably in addition to other excretory functions, is the active agent in the disposal of carbon dioxid, while the income of oxygen-holding water takes place by osmosis through the body walls, by ingulfing through the mouth, or by both methods.

C. Secretion and Excretion

Carbon dioxid is but one of the waste matters formed by the decomposition of proteids in vital activities. In the higher animals the final products are carbon dioxid and urea ($CO(NH_2)_2$), and it has long been assumed upon *a priori* grounds that a similar result follows combustion in Protozoa. Again, there have been but few observations to confirm this supposition. The contractile vacuole was regarded as an excretory vesicle (*Urinblase*) by Boeck ('47), Rood ('53), Stein ('56), Leydig ('57), Kölliker ('64), and more recently as an excretory organ by Maupas ('83), Rhumbler ('88), Griffiths ('89), Schewiakoff ('94), Delage and Hérouard ('96), besides many others. Bütschli[1] believed that it is a pure hypothesis to assume that the vacuole has an excretory function other than that of respiration, but Maupas ('83) insisted upon the physiological necessity of such an excretory organ, and cited as an argument the presence of contractile vacuoles in vegetable zoöspores, which, having chlorophyl, can presumably make use of all the carbon dioxid formed, and which, therefore, probably make use of the vacuole for secretion. Maupas's argument is offset by the fact of numerous Protozoa which have no contractile vacuoles, and it follows that if these can get rid of their waste organic matters by osmosis, it is quite possible that forms with vacuoles can do the same. Entz ('88) held that the crystals occasionally found in the vacuoles and reservoirs of different forms are uric acid (*Harnconcremente*), a supposition which was supported with direct evidence by Griffiths ('89). The latter determined the presence of uric acid in several different types of Protozoa, including the rhizopod *Amœba* and the ciliates *Paramœcium* and *Vorticella*. A number of animals were placed on a slide under a cover-glass, and killed with alcohol followed by nitric acid. The slide was then gently warmed, and ammonia was introduced. When the experiments were successful, a number of purple prismatic crystals of murexide appeared in the contractile vacuoles, showing that uric acid had been present. These results were repeatedly obtained, although the experiments were not always successful, showing, Griffiths says, that the vacuole may have some other functions besides secretion. Until this interesting series of experiments is confirmed, however, Griffiths's results must be inconclusive. If they are confirmed, on the other hand, the following reflection is warranted, and has a singular interest in the present-day problems of biology: "Through all the multitudinous changes," says Griffiths, "that have taken place during the lapse of ages in the development of the mammalian kidney, we find that the physiological functions are the same as occur in its original or primitive form, as represented in the Protozoa."[2]

The secretion from the protoplasmic body of definite particles of matter, which may or may not have been at some time a part of the animal protoplasm, and which play some further part in the life activi-

[1] ('88), p. 1452. [2] *Loc. cit.*, p. 135.

ties, is definitely established in a number of cases. The materials thus secreted vary in nature from purely inorganic solids, like calcium carbonate, silica, etc., to chitin, cellulose, fats, and jelly-like protoplasmic products. The simplest cases of secretion are seen in those Mastigophora and Sarcodina where the outer protoplasm becomes gelatinous, to form the jelly-like mantles of different types (many Flagellidia, Heliozoa, Radiolaria). In *Amœba* there is a secretion of such a substance which aids the animal in securing food by sticking it to larger objects, as well as by ensnaring the prey (Rhumbler, Verworn, Hofer). In *Euglena*, according to Klebs ('86), the protoplasm throws out a slimy mantle when the surrounding conditions are unsuitable. This mantle at first is not homogeneous, but is in the form of minute gelatinous threads which arise beneath the cuticle from the outer protoplasmic layer of the body. A network is then formed between the threads, which finally unite to form a homogeneous mantle about the animal. An identical process has been described by Schewiakoff ('94) and by Siedlecki ('99), in the movements of certain gregarines,[1] where a gelatinous layer beneath the membrane secretes filaments of slime-like material which harden outside the body.

In other instances the secreted material is in the form of granules which unite outside the body to form stalks or houses, or even shells. In many such cases the granules have not been a part of the body protoplasm, although they may have been created there. Such, for example, are the lime shells of the Reticulariida, or the silicious and acanthin skeletons of the Radiolaria. In other cases they appear to be a growth product of the organism, as in the branched stalks of many Flagellidia, although even here foreign particles may be used for this end. Thus, in freshly formed *Anthophysa* stalks, Kent and Bütschli observed that the excreted material was granular, and Ehrenberg had already noticed that, if *Anthophysa* colonies are fed with indigo, the colored particles collect at the base of the animal, while Kent ('81), repeating the experiment, observed that the granules were actually deposited to form a part of the new stalk material. The materials for the shells of Rhizopoda may be foreign particles analogous to the indigo granules, or they may be the result of chemical activity of the protoplasm. The various shells of *Difflugia*, *Centropyxis*, *Cyphoderia*, *Lithicolla*, etc., are examples of the first type, while *Euglypha*, *Quadrula*, most Heliozoa, Reticulariida, and Radiolaria are examples of the second.

In the Sarcodina, where the shells, as in Reticulariida, are formed by deposition of calcium carbonate, it has been shown by Carpenter, Kölliker, Wallich, and especially by Dreyer ('92), that the material

[1] Cf. p. 149.

is deposited from a well-defined portion of the protoplasm (*chitosarc*, Wallich), and not upon the outside of the body, but between two chitinous lamellæ situated in the ectoplasm (Fig. 149). The manufacture of the shell material of the Mollusca was explained by Steinmann as a purely chemical phenomenon, and not as an expression of vital activity. The early experiments of Harting ('73) in allowing carbonic acid alkalies to act upon albumin or other nitrogenous substances, thereby obtaining a precipitate of calcium carbonate in the form of granules similar to Coccoliths, gave Steinmann the clue to his theory that calcium chloride and other salts acting upon albumin in animal protoplasm give a similar precipitate. Applying this theory to the marine rhizopods, Dreyer argued that the protoplasmic body is saturated with dissolved calcium salts, and that albuminoid stuffs secreted by the living animal undergo fermentation through the agency of bacteria, and ammonium carbonate is formed, which, acting upon the calcium salts, results in the formation of calcium carbonate.[1] A similar explanation can be given for the diverse skeletons of the Radiolaria, the precipitate here collecting in the interstices of the protoplasmic alveoli, thus giving rise, first to the four-rayed type of spicule, and then to the compact skeletons (Fig. 41, p. 77).

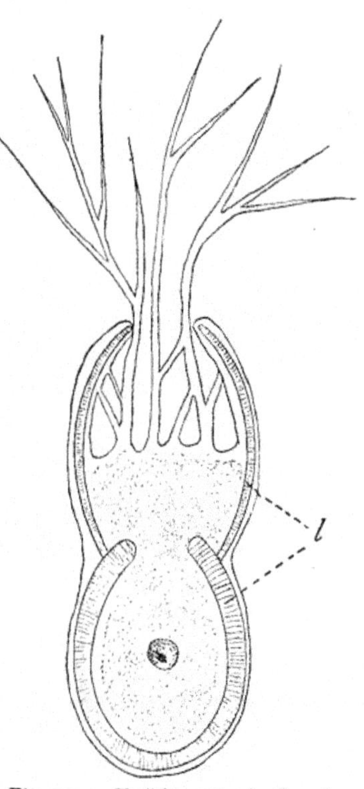

Fig. 149. — Shell formation in *Gromia fluviatilis* Duj. [DREYER.] *l*, the deposit between two lamella.

Among the most interesting materials secreted by the Protozoa are those which give rise, through rapid oxidation, to phosphorescence. A great many forms have this power, especially among the Dinoflagellidia and the Cystoflagellidia; and although the cause is not fully established, yet it is generally agreed, at the present time, that the light is due to the rapid oxidation of certain products of metabolism. One of the most noteworthy illustrations of phosphorescence is shown by *Noctiluca miliaris*, in which Quatrefages ('50) found that the light,

[1] Cf. Dreyer, p. 224.

which appears homogeneous under low magnification, is, in reality, caused by a vast number of minute light centres. These are closely grouped together in the centre of the light area, but are easily distinguished at the edges (Fig. 150). Radziszewski ('80) found that a number of different substances which are present in living organisms have this power of phosphorescence when in an alkaline medium.

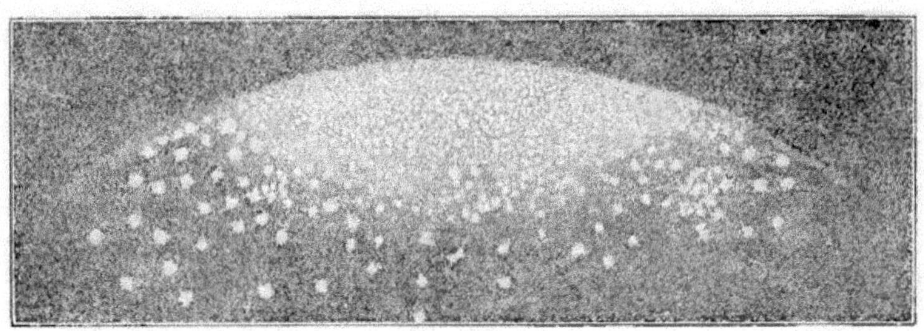

Fig. 150. — Phosphorescence in *Noctiluca miliaris* Sur. [QUATREFAGES.] A portion of the body is represented with numerous scintillating dots.

Amongst them are fat, lecithin, cholestearin, certain oils, grape sugar, etc. Watasé ('98) regarded the protrusion or secretion of such substances as due to protoplasmic contraction induced by any external stimulus, such as a blow or shock of any kind.

D. Irritability

When, by reason of an external stimulus, any normal movement of a protozoön is interrupted, or changed to another type of motion (which may be definite or indefinite in regard to the source of the stimulus), the change is brought about by protoplasmic reaction, which is expressed by the general term *irritability*.

The effect of heat upon the activities of Protozoa was early recognized by the students of the group. Spallanzani, for example, noted that increased warmth brought about increased activity in the contractile vacuole, as well as in movements of the animal. Subsequent observers have established the fact that movement is possible only within certain temperatures, and, further, that this limit in either direction varies with the organism considered. Between such limits, the amœboid movements of Rhizopoda and the vibratile movements of Flagellidia and Ciliata increase to a certain optimum, which again varies with the organism, but after this optimum is passed, movement gradually decreases until, at the point of maximum temperature, heat rigor sets in; and this temperature, if passed, results in death. On the other hand, with cooling, the movements become slower and slower, until a minimum temperature is reached and all motion ceases, after which, with further decrease in temperature, cold rigor and death ensue. In spite of these limitations, however, the Protozoa have the power of adaptation to high temperatures on the one hand, and low temperatures on the other. Ehrenberg cites the presence of *Nassula*, *Enchelys*, and *Amphileptus* in the waters of the Ischian hot springs, which reach a temperature far above the normal for most Ciliata and Flagellidia, while the presence of *Hæmatococcus* on high mountains and in polar regions, or of various Flagellidia in frozen pools, is common knowledge. The maximum temperature for common Flagellidia varies usually between 40° and 50° C. (Schultze, '63; Strasburger, '78; Klebs, '83), although Dallinger cites a number of common forms which live until 60° is reached, while the spores of these forms withstand temperatures far beyond the boiling point. With Rhizopoda, the maximum temperature seems to be somewhat less than with Flagellidia. Kühne ('64) found the maximum for *Amœba diffluens* and *Actinosphærium* at 45° C., and similar results were obtained by Schultze ('63) and Verworn ('89).

The Ciliata also have a generally lower maximum temperature than do the Flagellidia, although important exceptions are found where, by adaptation to new conditions, the organisms are able to withstand a considerable temperature. The common forms stand a temperature of from 38° to 42° C., and the activity of the cilia resembles that of ciliated epithelium described by Engelmann ('79). According to Rossbach's description of the increase of heat upon the activity of the ciliate *Stylonychia*, 4° marks the lower limit of motion when the animal is nearly at rest, above 25° the movements become violent, although still spontaneous and normal, while from 30° to 35° the motion becomes uncertain and the animals lose their power of direction, a condition which gives place to a violent rotation about the axis until death finally ensues. In connection with the general stimulation of the body by heat, there is frequently a definite movement of the organism toward or away from the source of the stimulus (*thermotaxis*). This seems to have been first observed by Stahl ('84) in connection with the plasmodia of *Æthalium septicum*. He placed a strip of filter paper, upon which a plasmodium had flowed, between two beakers in such a way that one end of the plasmodium lay in cold water (7°), while the other lay in warm water (30°). The streaming of the plasm began at once toward the warmer water until the plasmodium was completely immersed, although previously the movement had been in the opposite direction. Verworn ('89) observed a similar thermotactic phenomenon in *Amœba limax*, while, more recently, Mendelssohn ('95) has shown that, in the Ciliata, *Paramœcium* seeks warmer water in temperatures below 24° to 28°, while in temperatures above this the reverse reaction takes place.

Light as well as heat rays frequently have a similar directive effect upon Protozoa, a phenomenon called *phototaxis* by Strasburger, which is explained as follows by Verworn: "A ray of light extends through space from a source of light, in a straight direction, and diminishes in intensity with the distance, hence any two points in the line of the ray possess different intensities; the point that is nearer the source has the greater, that which is farther away has the less, intensity. A ray of light, therefore, fulfils very completely the conditions that are necessary to the appearance of unilateral stimulation; in fact, it is extremely difficult to establish conditions under which an organism is stimulated by light uniformly on all sides. As a result of this, stimulation by light calls out very pronounced directive effects."[1] A difference of opinion exists, however, as to the general effect of light upon organisms; Klebs ('85), on the one hand, asserted that all protoplasm is sensitive to light, while Verworn, after trying in vain to get reactions on the part of certain organisms, concluded that such is not the case. It appears from numerous investigations that the most phototactic forms are the flagellated cells, Strasburger ('78) observing that swarm-spores of different plants are attracted toward a light of a certain intensity, and move away from a light of greater intensity, while he and Cohn ('64) determined the fact that rays having a short wave-length, especially the blue and the violet, are more effective than those of the longer wave-length, such as the red. Engelmann ('82) saw *Euglena* gather in heaps along the region of the line *F* of a microspectrum, and he also established the fact that colorless forms, such as *Chilomonas paramæcium*, or the colorless swarm-spores of the Chytridiæ, are positively phototactic in weak, and negatively phototactic in strong, light.

Among the Infusoria, light reactions are much less marked than among the Mastigophora, and light seems to be quite unnecessary for their existence.[2] Nevertheless, in one form, at least, positive reactions have been demonstrated by Verworn ('89), who showed that *Pleuronema chrysalis* reacts vigorously to light stimuli through cobalt glass, while light through red glass gives no effects, an experiment which shows that it is not the longer but the shorter light vibrations that are effective.

The results with Rhizopoda have been somewhat more definite

[1] Lee, 1900, p. 447.

[2] The slight effect of light upon Ciliata is shown by the following table from Maupas ('88). The organisms were kept for one month in the light and one month in darkness, and during that time they increased by division as follows: —

Colpidium colpoda	in darkness		48	times;	in	light	46
Glaucoma scintillans	"	"	98	"	"	"	99
Paramœcium bursaria	"	"	9	"	"	"	9
Stylonychia pustulata	"	"	48	"	"	"	50

than with the Ciliata. *Pelomyxa palustris*, according to Engelmann ('79), moves energetically in darkness, but if a strong light be suddenly thrown upon it, a sudden contraction results and the organism rounds out into a spherical mass; but if light gradually increases, there is no response. Rhumbler ('98) came to a somewhat similar conclusion in regard to *Amœba proteus*, holding that food-taking is more energetic in the night than during the day, and is not frequently observed because the intense light of the microscope mirror causes sluggishness. The experiments of Leaming and Harrington ('99) upon the effect of light of different colors shows that *Amœba proteus* reacts vigorously in red lights, but protoplasmic movements almost cease in rays from the violet end of the spectrum.

Living protoplasm can also, within a certain limit, accommodate itself to chemical changes in the surrounding medium. Thus, *Amœbæ* may become accustomed to a 4 per cent solution of common salt, if it be slowly added, while a 1 per cent solution, if added suddenly, will kill them. Heliozoa can be transferred from salt water to fresh, and *vice versa*. *Æthalium septicum* will live in a 2 per cent solution of sugar if slowly added, and many other instances might be given. Such adaptation on the part of the organism signifies a gradual change in the chemical or physical make-up of its own substance, and in a variety of cases the rapidity of the change varies inversely as the distance from the source, hence as the intensity, of the stimulus. If a stimulus, therefore, comes from a certain direction, the effect, like that of a light stimulus, is a unilateral reaction, to which the general term *chemotaxis* has been given (Verworn). The greatest variety of substances may give these reactions, water, air, and other gases, alkalies, and acids of various kinds. Pfeffer ('88) found that, in various Flagellidia, including *Cryptomonas*, several species of *Bodo*, *Monas guttula*, *Trepomonas*, *Polytoma*, *Euglena*, etc., different substances cause very different reactions, — substances which cause positive reactions in some forms causing negative reactions in others. He also obtained the interesting result that substances which in weak solutions cause a positive reaction, in strong solutions cause a negative one, so that, as in light stimuli, a chemotactic optimum exists toward which the organism constantly strives. In Rhizopoda, although fewer instances of chemotaxis are known, the same general results have been obtained. Stahl's often cited experiments on the mycetozoan *Æthalium septicum* showed that this organism is positively chemotactic toward both harmful and beneficial stuffs. In all cases of harmful action upon the Rhizopoda, the reaction is expressed by the withdrawal of the pseudopodia, rounding out of the body, and final disintegration. With the Ciliata, according to Pfeffer, chemotactic reactions appear to be less delicate than with the Mastigophora.

Oxygen, as with most organisms, exercises a chemotactic effect, although the positive are much more rare than negative effects (Verworn and Jennings). Jennings, in a series of very careful experiments, found that alkalies give clearly marked negative chemotaxis, while substances "in which *Paramœcia* collect, giving the motor reaction only when they attempt to pass out of them, are substances having a weak acid reaction" ('99). The relative effect of the same stimulus upon different organisms is prettily shown by Massart's ('91) experiments upon *Spirillum* and *Anophrys*, in the presence of oxygen. Both organisms collect in a ring around a bubble of air, the ciliate nearer the source, the bacillus away from it. Carbon dioxid is said to give the same reaction as dilute acids, causing the aggregation of *Paramœcium* about it (Jennings), although other observers have been unable, thus far, to confirm these experiments.

The singularly interesting investigations which Pfeffer ('84, '88) made upon the chemotaxis of reproductive elements have thrown no little light upon the problems of fertilization, and have shown at the same time how minute a quantity of the stimulant is needed to give the required effect, too much causing a negative result. He found that antherozoids of the fern are attracted to the capillary tube, which holds a slight trace of malic acid, while a larger quantity turns them away. Carrying the experiment to other forms of life, it is probable that analogies to this may be found in almost all kinds of organisms, both animal and plant, and that positive chemotaxis is a necessary condition of the fertilization of many eggs by spermatocytes.

Jennings's ('99) interesting experiments and observations upon various forms of Infusoria and Mastigophora show that many hitherto supposedly directive chemical stimuli are effective, not because of their unilateral stimulation through the differences in density, but because they induce a natural *motor response* on the part of the organism. To illustrate with the case of *Paramœcium*, which affords the most satisfactory basis for this view, Jennings maintains that these organisms, which for a long time have been known to have an apparent affinity for weak acids, are not *attracted* by a drop of acid placed in the vessel which contains them, but wander into it in their aimless movements. The usual motor response of *Paramœcium* is to swim backward from the object struck, turn toward its side containing the peristome, and swim forward again. By this manœuvre an ordinary obstacle is avoided. The acid does not cause this reaction, but when the organism attempts to leave the drop, contact with the surrounding walls brings about the motor response, and the organism thus remains entrapped (Fig. 151). While this ingenious view is acceptable in the case of *Paramœcium*, it by no means follows that all so-called chemotropic or chemotactic effects can be similarly explained, and that chance brings the organism in the sphere of influence of the stimulant. On the other hand, it is equally probable that many of the so-called directive effects may be explained by this assumption, and extreme caution is needed in interpreting all such reactions. Garry ('00) has applied the distinction, first suggested by Loeb ('93), to be used in interpreting the responses of organisms to stimuli; if a directive effect is produced (that is, if the organism is distinctly oriented in respect to the centre of the stimulus), the effect is said to be chemotropic, but if an effect is produced which is not directive, it is said to be chemokinetic (*Unterschiedsempfindlich*, Loeb). According to Jennings, the reaction of *Paramœcium* would be neither chemokinetic nor equivalent to the unilateral stimuli induced by light or heat (phototaxis and thermotaxis).

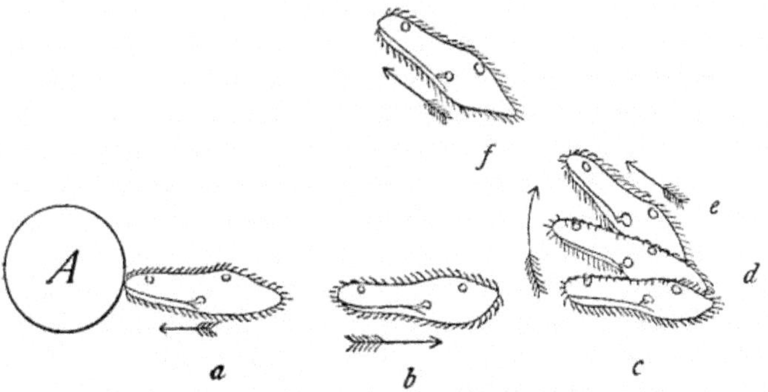

Fig. 151.—Motor response in *Paramœcium*. [JENNINGS.]
a–f. Successive positions after meeting with an obstruction *A*.

The various instances which have been described of positive and negative chemotaxis must be reinvestigated in the light of Jennings's conclusions, for it is obvious that the mere collection of organisms in a substance does not necessarily indicate the directive effect of that substance. Nevertheless, in some cases such an effect appears to be definitely established. Garry showed that certain organic acids have a positive directive effect upon the flagellate *Chilomonas paramœcium*, while various other acids, alkalies, and salts have merely a chemokinetic effect. He also obtained results which seem to indicate that it is the ions dissociated in certain solutions which are capable of stimulating the organism. These observations open out an unexplored field for investigation and promise extremely interesting results.

Irritability on the part of the Protozoa, in response to mechanical stimuli, has long been known. Rösel v. Rosenhof (1755) observed that *Amœba* becomes globular upon being shaken, an observation which was followed by numerous experiments by De Bary on Mycetozoa, by Haeckel ('70) on Rhizopoda and Radiolaria, and by Verworn ('89), who demonstrated on a number of Rhizopoda that stimuli of different grades of intensity cause different degrees of reaction, and that a stimulus applied at the end of a pseudopodium causes only a local irritation expressed by contraction at that point, and he expressed the belief that a regular graduated scale of irritability could be established for this group. Many forms show almost no reaction to slight stimuli, others a distinct contraction.

Among the Mastigophora and Infusoria, not only is the body more easily affected by mechanical stimuli, but transmission of the stimulus is far more rapid than in the Sarcodina. Impact of a ciliate with any object, possibly an organism much smaller than itself, causes a vigorous reaction expressed by sudden change of motion (Flagellidia, hypotrichous and holotrichous Ciliata), by contraction of myonemes (*Stentor*, Vorticellidæ, etc.), or by body contractions (*Euglena*, *Spirostomum*, etc.), all of which may be expressed by Jennings's term *motor response*.

A general result of mechanical stimulation is a motor response followed by the tendency to turn away from the source, and the general reaction, whether positive or negative, since it deals with the question of pressure in some form or other, is called *barotaxis* (Verworn).

In all forms of barotaxis, *e.g.* in *thigmotaxis*, the reactions may be either positive or negative. Negative thigmotaxis, the ordinary reaction to mechanical stimuli, such as pricking or crushing a local area, is expressed by movement away from the source of irritation. Positive thigmotaxis is expressed by movement toward the source of the stimulus. The latter is commonly seen in the attachment of pseudopodia of *Amœba* and other fresh-water Rhizopoda to solid bodies, and it affords an explanation of the ingulfing of solid bodies (*e.g.* food particles). Dewitz ('86) found positive thigmotaxis on the part of spermatozoa of the cockroach, a discovery of the greatest interest in view, again, of the question of fertilization.

The reactions to galvanic currents are, as a rule, quite definite, and as in iron filings over a magnet, there is usually a definite polar arrangement of the single-celled organisms, a reaction expressed by the term *galvanotaxis*. Pearl ('00), examining the reactions to galvanic stimuli in several ciliates and in the flagellate *Chilomonas paramœcium* in the light of Jennings's theory of motor responses, came to the conclusion that the usual reaction is distinctly affected by the current so that a "forced movement" in many cases is superimposed upon the normal motor response. The forced movement in ciliates, he states, is due to the action of certain cilia which are constrained by the current to lie in certain definite positions, — those on the cathode surface pointing toward the anterior end, those on the anode surface toward the posterior end.[1] In some forms, *e.g. Paramœcium* and the majority of Ciliata, and different species of *Amœba*, making the current results in active and direct movements toward the kathodic electrode; in others, *e.g.* the flagellate *Polytoma*, the reverse reaction takes place, the organisms swimming vigorously toward the anode, and still a third reaction may be observed in *Spirostomum ambiguum*, which turns at right angles to the direction of the current and remains so (Verworn, '89).

From the various experiments which have been made upon Protozoa, it appears that more or less similar motor reactions follow different kinds of stimuli, and that the organism reacts similarly to the same stimulus. With each type of stimulus there is, in most cases, an optimum of intensity below which a positive reaction is observed, but which, if exceeded, is followed by a negative reaction, while over-stimulation results in no response or in death. Furthermore, the experiments show that each type of organism behaves in its own way, a given stimulus violently affecting some species, while others are unaffected, although different individuals of the same species always react similarly. The directive effects of certain stimuli through operation in certain lines or on certain sides giving rise to unequal stimulation of the protoplasm, induce reactions which again are invariable. The thermotactic response of *Amœba*, or the phototactic reactions of *Euglena*, can be predicted with almost as much certainty as the reaction of mercury to heat. In some instances, notably in the more complicated forms, it appears that the organism as a whole is endowed with a set of motor responses which might be identified as instinctive. Stimulation at one point induces not a local response, as in Rhizopoda, but a reaction of the entire organism, which is poorly explained by the assumption of a machine-like organization of the cell, or by the statement that these responses are merely the expression of chemical and physical forces.

E. General Considerations

All of the normal functions of the protozoan cell show themselves second in complexity only to the analogous processes which take place in the Metazoa, and, as with the latter, many of them still remain unexplained. One point, nevertheless, has been very thoroughly demonstrated, *viz.* the importance of the nucleus in the maintenance of many of these functions. Brandt ('77) was the first to observe that non-nucleated fragments of *Actinosphærium* will die, while nucleated parts will live, and his observations were made the

[1] *Loc. cit.*, p. 123.

basis of a long series of experiments by Nussbaum ('84, '85), Gruber ('86), Balbiani ('88), Hofer ('89), and Verworn ('88, '89, '91), which have led to fruitful results. Amongst these, the most striking and suggestive fact is, that without the nucleus, the process of digestion cannot take place in any form of Protozoa. The beautiful and clear-cut experiments of Hofer ('89) and Verworn ('91) have demonstrated beyond a doubt that the digestive fluid is not prepared in the cytoplasm when the nucleus is absent. Hofer demonstrated that the slimy secretion which *Amœba* throws out to anchor itself before food-taking is never formed by the enucleated portions, and Verworn ('88) proved that enucleated pieces of *Polystomella* could not repair or regenerate the lost shell, while nucleated pieces quickly repaired it. Verworn came to the conclusion, which seems to be demonstrated, that enucleated protoplasmic masses cease entirely those chemical processes by which products of the normal cell are used or formed. Indeed, the generalization may now be made that no secretion takes place in enucleated fragments. On the other hand, the nucleus by itself, *i.e.* separated from the cytoplasm, has no longer the power to regenerate the lost parts, and like the enucleated cytoplasm, soon dies. "The nucleus needs the plasm, the plasm the nucleus," says Bütschli, "the activities of both are reciprocal, and one without the other cannot live,"[1] a generalization confirmed by Verworn's conclusive experiments ('91). The processes of secretion, therefore, whether for the purpose of digestion or for any other purpose in the life of the unicellular organism, are expressed by the constant chemical interchange which goes on between the cytoplasm and the nucleus.

Outside of secretion and reproduction, however, the usual functions of motion, and those of the contractile vacuole, are performed nearly as well by the enucleated as the nucleated fragments. Balbiani ('88) observed that non-nucleated pieces of infusoria, while unable to regenerate the lost parts, were nevertheless capable of limited motion, and of living and swimming about actively for several days, while the contractile vacuole continued its rhythmic pulsations. Hofer ('89) also observed that non-nucleated pieces of *Amœba proteus* will form pseudopodia and live for some time after the operation in active motion, a point which Verworn established in a striking manner in the case of *Thalassicolla nucleata*. Here the central capsule, after extirpation of the nuclei, forms a spherical mass which soon begins to assume the form of a typical radiolarian. The vacuolated portion, however, only begins to appear when degeneration sets in and the animal dies (Fig. 152). That the enucleated portions require oxygen and can use oxygen is shown by Verworn's negative experiments with enucleated Infusoria in

[1] ('88), p. 1642.

an oxygen-free medium. Enucleated pieces in the ordinary medium will live, but those in the oxygen-free medium quickly die, showing that oxygen-taking is possible without the nucleus. So, too, the contractile vacuole continues to pulsate in *Amœba* even nine days after the operation which separated it from the nucleus (Hofer).

The nucleus, furthermore, appears to have nothing to do directly with the functions of irritability, for non-nucleated fragments respond as readily to stimuli as do the nucleated parts. They show the same chemotactic, thermotactic, and galvanotactic reactions as the complete organisms, and Eimer's view that the nucleus is the seat of psychic activity as expressed by motion, is flatly contradicted (Verworn, '89).

Perty ('52), as well as Dujardin ('41), long since questioned the intel-

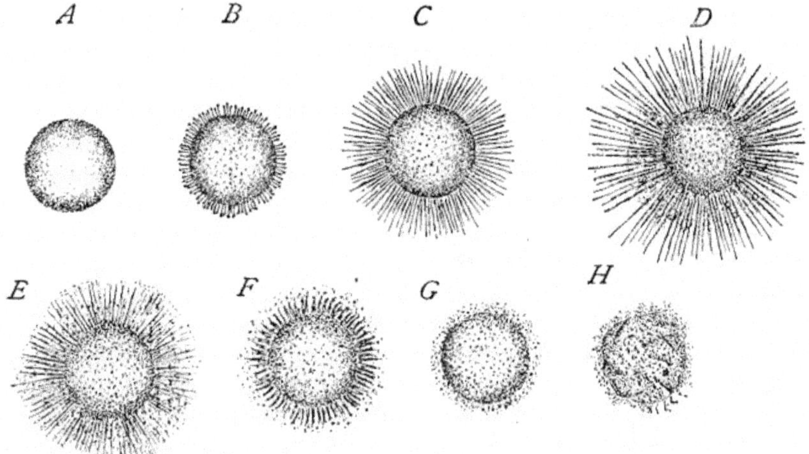

Fig. 152. — Isolated nucleus of *Thalassicolla nucleata* Hux. [VERWORN.]
A, B, C, D. Regenerative changes. *E, F, G, H.* Degenerative changes.

ligent power of the Infusoria in their reactions to stimuli, and more modern observers have generally accepted the view that the protoplasm of these animals responds automatically, in greater or less degree, to stimuli of various kinds. In the Sarcodina, the entire cell is equally irritable ; but higher in the scale, the outer plasm becomes more and more sensitive in relation to the entire body until, in the membranes of Infusoria, a generally sensitive skin is found which is sometimes drawn out into so-called sensory processes (feelers and tasters of hypotrichous forms), while in the Mastigophora special sensory or receptive spots are frequently present ("eye-spots"). This tendency to localization of the sensitive parts of the protoplasm is only in accord with our *a priori* conceptions of differentiation brought about by reaction to the environment or adaptation, and is strictly comparable to the differentiation of the diffuse ectodermal nervous system of the Cœlenterata, or to the ectodermal origin of the nervous system in all higher Metazoa.

Movement and reaction to stimuli have been the grounds usually adopted in support of the view that Protozoa are endowed with intelligence, and Eimer's view ('88) may be given as an example of this type of reasoning. He says: "The ciliated Infusoria behave so in regard to the outer world that one must ascribe to them the presence of a will. The simple consideration of their movement from place to place shows this. Change of position is entirely voluntary and is effected through the greatest variety of motions; sometimes all of the cilia, sometimes this one or that one, move more slowly or more rapidly or are kept quiet. The hypotrichously ciliated Infusoria, *e.g. Euplotes charon*, sometimes paddle rapidly through the water by means of all the cilia, sometimes run upon immersed objects, using their cilia as legs, with a motion like that of an Asellus," etc.[1] If Eimer could explain how he knows that such movement is voluntary or involuntary, the conclusions concerning intelligence would have some basis, but, as Haeckel says, the difference between voluntary and involuntary movement is as difficult to define as it is to fix the boundary between sensation (*Empfindung*) and irritability (*Reizbarkeit*), while the latter shows no more trace of intelligence than do the very sensitive reactions of *Mimosas*, *Dionæa muscipula*, and other higher plants. If by voluntary action is understood a movement brought about by an impulse or impulses, engendered more or less directly by external stimuli, or by changes in the inner condition of the organism, then, as Bütschli remarks, there is no reason for opposing the assumption; but if by voluntary action is meant that each stimulus or change in outer conditions induces a response brought about by the intelligent act of a will, then we are entirely without basis for the assumption of such action in the Protozoa.

While it is impossible to prove that movement is involuntary, observations upon living organisms under different conditions, such, for example, as the reactions of entire cells or the parts of cells to artificial stimuli of various kinds, make it reasonably certain.

Observations and experiments upon entire individuals are not conclusive, and the possibility of consciousness in perception and in reaction is not excluded. It might indeed be argued that acids "taste good" to *Paramœcium*, or that antherozoids of the fern "like" malic acid in small quantities as man likes alcohol, or that various Protozoa "know enough" to turn away from harmful substances, having acquired these characteristics through long ages of natural selection. But similar responses to similar stimuli are shown by the

[1] *Loc. cit.*, p. 340.

isolated muscle of a frog, and the question of consciousness resolves itself into the query, Does the single-celled organism compare with the isolated muscle or with the individual frog of which that muscle forms a part?

In the preceding chapters it has been shown that in many forms certain parts of the cell become differentiated for special functions, some of which are sensory, and in the whole group of Protozoa it is possible to arrange a scale of forms in which sensory differentiations of the plasm become more and more complex, *i.e.* certain regions of the cell become more irritable than others. The reactions are very much the same, however, whether external stimuli be applied to the least differentiated or the most complex of Protozoa. Verworn has further shown ('89) that enucleated and minute parts of different kinds of Protozoa also react to stimuli in exactly the same way as does the entire animal, observations which confirm Gruber's earlier view, that each protoplasmic element has its own "will expression" ('86), and justify Verworn's assertion that each protoplasmic part is the independent centre and source of its own movement, while the movement of a single-celled organism is only the synchronous movement of its many parts. Responses to stimuli are "reflex," therefore, and the conclusion forces itself that the protozoan organism has no more "consciousness" than the isolated muscle of the frog.

In addition to movement, however, other vital phenomena have been cited as evidence of intelligent action. Two of the most striking phenomena among the Protozoa are the apparent choice of food and the selection of certain materials for building the shell. Chlamydodontidæ live almost exclusively on diatoms and *Oscillaria*, although other food is abundant; the ciliates *Enchelys*, *Spathidium*, *Chœnia*, *Amphileptus*, *Lionotus*, *Dileptus*, and *Didinium* feed on ciliates alone (Bütschli), *Actinobolus* on *Halteria*, while numerous other forms of ciliates may be limited in other ways, or may be omnivorous. Among Mastigophora Cienkowsky ('65) observed *Colpodella pugnax* feeding exclusively on *Chlamydomonas*, while the rhizopod *Vampyrella spirogyræ* is limited to the cells of the alga *Spirogyra*. Carpenter, Romanes, and Brady ('84) all remarked upon the selective power of the marine rhizopods in building their shells, and the latter in particular ascribed a power of intelligent action on the part of certain forms, *e.g. Truncatulina lobatula*, which "protect themselves under certain circumstances with a covering of sand."

While these observations undoubtedly suggest willed acts on the part of the Protozoa in question, there is nevertheless room for an explanation along quite different lines. It is ever necessary to bear in mind, however, that all vital phenomena are exceedingly complex, even in the simplest of forms, and any explanation, whether based upon a comparison with our own consciousness or upon physical phenomena in the non-living world, must be considered only tentative. If protoplasm be regarded merely as a fluid substance, or better, a mixture of various, different fluids, forming an exceedingly complex chemical substance distinct in its individuality, and possessing certain recognizable properties, then it must be subject to the same laws which govern all fluids, and many of the so-called vital phenomena can be reduced to processes which in the inorganic world are familiar to physicists and chemists. Haeckel ('66) pointed out in connection with the supposed selective power of Protozoa, that an alum crystal selects only alum molecules from a mother liquid holding numerous salts in solution, and Berthold ('86) suggested that lifeless fluids will not absorb all kinds of stuff, but only such as have a certain chemical composition. Verworn ('89), experimenting with *Amœba* and *Pelomyxa*, found no selective action with the objects used; everything was surrounded by the pseudopodia and drawn into the body. On the other hand, certain Rhizopoda (as in the case of *Vampyrella spirogyræ*, which feeds solely on *Spirogyra*) eat only certain kinds of food, while Heliozoa, according to Meissner ('88), and Reticulariida, according to Verworn ('89), normally take in only living organisms as food. The latter, suspecting that in such cases it is only the mechanical irritation of moving organisms which causes the pseudopodial reaction, experimented with inorganic objects, such as a needle-point, a bit of paper, etc., and found that *Actinosphærium* among Heliozoa, and *Polystomella* among marine rhizopods, will absorb a moving needle-point or a bit of paper as readily as a struggling ciliate, and he concluded that the direct cause of food absorption, in these cases at least, was the mechanical stimulus. Again, in forms which draw food to the mouth opening by means of a vortex current due to flagella or cilia, Entz ('88) assumed a selective action from the fact that certain objects are passively thrown out of the current and do not reach the mouth opening at all. Stein accounted for the same occurrence by attributing a function of taste to certain of the adoral cilia. But Verworn called attention to the old experiments of Gleichen and Ehrenberg in feeding ciliates with powdered carmine and indigo, which was quickly ingested, while he also observed that digestible as well as indigestible particles are thrown out of the vortex current. Selection in such cases is not subjective, but depends upon the physical or chemical nature of the substances; the animal itself does not distinguish between good and bad.

Even more interesting in this respect are the observations and experiments of Rhumbler ('98) with organic and inorganic fluids. While corroborating the observations of Hofer and others, that objects are ingulfed at the posterior end of *Amœba*, he found that the ectoplasm is constantly moving inward to become changed again into endoplasm. The food particles accompany this inturning, and a filament

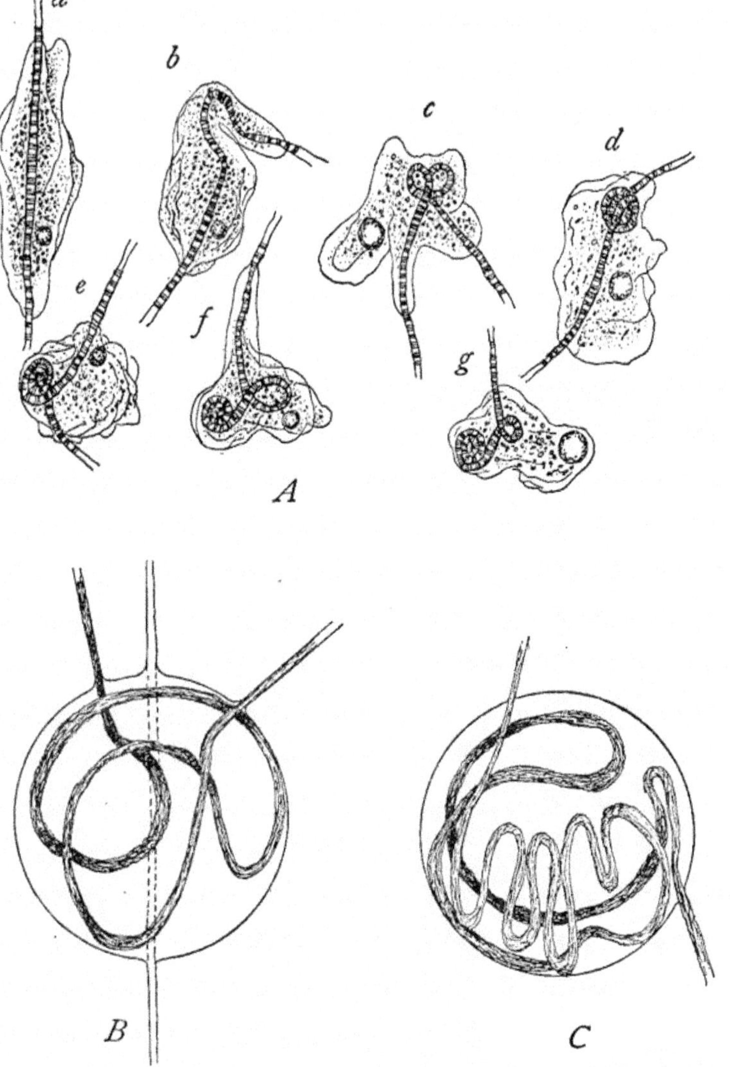

Fig. 153. — Reactions of *Amœba verrucosa* and of fluid substances. [RHUMBLER.]

A. Amœba verrucosa with a filament of *Oscillaria*. *B, C.* Chloroform drops with a filament of shellac. *a, b, c, d, e.* successive stages in the rolling of a filament of *Oscillaria*. *B, C.* Similar rolling of a shellac filament.

of *Oscillaria*, if pulled out after being partly ingested, leaves a channel in which it had lain (Fig. 153, *a–g*). After being drawn in, the filament of *Oscillaria* is rolled up into a small coil, and considerable space is thus saved. An *Amœba*, 90 μ in length, absorbed and coiled up a filament of *Oscillaria* 540 μ long. Quite similarly, a drop of water quickly draws into its substance a minute splinter of wood or glass, while all fluids show the same power in respect to certain substances. A drop of chloroform will draw in a shellac thread from the surrounding water, and will roll it up within its substance in exactly the same way that an *Amœba* rolls up a filament of *Oscillaria* (Fig. 153, *B*, *C*). Egg albumin and gum arabic in solution show the same phenomenon, the rapidity of ingestion depending upon the density of the medium. To a physicist such a process is explained by the phenomena of cohesion and adhesion; the coefficient of adhesion between the chloroform and the shellac filament is greater than the coefficient between shellac and water. But if a splinter of glass be mechanically inserted in a drop of chloroform suspended in water, the splinter will quickly leave the chloroform and seek the water. In this case the coefficient between glass and water is greater than that between glass and chloroform. Rhumbler tried the ingenious experiment of coating a glass splinter with a layer of shellac; when one end was placed against the chloroform drop, the splinter and shellac were quickly drawn into the drop. Here the shellac was soon dissolved by the chloroform, and the splinter was gradually left naked, whereupon it soon left the drop, being drawn into the surrounding water by reason of the greater coefficient of adhesion between glass and water. Here, according to Rhumbler, is an analogue of the process of feeding on the part of Rhizopoda. Bodies are ingested into the plasm because of the greater attraction to the fluid protoplasm than to water, then, through the chemical changes between protoplasm and the digestible parts of the foreign substances, the constituents of the foreign body are changed, and a corresponding change is wrought in the attractive force which keeps them together, that is, in the coefficient of adhesion, and defecation results.

Similarly with shell-formation, it has been shown by Verworn, Dreyer, Rhumbler, and others, that Rhizopoda pick up all sorts of foreign particles and excrete them at certain times upon the outside. Rhumbler found that the same process may be repeated by inorganic liquids, some of which (*e.g.* chloroform) show a selective tendency in picking up some objects and leaving others, while Verworn has shown that *Amœba* will not pick up certain objects unless the latter be made to irritate it so that a slimy secretion is poured out. In those forms of Rhizopoda which appear to select their building material, the selection is often due to the character of the material at their disposal, and partly to purely physical conditions, such as the exclusion of large sand grains because of the small mouth opening, or the inability of the protoplasm to hold objects above a certain weight.

In other cases where the shell material is deposited, such as the lime or silicious shells of Sarcodina, the substance itself is built up by a chemical process within the protoplasm, and the deposition may take place periodically or continuously.

The naked *Amœba* might be considered a complex chemical compound, endowed, like all compounds, with special properties — in this case with the power of motion, of irritability, of metabolism, and of growth and reproduction. Like many chemical compounds it is, during the living state, of unstable equilibrium, which involves a constant change in chemical composition. These several properties with which it is endowed, however, are not confined exclusively to the living organisms, for many inorganic compounds possess one or more of them. Thus, a drop of oil quite spontaneously assumes forms which simulate different species of *Amœba*, while a mixture of sugar or salt and olive oil simulates not only the movements, but even the structure, of living protoplasm (Bütschli). In these cases the motion is satisfactorily explained by the laws of surface tension, although the movements are almost as remarkable as those of a rhizopod. The motion of the protoplasm of a plant-cell is explained as the resultant of the chemical changes taking place within the body of the plant, and similarly the motion of *Amœba* has recently been interpreted by Berthold, Verworn, Bütschli, and Rhumbler, as a series of responses to changes in the chemical composition, with corresponding changes in density within the organism. The movements which engender the phenomena of phototaxis and thermotaxis, of chemotaxis and barotaxis, are also duplicated by inorganic substances. Thus the impact of ether waves upon various bodies brings about a rearrangement of molecules. Affinity, in chemistry, is the elective property by which one substance seeks another, and the mutual action is analogous to, if not the same as, chemotaxis; while the phenomena of cohesion and of adhesion come under the head of barotaxis. In a similar way it can be shown that irritability, or response to stimuli, has its analogue in the inorganic world; any compound substance in a state of unstable equilibrium, as in explosive compounds, will react to stimuli of various kinds, while even metabolism is simulated in inorganic objects, as shown in the excellent illustration cited by Verworn ('94) of nitric acid in the presence of sulphurous anhydride. Growth, too, is not confined to organic substances, as shown by the continual growth by accretion of various crystals or solids, while growth by intussusception takes place whenever a solid becomes dissolved in a liquid. The stimuli necessary to bring about protoplasmic reaction may be so delicate that we cannot perceive them, and we are thus led to assign some mystic cause under a term like "vital phenomena." Unexplained phenomena are daily seen in the laboratory, but the explanations are not regarded as hopeless; the sudden stoppage of a piece of camphor moving on the surface of pure water, by the introduction of the end of a finger, which gives to the water an exceedingly minute quantity of fatty matter, yet enough to equalize the surface tension, is understood through the laws of physics, although causes cannot be seen. At the present time, we can no more hope to understand the properties which we call "life" than we understand the ultimate causes of surface tension, of aquosity, or universal gravity; yet its manifestations can be examined and measured, and reduced to laws which are already applied in the domain of physical science.

The Protozoa, finally, must be regarded as forms in which the *organism* is less developed than in any other animal. In other words, the coördinating factors by means of which each protozoön acts as a unit are less complex than in other representatives of the animal kingdom, and this fact justifies the hope which lies at the bottom of modern "vitalism," that some day we may interpret these unknown factors in chemical and physical terms.

SPECIAL BIBLIOGRAPHY IX

Balbiani, G. — Recherches expérimentales sur la mérotomie des Infusoires ciliés. *Recueil zool. Suisse*, V., 1888.

Berthold, G. — Studien über Protoplasmamechanik. *Leipzig*, 1886.

Bütschli, O. — Untersuchungen über mikroskopische Schäume und das Protoplasma. *Leipzig*, 1892.

Greenwood, M. — On the Digestive Process in Some Rhizopoda. *Jour. of Physiol.* (English), VII. and VIII., 1886 and 1887.

Griffiths. — A Method of demonstrating the Presence of Uric Acid in the Contractile Vacuoles of Some Lower Organisms. *Proc. Roy. Soc. Edinburgh*, XVI., 1889.

Hofer, B. — Expérimentelle Untersuchungen über den Einfluss des Kerns auf das Protoplasma. *Jen. Zeit.*, 1889.

Jennings, H. S. — Studies on Reactions to Stimuli in Unicellular Organisms. *Amer. Jour. Phys.*, II., 1899.

Le Dantec, F. — Recherches sur la digestion intra-cellulaire chez les protozoaires. *Ann. d. l'Inst. Pasteur*, IV., 1890.

Meissner, M. — Beiträge zur Ernährungsphysiologie der Protozoen. *Z. w. Z.*, XLVI., 1888.

Rhumbler, L. — Physikalische Analyse von Lebenserscheinungen der Zelle. *Arch. f. Entwickelungsmechanik*, VII., 1898.

Verworn, M. — Psychophysiologische Protistenstudien. *Jena*, 1889.

Id. — Die physiologische Bedeutung des Zellkerns. *Arch. f. d. Ges. Physiologie*, LI., 1891.

Id. — Allgemeine Physiologie. *Leipzig*, 1894. Translated by F. S. Lee, New York, 1899.

BIBLIOGRAPHY

The titles included here are arranged in alphabetical order according to the system adopted in Minot's *Human Embryology*, in Wilson's *The Cell in Inheritance and Development*, and in other recent works. Each author's name is followed by the year of publication (abbreviated to the last two digits in all years between 1802–1901 inclusive); and where two or more publications appeared in the same year, they are referred to in the proper order. Thus, **BLOCHMANN, F., '94, 2,** refers to Blochmann's second paper in 1894.

ABBREVIATIONS

A. A.	Anatomischer Anzeiger.
A. B.	Archives de Biologie.
A. A. P.	Archiv für Anatomie und Physiologie.
A. M. N. H.	Annals and Magazine of Natural History.
A. m. A.	Archiv für mikroskopische Anatomie.
A. N.	Archiv für Naturgeschichte.
A. Entw.	Archiv für Entwicklungsgeschichte.
A. Z. E.	Archiv de zoologie expérimentale et générale.
B. C.	Biologisches Centralblatt.
C. R.	Comptes Rendus de l'Académie de Sciences.
C. R. S. B.	Comptes Rendus de la Société de Biologie.
J. M.	Journal of Morphology.
J. Z.	Jenaische Zeitschrift.
M. A.	Müller's Archiv.
M. J.	Morphologisches Jahrbuch.
Q. J.	Quarterly Journal of Microscopical Science.
Z. A.	Zoologischer Anzeiger.
Z. w. Z.	Zeitschrift für wissenschaftliche Zoologie.

AGARDH, '28. Icones Algar. Europ. — **Agassiz, L., '57.** Contribution to the Natural History of the United States of America. 1st Monograph: Essay on Classification, *Boston*. — **Alder, J., '51.** An Account of Three New Species of Animalcules: *A. M. N. H.* (2), VII, p. 426. — **Allman, G. J., '72.** Note on Noctiluca: *Q. J.*, XII (N. S.), pp. 326–32. — **Altmann, R., '94.** Die Elementarorganismen und ihre Beziehungen zu den Zellen: *2d Edition*. — **Archer, W., '69.** On Some Fresh-water Rhizopoda, new or little known: *Q. J.* (N. S.), IX. — **Id., '70.** On Some Fresh-water Rhizopoda, new or little known: *Q. J.* (N. S.), IX and X. — **Id., '75.** On Chlamydomyxa labyrinthuloides, nov. gen. et sp., a New Fresh-water Sarcodic Organism: *Q. J.*, XV (N. S.), p. 107. — **Id., '76.** Résumé of Recent Contributions to our Knowledge of Fresh-water Rhizopoda: *Q. J.* (N. S.), XVI, p. 347. — **Auerbach, L., '55.** Ueber die Einzelligkeit der Amoeben: *Z. w. Z.*, VII, pp. 365–430.

BALBIANI, G., '58. Note relative à l'existence d'une génération sexuelle chez les Infusoires: *Jour. de la Physiol.*, I, pp. 347–52. — **Id., '59.** Du rôle des organes générateurs dans la division spontanée des Infusoires ciliés: *C. R.*, XLVIII, pp. 266–9. — **Id., '60.** Note sur un cas de parasitisme improprement pris pour un mode de reproduction des Infusoires ciliés: *C. R.*, LI, pp. 319–22. — **Id., '61.** Recherches sur les phénomènes sexuelles des Infusoires: *Jour. de la Physiol.*, IV, pp. 102–30,

194-220, 431-48, 465-520. — **Id.**, **'63.** Sur l'organisation et la nature des Psorospermies : *C. R.*, LVII, pp. 157-61. —**Id.**, **'82.** Les Protozoaires : *Jour. de Micrographie*, V (1881), p. 63, VI (1882), p. 9. — **Id.**, **'88.** Recherches expérimentales sur la mérotomie des Infusoires ciliés : *Recueil Zoolog. Suisse*, V, pp. 1-72. — **Id.**, **'89.** Sur trois entophytes nouveaux du tube digestif des Myriapodes : *Jour. de l'Anatom. et de la Physiolog.*, XXV, pp. 5-45. — **Id.**, **'90.** Sur la structure intime du noyau du Loxophyllum meleagris : *Z. A.*, XIII, pp. 110-15 and 132-6. — **Id.**, **'95.** Sur la structure et la division du noyau chez le Spirochona gemmipara : *Ann. de Microg.*, VII, p. 289. **Baranetski, J.**, **'75.** — Influence de la lumière sur les plasmodia des Myxomycètes : *Mem. d. l. Soc. Nat. d. Cherbourg*, XIX, pp. 321-60. **Barbagello, P.** (with Cassagrandi, O.), **'95.** Ricerche biologiche e clinic sull' Amoeba coli (Lösch.) : *Bull. della sedute della Accademia Gioenia di Sci. Nat. in Catania*, XLI, pp. 7-19. — **De Bary, '64.** Die Mycetozoen : *2d Ed.*, *Leipzig*. — **Id.**, **'87.** Comparative Morphology and Biology of the Fungi, Mycetozoa, and Bacteria. Translated from German Edition of 1884 : *Oxford*. — **Behla, Robt.**, **'98.** Die Amöben insbesondere vom parasitären und culturellen Standpunkt : *Berlin*. — **Van Beneden, E.**, **'83.** Recherches sur la maturation de l'œuf et la fécondation : *A. B.*, IV, pp. 265-637. — **Van Beneden et Neyt, '87.** Nouvelle recherches sur la fécondation et la division mitosique chez l'Ascaris megalocephalus : *Bull. Acad. Roy. d. Belgique*, XIV, pp. 215-95. — **Bergh, R. S.**, **'79.** Tiarina fusus Cl. and Lach. : *Vidensk. Meddel. fr. d. Naturh. Foren. i. Kjobenhavn*, 1879-80, pp. 265-70. — **Id.**, **'81.** Der Organismus der Cilioflagellaten : *M. J.*, VII, pp. 177-288. — **Id.**, **'89.** Recherches sur les noyaux de l'Urostyla grandis et de l'Urostyla intermedia n. sp. : *A. B.*, IX, pp. 497-513. **Berthold, G.**, **'86.** Studien über Protoplasmamechanik : *Leipzig*, 1886. — **Beyerinck, —**, **'90.** Culturversuche mit Zoochlorellen, Lichengonidien und andere Algen : *Bot. Zeit.* — **Blainville, H. de, '22.** Article "Infusoire," in *Dictionnaire des Sciences Naturelles*, T. 23, pp. 416-20. — **Blochmann, F.**, **'88.** Zur Kenntniss der Fortpflanzung von Euglypha alveolata : *M. J.*, XIII, pp. 173-83. — **Id.**, **'94, I.** Ueber die Kerntheilung bei Euglena : *B. C.*, XIV, pp. 194-7. — **Id.**, **'94, 2.** Zur Kenntniss von Dimorpha nutans (Gruber) : *B. C.*, XIV, pp. 197-201. — **Boeck, C. P. B.**, **'47.** Nogle Forhold af Bygningen og Udviklingem af Polygastrica : *Forh. Skandin. Naturfor. Christiania*, p. 270. — **Borgert, Adolf, '91.** Ueber die Dictyochiden, insbesondere über Distephanus speculum ; so wie Studien an Phaeodarien : *Z. w. Z.*, LI, pp. 629-76. — **Id.**, **'96.** Fortpflanzungsverhältnisse bei tripyleen Radiolarien (Phaeodarien) : *Verh. d. Deutsch. Zool. Ges.* — **Bory de Saint Vincent, '24.** Classification des animaux domestiques : *Encycl. Méthode*, II, *Paris*. — **Bosc, F. J.**, **'98.** Le Cancer maladie infectieuse a Sporozoaires : *Arch. de Phys. norm. et path.*, *Paris*, XXX, pp. 458-71 and 484-93. — **Bourne, A. G.**, **'91.** On Pelomyxa viridis sp. n. and on the vesicular nature of Protoplasm : *Q. J.* (N. S), XXXII, pp. 357-74. — **Boveri, Th.**, **'88.** Zellenstudien I, II : *J. Z.*, XXII. — **Id.**, **'90.** Zellenstudien, III, *etc.* : *J. Z.*, XXIV., pp. 314-401. — **Id.**, **'95.** Ueber das Verhalten der Centrosomen bei der Befruchtung des Seeigeleies, nebst allgem. Bemerkungen über Centrosomen u. Verwandtes : *Verh. d. Phys. Med. Ges. z. Würzburg* (N. F.), XXIX, 1, p. 41. — **Id.**, **'01.** Zellenstudien, IV, *Jena*, 1901. — **Brady, H. B.**, **'79.** Notes on some of the reticularian Rhizopoda of the "Challenger" Expedition. 1. On new or little-known Arenaceous types : *Q. J.*, XIX, p. 28. — **Id.**, **'82, '84.** Report on the Foraminifera dredged by H. M. S. "Challenger" during the years 1873-6 : *Challenger Reports Zoology*, IX. — **Id.**, **'84.** On Keramosphæra, a new type of Porcellanous Foraminifera : *A. M. N. H.* (5) X, pp. 242-5. — **Brandt, H.**, **'77.** Ueber Actinosphaerium Eich : *Dissertation*, *Halle*. — **Brandt, K.**, **'81.** Färbung lebender einzelligen Organismen : *B. C.*, I, pp. 202-5. — **Id.**, **'82.** Ueber die morphologische und physiologische Bedeutung des Chlorophyls bei Thieren : *Arch. f. Anat. u. Phys.* (*Phys. Abth.*), pp. 125-51. — **Id.**, **'85.** Koloniebildenden Radiolarien (Sphaerozoeën) des Golfes von Neapel: *Fauna and Flora, G. v. N.* — **Brauer, Aug., '93.** Zur Kenntniss d. Spermatogenese v. Ascaris megalocephalus: *A. m. A.*, XLII, p. 153. — **Id., '94.** Ueber die Encystierung von Actinosphaerium Eich: *Z. w. Z.*, LVIII, pp. 189–221. — **Brücke, E., '61.** Die Elementarorganismen: *Wiener Sitzungsber.* XLIV, *Abth.* II, pp. 381–407. — **Bruch, C., '50.** Einige Bemerkungen über die Gregarinen: *Z. w. Z.*, II, pp. 110–14. — **Buck, E., '77.** Einige Rhizopodenstudien: *Z. w. Z.*, XXX, pp. 1–49. — **Buehler, A., '95.** Protoplasma-Structur in Vorderhirnzellen der Eidechsen: *Verh. Phys. Med. Ges. Würzburg* (2), XXIX, pp. 209–52. — **Buffon, G. L. de, 1749.** Histoire Naturelle générale et particulier: *T.* I–III, *Edition* 1812. — **Burnett, W. J., '54.** On the zoölogical nature of Infusoria: *Proc. Boston Soc. Nat. Hist.*, IV, pp. 331–5. — **Bütschli, O., '73.** Einiges über Infusorien: *A. m. A.*, IX, pp. 657–78. — **Id., '74.** Zur Kenntniss der Fortpflanzung bei Arcella vulgaris Ehr., *A. m. A.*, XI, pp. 460–6. — **Id., '76.** Studien über d. erst. Entwicklungsvorg. d. Eizelle, d. Zelltheilung u. die Conjugation der Infusorien: *Abhand. d. Senkenberg. naturfor. Gesell. Fr. a. M.*, X, pp. 213–452. (Preliminary in *Z. w. Z.*, XXV, 1875, p. 426.) — **Id., '77.** Ueber den Dendrocometes paradoxus Stein, nebst einigen Bemerkungen, *etc.*: *Z. w. Z.*, XXVIII, pp. 49–67. — **Id., '78.** Beiträge zur Kenntniss der Flagellaten und einiger verwandten Organismen: *Z. w. Z.*, XXX, pp. 205–81. — **Id., '82.** Beiträge zur Kenntniss der Fischpsorospermien: *Z. w. Z.*, XXV, pp. 629–51. — **Id., '83.** Protozoa in *Bronn's Klassen und Ordnungen des Thierreichs*, 1883–88. — **Id., '85.** Einige Bermerkungen über gewisse Organisationverhältnisse der Cilioflagellaten und der Noctiluca: *M. J.*, X, pp. 529–77. — **Id., '92, 1.** Ueber die so-genannten Centralkörper der Zelle und ihre Bedeutung: *Verh. d. Nat. Med. Verh. z. Heidelberg* (N. F.), IV, pp. 535–8. — **Id., '92, 2.** Untersuchungen über mikroskopische Schäume und das Protoplasma: *Leipzig*, 1892. (Translated by E. A. Minchin, *London*, '94.) — **Id., '96.** Weitere Ausführungen über den Bau der Cyanophyceen und Bakterien: *Leipzig.*

CALENDRUCCIO, S., '90. Animali parasiti dell' uomo in Sicilia: *Att. d. Accad. Gioenia d. Sci. Nat. in Catania* (4), II, p. 95. — **Calkins, G. N., '91.** On Uroglena, a Genus of Colony-building Infusoria observed in Certain Water Supplies of Massachusetts: 23*d Annual Report of the Massachusetts State Board of Health.* — **Id., '92, 1.** The Microscopical Examination of Water: *Report Massachusetts State Board of Health*, 1892, pp. 397–421. — **Id., '92, 2.** A Study of Odors Observed in the Drinking Waters of Massachusetts: *Report Massachusetts State Board of Health*, 1892, pp. 353–90. — **Id., '98, 1.** The Phylogenetic Significance of Certain Protozoan Nuclei: *Ann. N. Y. Acad. of Sciences*, XI, pp. 379–400. — **Id., '98, 2.** Mitosis in Noctiluca miliaris, and its Bearing on the Nuclear Relations of the Protozoa and the Metazoa: *Ginn & Co., Boston*, 1898 (also in *J. M.*, XV, 1899, Part III). — **Id., '99, 1.** Nuclear Division in Protozoa: *Woods Hole Biological Lectures*, 1899, No. 13. — **Id., '99, 2.** Report upon the recent Epidemic among Brook Trout (Salvelinus fontinalis) on Long Island: *Fourth Ann. Rep. Comm. of Fisheries, Game, and Forests of the State of New York.* — **Carpenter, W. B., '56.** Researches on the Foraminifera: 2*d Series Phil. Trans.*, p. 547. — **Id., '62.** Introduction to the Study of the Foraminifera: *Ray Society.* — **Id., '65.** Additional Note on the Structure of Eozoön Canadense: *Q. J. Geol. Soc.*, XXI, pp. 59–66. — **Id., '66.** Supplemental Notes on the Structure and Affinities of Eozoön Canadense: *Q. J. Geol. Soc.*, XXII, pp. 219–28. — **Id., '70, 1.** Descriptive Catalogue of Objects from the Deep Sea: *London.* — **Id., '70, 2.** On the Rhizopodal Fauna of the Deep Sea: *Proc. Roy. Soc.*, XVIII, p. 59. — **Carter, H. J., '58.** On Fecondation in Eudorina elegans and Cryptoglena: *A. M. N. H.* (III) II, pp. 237–53. — **Id., '63.** On the Presence of Chlorophyl-cells and Starch Granules as Normal Parts of the Organism, and on the Reproductive Process, etc.: *A. M. N. H.* (III), XII, pp. 249–64. — **Id., '64.** On Fresh-water Rhizopoda of England and India: *A. M. N. H.* (III), XIII, pp. 13–38. — **Id., '65.** On the Fresh- and Salt-water Rhizopoda of England and India: *A. M. N. H.* (III), XV, pp. 277-93. — **Id., '77, 1.** Description of Bdelloidina aggregata, a new Genus and Species of Arenaceous Foraminifera, etc.: *A. M. N. H.* (IV), XIX, pp. 201-9. — **Id., '77, 2.** On a Melobesian Form of Foraminifera (Gypsina melobesioides), etc.: *A. M. N. H.* (IV), XX, p. 172. — **Carus, C. G., '23.** Beitrag z. Gesch. d. unter Wasser an ver. Thierkörpern sich erzeug. Schimmel- u. Algen-gattungen: *Nov. Act. Ac. Leop. C.*, XI, II, p. 506. — **Id., '32.** Neue Unters. über die Entwicklungsgeschichte unserer Flussmuscheln: *Nov. Act. Ac. Leop. C.*, XVI, pp. 70-80. — **Carus, J. V., and Gerstaecker, '63.** Text-book of Zoology, vol. 2, p. 570. —**Cassagrandi** (see Barbagello). — **Cattaneo, G., '78.** Intorno all' ontogenesi dell' Arcella vulgaris Ehr.: *Att. Soc. Ital. d. Sci. Nat.*, XXI, pp. 331-43. — **Celli, A.**, and **Fiocca, R., '95.** Ueber die Aetiologie der Dysenterie: *Cent. f. Bakt. u. Parasitenkunde*, XVII, pp. 309-10. — **Certes, A., '85.** De l'emploi des matières colorantes de l'étude physiologique et histiologique des infusoires vivants: *C. R. S. B., Paris* (8), II, pp. 197-202. — **Cienkowsky, L., '55.** Bemerkungen über Stein's Acinetenlehre: *Bull. Phys. Math. Acad. d. St. Petersburg*, XIII, p. 297. Also in *Q. J.*, V, pp. 96-103. — **Id., '56.** Zur Kenntniss eines einzelligen Organismus: *Bull. Phys. Math. Ac. St. Petersburg*, XIV, pp. 261-7. — **Id., '63.** Zur Entwicklungsgeschichte der Myxomyceten: *Jahrb. f. Wiss. Bot.*, III, pp. 325-37. — **Id., '65, 1.** Beiträge zur Kenntniss der Monaden: *A. m. A.*, I, pp. 203-32. — **Id., 65, 2.** Die Chlorophyllhaltigen Glœocapsen: *Bot. Zeit.*, XXIII, pp. 21-7. — **Id., '67, 1.** Ueber die Clathrulina, eine neue Actinophryengattung: *A. m. A.*, III, pp. 311-16. — **Id., '67, 2.** Ueber den Bau und die Entwicklung der Labyrinthuleen: *A. m. A.*, III, pp. 274-310. — **Id., '70.** Ueber Palmellaceen und einige Flagellaten: *A. m. A.*, VI, pp. 421-38. — **Id., '73.** Ueber Noctiluca miliaris: *A. m. A.*, IX, pp. 47-61. — **Id., '76.** Ueber einige Rhizopoden und verwandten Organismen: *A. m. A.*, XII, pp. 15-50. — **Claparède, E., '58** (see Claparède and Lachmann). — **Id., '67.** Miscellanées zoologiques: *Ann. d. Sc. Natur.* (5), VIII, p. 30. — **Claparède and Lachmann, E., '58-'60.** Études sur les Infusoires et les rhizopodes: *Mém. Inst. Genèvoise*, V, VI, and VII, 1850 and 1860. — **Clark, H. J., '65.** Proofs of the Animal Nature of the Cilio-flagellate Infusoria, as based upon Investigations of the Structure and Physiology of one of the Peridiniæ: *Proc. Amer. Acad. Arts and Sc.*, pp. 393-402. — **Id., '66.** On the Spongiæ ciliatæ, as Infusoria flagellata, or Observations on the Structure, Animality, and Relationships of Leucosolenia botryoides Bow. *Mem. Bos. Soc. Nat. Hist.*, I, pt. 3, pp. 1-36. — **Clarke, J. J., '95.** Observations on Various Sporozoa: *Q. J.* (N. S.), XXXVII, pp. 277-87. — **Cohn, F., '62.** Ueber die Contractilen Staubfäden der Disteln: *Z. w. Z.*, XII, pp. 366-71. — **Id., '64.** Ueber die Gesetze der Bewegung Mikroskopischer Thiere und Pflanzen unter Einfluss des Lichtes: *42d. Jahr. Ber. und Abhandl. der Schlesw. Gesell. f. Vaterl. Cult.*, pp. 35-6. — **Id., '66.** Neue Infusorien im Seeaquarium: *Z. w. Z.*, XVI, pp. 253-302. — **Id., '75.** Die Entwicklungsgeschichte der Gattung Volvox: *Beit. z. Biol. d. Pflanzen*, I Heft, III, pp. 93-114. — **Id., '76.** Bemerkungen über die Organisation einiger Schwärmzellen: *Beit. z. Biol. d. Pflanzen*, II, pp. 101-21. — **Cohn, L., '96.** Ueber die Myxosporidien von Esox lucius und Perca fluviatilis: *Zool. Jahr.* (*Anat.*), IX, pp. 227-72. — **Corti, B., 1774.** Osservazioni microscopiche sulla Tremella, *etc.*, *Lucca*, 1774. — **Councilman, W. T., '91.** The Form of Dysentery produced by the Amœba coli: *Jour. Amer. Med. Ass.* — **Cuénot, L., '97.** Évolution des Grégarines cœlomiques du Grillon domestique: *C. R.*, CXXV, pp. 52-4. — **Cunningham, D. D., '81.** On the Development of Certain Microscopic Organisms occurring in the Intestinal Canal: *Q. J.*, XXI, pp. 234-90.

DALLINGER and DRYSDALE, '73. Researches on the Life History of the Monads: *Monthly Mic. Jour.*, '73, X, pp. 53-58; '74, XI, pp. 7-10, 69-72, 97-103; 75, XII and XIII, pp. 261-9, 185-97. — **Dangeard, P. A., '89.** On the formation of the Antherozoids in Eudorina elegans: *A. M. N. H.* (6), V, pp. 343-4. — **Id.**, **'90**. Contribution à l'étude des organismes inférieurs: *Le Botaniste*, II, pp. 14-29. — **Danilewsky, B.**, **'85**. Die Haematozoen der Kaltblüter: *A. m. A.*, XXIV, pp. 588-98. — **Id.**, **'90, 1**. Developpement des parasites malarique dans les leucocytes des oiseaux: *Ann. de l'Institut Pasteur*, IV, pp. 427-45. — **Id.**, **'90, 2**. Sur les microbes de l'infection malarique aigue et chronique chez les oiseaux et chez l'homme: *Ann. d. l'Inst. Pasteur*, IV, p. 753. — **Id.**, **'91**. Contribution à l'étude de la microbiosis malarique: *Ann. d. l'Inst. Pasteur*, V, p. 758. — **Le Dantec, F.**, **'90**. Recherches sur la digestion intra-cellulaire chez les protozoaires: *Ann. d. l'Inst. Pasteur*, IV, pp. 777-91, and V, pp. 163-70. — **Id.**, **'92**. Recherches sur la symbiose des algues et des Protozoaires: *Ann. d. l'Inst. Pasteur*, VI, pp. 190-8. — **Id.**, **'94**. Études biologiques comparatives sur les Rhizopodes lobés et reticulés d'eau douce: *Bull. Sci. d. la France et de la Belg.*, XXVI, pp. 56-99. — **Id.**, **'95**. Note sur quelques phénomènes intracellulaires: *Bull. Sci. d. l. Fr. et d. l. Belg.*, XXV, pp. 398-416. — **Dawson, J. W.**, **'65**. On the Structure of Certain Organic Remains in the Laurentian Limestone of Canada: *Q. J. Geol. Soc.*, XXI, pp. 51-9. — **Id.**, **'75**. The Dawn of Life, *etc.*, *London*. — **Delage et Hérouard, '96**. Traité de zoologie concrète; I, La Cellule et les Protozoaires, *Paris*. — **Dewitz, J.**, **'86**. Ueber Gesetzmässigkeit in der Ortsveränderung der Spermatozoen, *etc.*: *Arch. f. Ges. Physiol.*, XXXVIII, p. 358. — **Diesing, K. M.**, **'65**. Revision der Prothelminthen: Abth. Mastigophoren. *Sitz. Ber. d. math.-nat.-Kl. d. Akad. z. Wien*: LII, pp. 287-402. — **Doflein, F.**, **'96**. Ueber Kentrochona nebaliæ: *Z. A.*, XIX, pp. 362-6. — **Id.**, **'98**. Studien zur Naturgeschichte der Protozoen, III, Ueber Myxosporidien: *Zool. Jahrb.* (*Anat.*), XL, pp. 281-350. — **Id.**, **'00**. Studien zur Naturgeschichte der Protozoen, IV, Zur Morphologie und Physiologie der Kern- und Zelltheilung: *Z. J.* (*Anat.*), XIV, pp. 1-59. — **Donné, A.**, **'37**. Recherche microscopique sur la nature du mucus: *Paris*. — **Dreyer, F.**, **'92**. Die Gerüstbildung bei Rhizopoden, *etc.*: *J. Z.*, XIX, pp. 204-469. — **Drüner, L.**, **'95**. Studien über den Mechanismus der Zelltheilung: *J. Z.*, XXIX, pp. 271-345. — **Dufour, L.**, **'28**. Note sur la Grégarine, nouveau genre de ver qui vit en troupeau dans les intestins de divers insectes: *Ann. d. Sc. Nat.* (1), XIII, pp. 366-9. — **Dujardin, F.**, **'35**. Recherches sur les organismes inférieures: *Ann. d. Sc. Nat.* (2), IV, pp. 343-76. — **Id.**, **'38**. Sur les monads à filament multiple: *Ann. d. Sc. Nat.* (2), X, pp. 17-20. — **Id.**, **'41**. Histoire naturelle des zoophytes Infusoires, comprenant la physiologie et la classification de ces animaux, *etc.*: *Paris*. — **Dunal, F.**, **'38**. Sur les Algues qui colorent en rouge certaines eaux des marais salans méditerranéens: *Ann. d. Sc. Nat.* (2), IX, pp. 172-5.

EHRENBERG, C. G., **'30**. Beiträge zur Kenntniss der Organisation der Infusorien u. ihrer geog. Verbreitung besonders in Sibirien: *Abh. d. Berlin Akad.*, pp. 1-88. — **Id.**, **'31**. Ueber die Entwickelung und Lebensdauer d. Infusoriensthiere, *etc.*: *Abh. d. Berlin Akad.*, pp. 1-154. — **Id.**, **'33**. Dritter Beiträge zur Erkenntniss grosser Organisation in der Richtung des Kleinster Raumes: *Abh. d. Berlin Akad.*, pp. 145-336. — **Id.**, **'35**. Zusätze zur Erkenntniss d. grosser organ. Ausbildung in den kleinster thierischen Organismen: *Abh. d. Berlin Akad.*, pp. 151-80. — **Id.**, **'38**. Die Infusionsthierchen als vollkommene Organismen: *Leipzig*. — **Id.**, **'41**. Verbreitung und Einfluss des mikroskopischen Lebens in Sud- und Nord-Amerika: *Abh. d. Berlin Akad.*, pp. 291-438. — **Id.**, **'48**. Beobachtung zwei generisch neuer Formen des Frühlingsgewässers bei Berlin als lebhaft grüner Wasserfärbung: *Monats. ber. d. Berlin Akad.*, pp. 233-7. — **Id.**, **'54, 1**. Mikrogeologie: *Berlin*. — **Id.**, **'54, 2**. Nova Genera maris profundi: *Monatsber. d. Berlin Akad.*, pp. 236-9. — **Eimer, G. H. T.**, **'88**. Die Entstehung der Arten auf Grund von Vererben erworbener Eigenschaften, *etc.*: *Jena*. — **Eismond, J.**, **'90**. Zur Frage über den

Saugmechanismus bei Suctorien: *Z. A.*, XIII, pp. 721-3.—**Ellis, J., 1769.** Observations on a particular manner of increase in the Animalcula of Vegetable Infusions, *etc.: Phil. Trans.*, LIX, pp. 138-52.—**Engelmann, Th. W., '62.** Zur Naturgeschichte der Infusionsthiere: *Z. w. Z.*, XI, 347-93.—**Id., '69.** Beiträge zur Physiologie des Protoplasmas: *Arch. f. d. Ges. Physiol.*, II.—**Id., '75.** Contractilität und Doppellbrechung: *Arch. f. d. Ges. Physiol.*, XI.—**Id., '76.** Ueber Entwickelung und Fortpflanzung von Infusorien: *M. J.*, I, pp. 573-635.—**Id., '78, 1.** Zur Physiologie der Contractilen Vacuolen der Infusionsthiere: *Z. A.*, I, pp. 121-2.—**Id., '78, 2.** .Ueber Gasentwickelung im Protoplasma lebender Protozoen: *Z. A.*, I, pp. 152-3.—**Id., '79.** Physiologie der Protoplasma und Flimmerbewegung: *Hermann's Handbuch d. Physiol.*, I, *Leipzig*.—**Id., '82.** Ueber Licht- und Farbenperception niederster Organismen (Euglena): *Arch. f. d. Ges. Physiol.*, XXIX, p. 387.—**Id., '83.** Ueber Thierisches Chlorophyl: *Arch. f. d. Ges. Physiol.*, XXXII, pp. 80-96.—**Entz, G., '78.** Zur Gasentwickelung im Protoplasma lebender Protozoen: *Z. A.*, I, pp. 248-56.—**Id., '79.** Ueber einige Infusorien des Salzteiches zu Szammosfalva: *Termés. Füzetek.*, III, p. 40.—**Id., '82.** Das Konsortialverhältniss von Algen u. Tieren: *B. C.*, II, pp. 451-64.—**Id., '83, 1.** A tordai és szamosfalva sóstavak ostorosai (Flagellata). Die Flagellaten des Kochsalzteiche zu Torda, *etc.: Term. Füzet.*, VII, p. 139.—**Id., '83, 2.** Beiträge zur Kenntniss der Infusorien: *Z. w. Z.*, XXXVIII, pp. 167-89.—**Id., '84.** Ueber Infusorien des Golfes von Neapel: *Mitt. d. zool. Stat. z. Neap.*, V, pp. 289-444.—**Id., '88.** Studien über Protisten: *Budapesth*.—**Id., '91.** Die elastischen und contractilen Elemente der Vorticellinen: *Math. u. Naturwis. Berichte a. Ungarn*, X, pp. 1-48.

FABRE-DUMERGUE, P., '85. Note sur les infusoires ciliés de la baie de Concarneau: *Jour. d. l'Anat. et de la Physiol.*, XXI, pp. 554-68.—**Id., '88.** Recherches anatomiques et physiologiques sur les infusoires ciliés: *Ann. d. Sci. Nat.* (7), V, pp. 1-140.—**Id., '90.** Sur le système vasculaire contractile des infusoires ciliés: *C. R. S. B., Paris* (9), II, pp. 391-3.—**Fisch, C., '85.** Untersuchungen über einige Flagellaten und verwandte Organismen: *Z. w. Z.*, XLII, pp. 47-125.—**Fischer, Alf., '94.** Ueber die Geisseln einiger Flagellaten: *Jahr. Wiss. Bot.*, XXVII, pp. 187-235.—**Fiszer, Z., '85.** Forschungen über die pulsirende Vacuole der Infusorien: *Wszechswiat*, IV, pp. 691-4, *etc.* (see Bütschli).—**Flemming, W., '82.** Zellsubstanz, Kern und Zelltheilung: *Leipzig*.—**Id., '91.** Attraktionssphären und Centralkörper in Gewebszellen und Wanderzellen: *A. A.*, 1891, pp. 78-81.—**Focke, G. W., '36.** Ueber einige Organisationsverhältnisse bei polygastrischen Infusorien u. Räderthieren: *Isis*, 1836, pp. 785-7.—**Foettinger, A., '81.** Recherches sur quelque infusoires nouveaux, parasites des Céphalopodes: *A. B.*, II, pp. 345-78.—**Foulke, S. G., '83.** Observations on Actinosphærium Eich: *A. M. N. H.* (5), XII, p. 206.—**Francé, R., '93.** Ueber die Organisation der Choanoflagellaten: *Z. A.*, 1893, pp. 44-5.—**Id., '97.** Der Organismus der Craspedomonaden: *Budapest* (German text, pp. 115-248).—**Frenzel, J., '85.** Ueber einige in Seethieren lebende Gregarinen: *A. m. A.*, XXIV, pp. 545-88.—**Id., '92.** Ueber einige merkwürdige Protozoen Argentiniens: *Z. w. Z.*, LIII, pp. 334-60.—**Fresenius, G., '58.** Beiträge z. Kenntniss mikrosk. Organismen: *Abh. d. Senkenb. Naturfors. Gesell. Frankfurt*, II, pp. 211-42.—**Fromentel, E. de, '74.** Études s. l. microzoaires ou infusoires proprement dits: *Paris*.

GABRIEL, B., '75. Ueber Entwicklungsgeschichte der Gregarinen: *Jahres Ber. d. Schles. Ges. f. Vaterl. Cultur. Breslau*, 1875, pp. 44-6.—**Id., '80.** Zur Classification der Gregarinen: *Z. A.*, III, pp. 569-72.—**Garry, W. E., '00.** The Effects of Ions upon the Aggregation of Flagellated Infusoria: *Am. Jour. Phys.*, III, pp. 291-315.—**Gasser, J., '95.** Note sur les causes de la dysentérie: *Arch. de méd. expér. et d'Anat. Path.*, II, p. 198.—**Geddes and Thompson, '89.** The Evolution of Sex: *London*. — **Giard, A., '76.** Sur une nouvelle espèce de Psorospermie (Lithocystis Schneideri) parasite de l'Echinocardium cordatum: *C. R.*, LXXXII, pp. 1208–10. — **Gleichen, W. F., 1778.** Abhand. über d. Sämen- u. Infusionsthierchen und über die Erzeugung, *etc.*: *Nürmberg* (see Bütschli). — **Goetze, J. A. E., 1777.** Infusionsthierchen die andre fressen: *Beschäft. d. Berliner Ges. naturf. Freunde*, III, pp. 375–84. — **Goldfuss, G. A., '20.** Handbuch der Zoologie: Bd. I. — **Goroschankin, T., '75.** Genesis in Typus der palmellenartigen Algen, *etc.*: *Mitt. Kais. Ges. Naturf. Freunde Moskau*, XVI (see Bütschli). — **Id., '90.** Beit. z. Kenntniss d. Morphologie und Systemat. d. Chlamydomonadinen: *Bull. d. l. Soc. Nat. Moscou*, II, p. 498. — **Gould, L. J., '94.** Notes on the Minute Structure of Pelomyxa palustris Greeff: *Q. J.*, XXXVI, p. 295. — **Grassi, B., '81.** Intorno ad alcuni protisti endoparassitici, *etc.*: *Att. Soc. Italiana d. Sci. Nat.*, XXIV, pp. 135–224. — **Id., '82.** Intorno ad alcuni protisti endoparassitici, *etc.*: *Att. Soc. Italiana Sci. Nat.*, XXIV, pp. 1–94. — **Id., '85.** Intorno ad alcuni protozoi dar. d. Termiti: *Att. Gioenia Sci. Nat.*, XVIII, pp. 464–5. — **Grassi, B., and Feletti, R., '92.** Contribuzione allo studio dei parassiti malarici: *Att. Acc. Catania* (4), V, pp. 1–80. — **Gray, J. E., '58.** On Carpenteria and Dujardinia, two genera of a new form of Protozoa: *Proc. Zool. Soc., London*, XXVI, p. 266. — **Greeff, R., '66.** Ueber einige in der Erde lebende Amoeben und andere Rhizopoden: *A. m. A.*, II, pp. 299–331. — **Id., '67.** Ueber Actinophrys Eichhornii und einen neuen Süsswasserrhizopoden: *A. m. A.*, III, p. 396. — **Id., '68.** Beobachtungen über die Fortpflanzung v. Infusorien: *Sitz. Ber. d. Niederrh. Ges. z. Bonn.*, pp. 90–2. — **Id., '69.** Ueber Radiolarien und Radiolarien-artige Rhizopoden des süssen Wassers: *A. m. A.*, V, pp. 464–502. — **Id., '71, 1.** Untersuchungen über den Bau und die Naturgesch. d. Vorticellen: *A. f. Naturg.*, XXXVI and XXXVII, pp. 353–84 and 185–221. — **Id., '71, 2.** Ueber die Actinophryen oder Sonnenthierchen des süssen Wassers als echte Radiolarien, *etc.*: *Sitz. Ber. der niederrh. Ges. in Bonn.*, XXVIII, pp. 4–6. — **Id., '73.** Ueber Radiolarien und Radiolarienartige Rhizopoden des Süss-wassers: *Sitz. Ber. d. Ges. z. Beford. d. Ges. Naturw. z. Marburg*, 1873, pp. 47–64. — **Id., '74.** Pelomyxa palustris (Pelobius), ein amöbenartigen Organismus des süssen Wassers: *A. m. A.*, X, pp. 51–74. — **Id., '91.** Ueber Erd-Amoeben: *B. C.*, XI, pp. 601–8 and 633–40. — **Greenwood, M., '86.** On the Digestive Process in Some Rhizopods, I: *Jour. of Phys*, VII, p. 253. — **Id., '87.** On the Digestive Process, *etc.*: *Jour. of Phys.*, VIII, p. 263. — **Id., '94.** On the Constitution and Mode of Formation of "Food Vacuoles" in Infusoria, as Illustrated by the History and Processes of Digestion in Carchesium: *Phil. Trans.*, 1894, p. 353. — **Griffiths, '89.** A Method of Demonstrating the Presence of Uric Acid in the Contractile Vacuoles of Some Lower Organisms: *Proc. Roy. Soc., Edinburgh*, XVI, pp. 131–5. — **Grenacher, H., '69.** Bemerkungen über Acanthocystis viridis Ehbg. Sp.: *Z. w. Z.*, XIX, pp. 289–96. — **Gruber, A., '80.** Kleine Beiträge z. Kenntniss der Protozoën: *Ber. d. Naturf. Gesell. z. Freiburg*, VII, pp. 533–55. — **Id., '81, 1.** Beiträge zur Kenntniss der Amöben: *Z. w. Z.*, XXXVI, pp. 459–70. — **Id., '81, 2.** Dimorpha nutans. Eine Mischform von Flagellaten und Heliozoen: *Z. w. Z.*, XXXVI, pp. 445–70. — **Id., '83, 1.** Untersuchungen über einige Protozoen: *Z. w. Z.*, XXXVIII, p. 45. — **Id., '83, 2.** Ueber Kerntheilung bei einige Protozoen: *Z. w. Z.*, XXXVIII, p. 372. — **Id., '84, 1.** Studien über Amöben: *Z. w. Z.*, XLI, p. 222. — **Id., 84, 2.** Ueber Kern- und Kerntheilung bei den Protozoen: *Z. w. Z.*, XL, pp. 121–53. — **Id., '84, 3.** Die Protozoen des Hafens von Genua: *Nov. Act. d. Kais. Leop.-Carol. Deut. Acad. d. Naturfor.*, XLVI, pp. 475–539. — **Id., '85.** Ueber Künstliche Teilung bei Infusorien: *B. C.*, V, pp. 137–41. — **Id., '86.** Beiträge z. Kenntniss der Physiologie und Biologie der Protozoën: *Ber. d. Naturf. Ges. z. Freiburg*, 1886, I, p. 33. — **Id., '87, 1.** Ueber die Bedeutung der Conjugation bei den Infusorien: *Ber. d. Naturf. Ges. z. Freiburg*, II, p. 31. — **Id., '87, 2.** Der Conjugationsprocess bei Paramœcium aurelia: *Ber. d. Nat. Ges. z. Freiburg*, II, p. 43. — **Id.**, **'94**. Amöben-studien: *Ber. d. Naturf. Ges. z. Freiburg*, VIII, pp. 24-34. — **Gruby, M.**, **'43**. Recherches et observations sur une nouv. espèce d'hematozoaire, Trypanosoma sanguinis: *C. R.*, XVII, pp. 1134-36. — **Gurley, R. R.**, **'93**. The Myxosporidia, or Psorosperms, of Fishes and the Epidemics produced by them: *Bull. U. S. Fish Commission*, XI, pp. 65-304.

HAECKEL, E., **'62**. Die Radiolarien (Rhizopoda radiaria): *Berlin*. — **Id.**, **'66**. Generelle Morphologie: *Berlin*. — **Id.**, **'68**. Monographie der Moneren: *J. Z.*, IV, pp. 64-145. — **Id.**, **'70**. Studien über Moneren und andere Protisten: in *Biol. Studien*, Heft I, *Leipzig*. — **Id.**, **'73**. Zur Morphologie der Infusorien: *J. Z.*, VII, pp. 516-60. — **Id.**, **'88**. The Radiolaria: *Challenger Reports, Zoology*, Vols. 17 and 18. — **Id.**, **'94**. Systematische Phylogenie des Thierreichs: I, *Phylog. d. Protisten und Pflanzen*. — **Harrington, N. R., and Leaming, E.**, **'98**. The Reaction of Amœba to Lights of Different Colors: *Amer. Jour. of Phys.*, III, pp. 9-18. — **Harting, P.**, **'73**. Recherches de morphologie synthétique sur la production artificielle de quelques formations calcaires organiques: *Verh. d. K. Ak. v. Wetenschappen*, XIII (see Dreyer, '92). — **Hartog, M. M.**, **'91**. Some Problems of Reproduction, *etc.*: *Q. J.*, XXXIII, pp. 1-81. — **Heidenhain, M.**, **'94**. Neue Untersuchungen über die Centralkörper und ihre Beziehungen z. Kern- und Zellen-protoplasma: *A. m. A.*, XLIII, pp. 423-758. — **Heitzmann, J.**, **'73**. Untersuchungen über das Protoplasma: *Sitz. Ber. d. K. Akad. Wiss. z. Wien*, LXVII, pp. 100-15. — **Henle, J.**, **'45**. Ueber die Gattung Gregarina: *Arch. f. Anat. und Phys.*, 1845, pp. 369-74. — **Henneguy, L. F.**, **'92**. (See Thélohan.) — **Hermann, F.**, **'91**. Beitrag zur Lehre von der Entstehung der Karyokinetischen Spindel: *A. m. A.*, XXXVII, pp. 569-86. — **Hertwig, O.**, **'75**. Beiträge zur Kenntniss der Bildung, Befruchtung und Theilung des Thierischen Eies: *M. J.*, I, p. 347. (See also Vols. III and IV.) — **Id.**, **'93**. Die Zelle und die Gewebe: *Jena*. — **Hertwig, R.**, **'74**. Ueber Mikrogromia socialis, eine Coloniebildende Monothalamie des süssen Wassers: *A. m. A.*, X, Suppl., pp. 1-34. — **Id.**, **'76, 1**. Ueber Podophrya gemmipara, nebst Bemerkungen zum Bau und zur Syst. Stellung d. Acineten: *M. J.*, I, pp. 20-82. — **Id.**, **'76, 2**. Beiträge zu einer einheitlichen Auffassung der verschiedenen Kernformen: *M. J.*, II, pp. 63-80. — **Id.**, **'77, 1**. Ueber den Bau und die Entwicklung der Spirochona gemmipara: *J. Z.*, XI, pp. 149-87. — **Id.**, **'77, 2**. Studien über Rhizopoden: *J. Z.*, XI, pp. 324-48. — **Id.**, **'77, 3**. Ueber Leptodiscus medusoides, eine neue, den Noctilucen verwandte Flagellate: *J. Z.*, XI, pp. 307-322. — **Id.**, **'79**. Der Organismus der Radiolarien: *Jena. Denkschriften*, II, pp. 129-277. — **Id.**, **'84**. Ueber die Kerntheilung bei Actinosphaerium Eich: *J. Z.*, 1884, pp. 490-518. — **Id.**, **'89**. Ueber die Konjugation der Infusorien: *Abh. d. bayr. Akad. d. Wiss., München*, Cl. II, XVII. — **Id.**, **'92**. Ueber Befruchtung und Conjugation: *Verh. d. deut. Zool. Ges.*, 1892, pp. 95-113. — **Id.**, **'95**. Ueber Centrosoma und Centralspindel: *Sitz. Ber. d. Ges. Morph. u. Phys. z. München*, I, 1895. — **Id.**, **'98, 1**. Ueber die Bedeutung der Nucleolen: *Sitz. Ber. d. Ges. f. Morph. u. Phys., München*, 1898, I, pp. 1-6. — **Id.**, **'98, 2**. Ueber Kerntheilung, Richtungskörperbildung und Befruchtung von Actinosphaerium Eich: *Abh. d. K. bay. Akad. d. Wiss., München*, II Kl., XIX, pp. 1-104. — **Id.**, **'99, 1**. Was veranlasst die Befruchtung der Protozoen? *Sitz. Ber. d. Ges. f. Morph. u. Phys., München*, 1899, I, pp. 1-8. — **Id.**, **'99, 2**. Mit welchem Recht unterscheidet man geschlechtliche und ungeschlechtliche Fortpflanzung? *Sitz. Ber. d. Ges. f. Morph. u. Phys., München*, 1899, II. — **Id.** **'99, 3**. Ueber Encystierung und Kern vermehrung bei Arcella vulg.: *Fest. v. Kupffer.*, pp. 367-82. — **Hertwig und Lesser, '74**. Ueber Rhizopoden und denselben nahe stehende Organismen: *A. m. A.*, X. Suppl., pp. 35-243. — **Hill, J.**, 1752. History of Animals. A General Natural History. Vol. III: *London*, 1748-52. — **Hofer, B.**, **'89**. Experimentelle untersuchungen über den Einfluss des Kerns auf das Protoplasma: *J. Z.*, XVII, p. 105. — **Hofmeister, B., '67.** Die Lehre v. d. Pflanzenzelle: *Leipzig.* — **Holman, L. E., '86.** Multiplication of Amœba: *Proc. Ac. Sci. Philadelphia*, 1886, pp. 346–8. — **Hooke, R., 1665.** Micrographia, or some physiological Descriptions of minute Bodies made by magnifying Glasses, *etc.: London*, 1665. — **Howell, Wm. H., '97.** An American Text-book of Physiology: *Philadelphia.* — **Hoyer, H., '99.** Ueber das Verhalten der Kerne bei der Conjugation des Infusors Colpidium colpoda St.: *A. m. A.*, LIV, pp. 95–134. — **Huxley, T. H., '51.** Zoological Notes and Observations made on Board H. M. S. "Rattlesnake"; III, Upon Thalassicolla, a new Zoophyte: *A. M. N. H.* (2), VIII, pp. 433–42. — **Id., '57.** On Dysteria, a new Genus of Infusoria: *Q. J.*, V, pp. 78–82. — **Id., '70.** Spontaneous Generation: *Nature.* (Also Lay Sermons and Addresses, Ed. 1878, p. 345.) — **Id., '78.** Article "Biology" in *Encyclop. Brit.*

ISHIKAWA, C., '91. Vorläufige Mittheilungen über die Conjugationverscheinungen bei den Noctiluceen: *Z. A.*, 1891, pp. 12–14. — **Id., '93.** Ueber die Kerntheilung bei Noctiluca miliaris: *Ber. d. Naturf. Ges. z. Freiburg*, VIII, pp. 54–68. — **Id., '94.** Studies on Reproductive Elements. I. Noctiluca miliaris Sur.: Its Division and Spore-formation: *Jour. Coll. Sci. Imper. Univ., Japan*, VI, pp. 297–334. — **Id., '96.** Ueber eine in Misaki vorkommende Art von Ephelota und über ihre Sporenbildung: *Jour. Coll. Sci. Imper. Univ., Japan*, X, pt. II, pp. 119–37. — **Id., '99.** Further Observations on the Nuclear Division of Noctiluca: *Jour. Coll. Sci., Japan*, XII, pt. IV, pp. 243–62. — **Israel, O., '94.** Ueber eine eigenartige Contractionserscheinungen bei Pelomyxa palustris Greeff: *A. m. A.*, pp. 228–36.

JACKSON, W. H., '75. On a New Peritrichous Infusorian, Cyclochæta spongillæ: *Q. J.*, XV, pp. 243–9. — **Jennings, H. S., '99, 1.** Studies on Reactions to Stimuli in Unicellular Organisms. II. The Mechanism of the Motor Reactions of Paramœcium: *Amer. Jour. Physiol.*, II, pp. 311–41. — **Id., '99, 2.** Studies on Reactions to Stimuli in Unicellular Organisms. III. Reactions to Localized Stimuli in Spirostomum and Stentor: *Amer. Naturalist*, XXXIII, pp. 373–89. — **Id., '99, 3.** Studies on Reactions to Stimuli in Unicellular Organisms. IV. Laws of Chemotaxis in Paramœcium: *Amer. Jour. Physiol.*, II, pp. 355–79. — **Id., '00, 1.** Studies on Reactions to Stimuli in Unicellular Organisms. V. On the Movements and Motor Reflexes of the Flagellata and Ciliata. — **Id., '00, 2.** Studies on Reactions to Stimuli in Unicellular Organisms. VI. On the Reactions of Chilomonas to Organic Acids: *Amer. Jour. Physiol.*, III, No. 9. — **Id., '00, 3.** Reactions of Infusoria to Chemicals; a Criticism: *Amer. Naturalist*, XXXIV, pp. 259–65. — **Id., '00, 4.** The Behavior of Unicellular Organisms. — *Woods Hole Biol. Lectures*, '99, pp. 93–112. — **Jickeli, C. F., '84.** Ueber die Kernverhältnisse der Infusorien, I: *Z. A.*, VII, pp. 468–73 and 491–7. — **Joblot, L., 1754.** Observations d'histoire natur., faites avec le microscope, *etc.: Paris.* — **Johnson, H. P., '93.** A Contribution to the Morphology and Biology of the Stentors: *J. M.*, VIII, pp. 468–552. — **Joly, N., '40.** Histoire d'un petit crustacé (Artemia salina Leach) auquel on a faussement attribué la colorat. en rouge des marais, *etc.: Ann. Sc. Nat.* (Zool.) (2), XIII, pp. 225–90. — **Joukowsky, D.** Beiträge zur Frage nach den Bedingungen der Vermehrung und des Eintrittes der Conjugation bei den Ciliaten: *Verh. Nat. Med. Ver., Heidelberg*, XXVI, pp. 17–42. — **Julin, Ch., '93.** Le corps vitellin de Balbiani et les éléments de la cellule des Métazoaires qui correspondent au macronucleus des Infusoires ciliés: *Bull. Sc. de la Fr. et de la Belg.*, XXV, pp. 294–345.

KENT, W. S., '81. Actinomonas: *Manual of the Infusoria, London.* — **Keuten, J., '95.** Die Kerntheilung von Euglena viridis Ehr.: *Z. w. Z.*, LX, pp. 215–35. — **King and Rowney, '66.** On the so-called "Eozoonal Rock": *Q. J. Geol. Soc.*, XXII, pp. 185–218 (also A. M. N. H., 4, XIII and XIV). — **Kirschner,**

O., '79. Zur Entwicklungsgesch. von Volvox minor (Stein): *Beit. z. Biol. d. Pflanzen, Cohn*, III, pp. 95-102. — **Klebs, G., '84.** Ein kleiner Beitrag zur Kenntniss der Peridineen: *Bot. Zeit.*, XLII, pp. 721-33 and 737-45. — **Id., '85.** Ueber Bewegung und Schleimbildung der Desmidiaceen: *B. C.*, V, p. 353. — **Id., '86.** Ueber die Organisation der Gallerte bei einigen Algen und Flagellaten: *Unters. Tübing. Institute*, II. — **Id., '92.** Flagellatenstudien, I: *Z. w. Z.*, LV, pp. 265-445. — **Kofoid, C. A., '98.** Plankton Studies. II. On Pleodorina Illinoisensis, a New Species from the Plankton of the Illinois River: *Bull. Ill. State Lab. of Nat. Hist.*, V, pp. 273-93. — **Id., '99.** Plankton Studies. III. On Platydorina, a new genus of the Family Volvocidæ, from the Plankton of the Illinois River: *Ill. State Lab. Nat. Hist.*, pp. 421-39. — **Kölliker, A., '45.** Die Lehre von den thierischen Zelle und den einfachen thierischen Formelementen, nach neuesten Fortschritten dargestellt: *Z. W. Bot.*, II, pp. 46-102. — **Id., '49.** Beiträge zur Kenntniss niederer Thiere: *Z. w. Z.*, I, pp. 1-37. — **Id., '64.** Icones Histologicæ, I: *Leipzig*. — **Kostanecki, K., '97.** Ueber die Bedeutung der Polstrahlung während der Mitose. *etc.*: *A. m. A.*, XLIX, pp. 651-706. — **Krassilstschik, J., '82.** Zur Entwicklungsgeschichte und Systematik der Gattung Polytoma Ehr.: *Z. A.*, pp. 426-9. — **Kruse, W., '90.** I. Ueber Blutparasiten: *Arch. Path. Anat.*, CXXI, p. 359. — **Kudelski, A., '98.** Note sur la Metamorphose partielle des Noyaux chez les Paramœcium: *Bibl. Anat., Fasc.* V, pp. 270-2. — **Kühne, W., '64.** Untersuchungen über das Protoplasma und die Contractilität: *Leipzig*. — **Kunstler, J., '89.** Recherches sur la morphologie des Flagellés. — *Bull. Sci. de la Fr. et de la Belg.*, 1889, pp. 399-515.

LABBÉ, A., '93. Coccidium Delagei, *etc.*: *A. Z. E.* (3), I, pp. 267-86. — **Id., '94.** Recherches zoologiques et biologiques sur les parasites endoglobulaires du sang des Vertébrés: *A. Z. E.* (3), II, pp. 55-258. — **Id., '95.** Sur le noyau et la division nucléaire chez les Benedenia: *C. R.*, 1895, pp. 381-5. — **Id., '96.** Recherches zoologiques, cytologiques et biologiques sur les Coccidiés: *A. Z. E.* (3), IV, pp. 517-654. — **Id., '99.** Sporozoa, in *Das Tierreich, Berlin*. — **Lachmann, J., '56.** Ueber die Organisation der Infusorien, besonders d. Vorticellen: *Arch. f. Anat. u. Phys.*, 1856, pp. 340-98. — **Lamarck, J. P. B. A. de, '04.** Coquilles fossils: *Ann. de Museum*, V, p. 179. — **Id., '15.** Histoire Naturelle des Animaux sans Vertébrés: I, II, *Paris*. — **Lambl, W., '60.** Beobachtungen und Studien aus dem Gebiete der patholog. Anatomie und Histologie: *Franz-Josef Kinderhospital in Prag*, 1860, p. 362. — **Lameere, A., '91.** Prolégomènes de Zoogénie: *Bull. Sc. de la France et de la Belg.*, XXIII, pp. 399-411. — **Langmann, G., '99.** On Hæmosporidia in American Reptiles and Batrachians: *N. Y. Med. Journ.*, Jan., 1899. Reprint pp. 1-17. — **Lankester, E. R., '79.** Lithamœba discus.: *Q. J.* (N. S.), XIX, pp. 484-7. — **Id., '82.** On Drepanidium ranarum, *etc.*: *Q. J.*, XXII, pp. 53-65. — **Id., '91.** Protozoa, in *Zoological Articles* from *Encyclopedia Brit.*, *Edinburgh*. — **Lauterborn, R., '90.** Protozoenstudien, III, Multicilia: *Z. w. Z.*, LX, pp. 236-48. — **Id., '94.** Ueber die Winterfauna, *etc*: *B. C.*, XIV, pp. 390-98. — **Id., '95.** Protozoenstudien, Kern- und Zelltheilung von Ceratium hirundinella O. F. M.: *Z. w. Z.*, LIX, pp. 167-91. — **Laveran, '93.** Etiologie de la dysenterie: *Sem. Méd.*, 1893, p. 508. — **Id., '98.** Sur les modes de reproduction d'Isospora Lacazei: *C. R. S. B., Paris* (10), V, pp. 1139-42. — **Ledermüller, M. F., 1760-63.** Mikroskopische Gemüths und Augen ergötzungen: *Nürnberg*. — **Lee, F. S., '98.** Translation of Verworn's Allgemeine Physiologie: *N. Y.* — **Leeuwenhoek, A., 1677.** Letter translated in Philosophical Transactions, *London*, XII, p. 821. — **Léger, L., '92.** Recherches sur les Grégarines: *Tabl. Zool.*, III, pp. 1-182. — **Id., '93.** Sur une nouvelle grégarine terrestre des larves de Mélolonthides de Provence: *C. R.*, CXVII, p. 129. — **Leidy, J., '50.** Nyctotherus, a new genus of Polygastrica allied to Plœsconia: *A. M. N. H.* (2), V, pp. 158-9. — **Id., '77, 1.** The Parasites of the Termites: *Journ. Ac.*

Nat. Sc. Phil. (N. S.). VIII, pp. 425-47. — **Id.**, **'77, 2.** Remarks upon Rhizopods, and Notice of a New Form: *Proc. Ac. Sci. Phil.*, 1877, p. 293. — **Id.**, **'80.** Rhizopods in the Mosses of the Summit of Roan Mountain, N. Carolina: *Proc. Ac. Nat. Sci. Phil.*, 1880, pp. 333-40. — **Leuckart, R.**, **'52.** Parasitismus und Parasiten: *Arch. f. Phys. Heilk.*, XI, pp. 429-36. — **Id.**, **'79.** Die Parasiten des Menschen: *Leipzig.* — **Leydig, F.**, **'51.** Ueber Psorospermien und Gregarinen: *Arch. f. Anat. u. Phys.*, 1851, pp. 221-33. — **Id.**, **'57.** Lehrbuch d. Histologie: *Frankfurt*, pp. 15-18, 133, 329, 344, 395. — **Limbach, J.**, **'80.** Einige Bemerk. über d. contract. Behälte, der Vorticellen: *Kosmos, Lemburg*, pp. 213-21. — **Linnæus, C. V.**, **1758-66.** Systema naturæ: *Ed. X, Ed. XIII.* — **Lister, J. J.**, **'95.** Contribution to the Life History of the Foraminifera: *Phil. Trans.*, CLXXXVI, pp. 401-53. — **Loeb, J.**, **'93.** Ueber künstliche Umwandlung positiv heliotropischen Thiere in negativ heliotropische und umgekehrt: *Arch. f. d. Ges. Phys.*, LIV, pp. 81-107. — **Loeb and Hardesty**, **'95.** Ueber die Localisation der Athmung in der Zelle: *Arch. f. d. Gesam. Phys.*, LXI, pp. 583-94. — **Lösch, F.**, **'75.** Massenhaft Entwickelung von Amoeben im Dickdarm: *Virchow's Arch.*, XLV, p. 196. — **Lutz, A.**, **'89.** Ueber ein Myxosporidium aus der Gallenblase brasilianischer Batrachier: *Centr. Bakt.*, V, pp. 84-88.

MACBRIDE, T. H., **'99.** The North American Slime Moulds: *N. Y.* — **Maccallum, W. G.**, **'97.** On the Flagellated Form of the Malarial Parasite: *Lancet*, Nov. 13, pp. 1240-1. — **Macfarland, F. M.**, **'97.** Celluläre Studien an Mollusken Eiern: *Zool. Jahr.* (*Anat.*), X, pp. 227-65. — **Manson, P.**, **'96.** The Life-history of the Malaria Germ: *Lancet*, 1896. — **Id.**, **'99.** Tropical Diseases: *London, New York, and Paris.* — **Marshall**, **'93.** — Beiträge zur Kenntniss der Gregarinen: *Arch. f. Naturg.*, LXIX, pp. 29-44. — **Mass. State Board of Health**: *Annual Reports*, 1891, 1892, and 1893. — **Massart, J.**, **'91.** Recherches sur les organismes inférieures: *Bull. d. l'Acad. Roy. Belgique* (3), XXII, pp. 148-67. — **Maupas, E.**, **'81.** Contributions à l'étude des Acinétiens: *A. Z. E.* (1), pp. 299-368. — **Id.**, **'83, 1.** Contributions à l'étude morphologique et anatomique des Infusoires ciliés: *A. Z. E.* (2), I, pp. 427-664. — **Id.**, **'83, 2.** Les Suctociliés de M. Mereschkowsky: *C. R.*, XCV, pp. 1381-4. — **Id.**, **'88.** Recherches expérimentales sur la multiplication des Infusoires ciliés: *A. Z. E.* (2), VI, pp. 165-277. — **Id.**, **'89.** Le rejeunissement karyogamique chez les ciliés: *A. Z. E.* (2), VII, pp. 149-517. — **Meissner, M.**, **'88.** Beiträge zur Ernährungsphysiologie der Protozoen: *Z. w. Z.*, XLVI, pp. 498-516. — **Mendelssohn, M.**, **'95.** Ueber den Thermotropismus einzelliger Organismen: *Arch. f. d. Ges. Phys.*, LX, pp. 1-27. — **Mereschkowsky, C.**, **'81.** Note on Wagnerella borealis, a Protozoan: *A. M. N. H.* (5), VIII, pp. 288-90. — **Id.**, **'82.** Les Suctociliés, nouv. groupe d'Infusoires, intermédiares entre les Ciliés et les Acinétiens: *C. R.*, XCV, pp. 1232-4. — **Mesnil, F.**, **'00.** Essai sur la Classification et l'origine des Sporozoaires: *Cinquantenaire de la Soc. d. Biol.*, pp. 1-17. — **Metschnikoff, E.**, **'63.** Untersuchungen über den Stiel der Vorticellinen: *Arch. f. Anat. u. Phys.*, 1863, pp. 180-6. — **Id.**, **'64.** Ueber die Gattung Sphærophrya: *Arch. f. Anat. u. Phys.*, 1864, pp. 258-61. — **Id.**, **'83.** Untersuchungen über die intracelluläre Verdauung bei Wirbellosen: *Arb. a. d. Zool. Inst. Wien*, V, pp. 141-68. — **Id.**, **'89.** Recherches sur la digestion intracellulaire: *Ann. de l'Inst. Pasteur*, III, pp. 25-9. — **Meyen, J.**, **'39.** Einige Bemerkungen über den Verdauungsapparat der Infusorien: *Arch. f. Anat. u. Phys.*, 1839, pp. 74-9. — **Meyer, H.**, **'97.** Ochromonas granulosa, Untersuchungen über einige Flagellaten: *Rev. Suisse de Zool.*, V. — **Mingazzini, P.**, **'90.** Sullo svillupo dei Myxosporidi: *Boll. Soc. Napoli* (1), IV, p. 160. — **Id.**, **'91, 1.** Le gregarine monocistidee dei Tunicati e della Capitella: *Att. Acc. Lincei* (4), VII, p. 407. — **Id.**, **'91, 2.** Gregarine monocistidee nuovo o poco conosciente del Golfo di Napoli: *Att. Acc. Lincei* (4), VII, pp. 229-35. — **Id.**, **'91, 3.**

Le Gregarine delle Oloturie: *Att. Acc. Lincei* (4), VII, pp. 313-19. — **Id.**, **'92, 1.** Contributo alla conoscenza dei Coccidi: *Rend. d. Acc. d. Lincei* (5), I, pp. 175-81. — **Id., '92, 2.** Nuovo specie di Sporozoi: *Att. d. Acc. Lincei* (5), I, pp. 396-402. — **Minot, C. S., '77.** Recent Investigations of Embryologists: *Proc. Boston Soc. Nat. Hist.*, XIX. — **Id., '79.** Growth as a Function of Cells: *Proc. Boston Soc. Nat. Hist.*, XX. — **Möller, V. von, '78.** Die Spiralgewundnen Foraminiferen des russischen Kohlenkalks: *Mém. d. l'Acad. Impér. d. St. Petersbourg* (7), XXV, pp. 1-140. — **Montfort, D. de., '10.** Conchyliologie systématique et classification méthodique des Coquilles: *Paris.* — **Moore, J. E. S., '93.** On the Structural Differentiation of the Protozoa as seen in Microscopical Sections: *J. Linn. Soc.*, XXIV. — **Mottier, D. M., '97.** Beiträge zur Kenntniss der Kerntheilung in den Pollenmutterzellen: *Jahrb. Wiss. Botan.*, XXX. — **Müller, J. O. H., '41.** Ueber einen krankhaften Hautaufschlag mit specifisch. Organisirten Sämenkörperchen: *Mon. Ber. d. Berlin Akad.*, 1841, pp. 212-22, 246-50. — **Id., '56.** Einige Beobachtungen an Infusorien: *Mon. Ber. d. Berlin Akad.*, 1856, pp. 390-2. — **Id., '55-'58.** Four Contributions from 1855 to 1858. Ueber Sphærozoum u. Thalassicolla, 1855, pp. 229-53: *Mon. Ber. d. Berlin Akad.*, 1855, 1856, 1858. — **Müller, O. F., 1773.** Vermium terrest. et fluviatil. sen animal. infusor., *etc.*: *Leipzig*, 1773. — **Id., 1785.** Animalcula infusoria, fluviat. et marina, *etc.*; Op. posth. cura O. Fabricii: *Leipzig*, 1786. — **Munier-Chalmas et Schlumberger, '83.** Nouvelles Observations sur le dimorphisme des foraminifères: *C. R.*, LXXXIII, pp. 862-6, 1598-1601.

NÄGELI, C., '84. Mechanisch-physiologische Theorie der Abstammungslehre: *München u. Leipzig*, 1884. — **Needham, T., 1784.** A Summary of Some Late Observations upon the Generation, Composition, and Decomposition of Animal and Vegetable Substances: *Phil. Trans. London*, XLV, pp. 615-66. — **Nitzsch, C. L., '27.** Article "Cercaria," in Ersch und Gruber's *Allgem. Encycl. d. Wissenschäften u. Künste.*, Th. 16, p. 68. — **Norman, A. M., '78.** On the Genus Haliphysema, with Description of Several Forms apparently allied to it: *A. M. N. H.* (5), I, pp. 265-84. **Nussbaum, M., '84.** Ueber Spontane und Künstliche Zelltheilung: *Sitz. ber. d. niederrheinischen Gesellschaft für Natur und Heilkunde in Bonn*, XLI, p. 259. — **Id., '85.** Ueber die Theilbarkeit der lebendigen Materie: *A. m. A.*, XXVI, pp. 484-538.

OERSTED, '73. System der Pilze, Lichnen, und Algen: *Leipzig*, p. 141. — **Oken, L. von, '05.** Ueber der Zeugung: *Bamberg.* — **Id., '15.** Lehrbuch der Naturgeschichte (See Elements of Physiophilosophy. Translation by Alfred Tulk. *Ray Soc.*, p. 189, 1847.): *Leipzig.* — **D'Orbigny, A., '26.** Tableau méthodique de la classe des Céphalopods, *etc.*: *Ann. d. Sc. Nat.*, VII, pp. 96, 245. — **Id., '39.** Voyage dans l'Amérique Méridionale: *Paris*, V, p. 5, Foraminifères. — **Id., '40.** Mémoire sur les Foraminifères de la Craie blanche du bassin de Paris: *Mém. Soc. Géol. de France*, IV, p. 1. — **Id., '46.** Foraminifères fossiles du bassin tertiaire d. Vienne: *Paris.* — **Osterhout, W. J. V., '97.** Ueber Entstehung der Karyokinetischen Spindel bei Equisetum: *Jahrb. Wiss. Bot.*, XXX. — **Owen, R., '43.** On the Generation of the Polygastric Infusoria: *Edin. n. Phil. Journ.*, XXXV, pp. 185-90.

PARKER, W. K., Brady and Jones, '70. A Monograph of the Genus Polymorphina: *Trans. Linn. Soc., London*, XXVII, p. 197. — **Parker and Jones, '59.** On the Nomenclature of the Foraminifera: *A. M. N. H.* (3), III, IV. — **Pearl, R., '00.** Studies on Electrotaxis. 1. On the Reactions of Certain Infusoria to the Electric Current: *Amer. Jour. Phys.*, IV, pp. 96-123. — **Pénard, E., '88.** Contributions à l'étude des dinoflagellés: *Dissertation Fac. Genève.* — **Id., '90, 1.** Die Heliozoën der Umgegend von Wiesbaden: *Jahrb. d. Nassau. Ver. f. Nat. K. Wiesbaden.* — **Id., '90, 2.** Études sur les rhizopodes d'eau douce: *Mém. Soc. Phy. Hist. Nat. Genève*, XXXI, No. 2. — **Id., '90, 3.** Ueber einige neue oder wenig bekannte Protozoën: *Jahrb. d. Nassau. Ver. Naturk. Wiesbaden*, LXIII.—**Perty, M., '49, 1.** Mikroskopische Organismen der Alpen und der italienischen Schweiz: *Mitth. d. Naturf. Ges. in Bern*, 1849, Nos. 164-5, pp. 153-76.—**Id., '49, 2.** Ueber verticale Verbreitung mikroskopischer Lebensformen: *Mitth. d. Naturf. Ges. in Bern*, 1849, Nos. 146-49, pp. 17-45.—**Id., '49, 3.** Eine physiologische Eigenthümlichkeit der Rhizopodensippe Arcella Ehr.: *Mitt. d. Naturf. Ges. in Bern*, 1849, Nos. 158-9, pp. 124-6.—**Id., '52.** Zur Kenntniss kleinster Lebensformen, *etc.*: *Bern*.—**Petridis, '98.** Recherches bactériologiques sur la pathogénie de la dysenterie et de l'abscès du foie d'Egypt: *Ann. d. Mic., Paris*, X, p. 192.—**Pfeiffer, L., '95.** Protozoën als Krankheitserreger; *1st ed.*, 1890, *2d ed.*, 1891, with Supplement, 1895.—**Pfitzner, W., '86.** Zur Kenntniss der Kerntheilung bei den Protozoën: *M. J.*, XI, pp. 454-67.—**Pflüger, E., '75.** Ueber die physiologische Verbrennung in den lebendigen Organismen: *Arch. f. d. Ges. Physiol.*, X, p. 251.—**Phillips, J., '45.** On the Remains of Microscopic Animals in the Rocks of Yorkshire: *Proc. Geol. and Polytech. Soc. W. R., Yorkshire, Leeds*, II, p. 277.—**Id., '82.** Note on a New Ciliate Infusorian allied to Pleuronema (Calyptotiica n. g.): *Jour. Linn. Soc., London* (Zool.), XVI, pp. 47-68.—**Pineau, F., '45.** Recherche sur le développement des animalcules infus. et des moisissures: *Ann. d. Sc. Nat.* (III.), *Zool.*, III, pp. 182-9, and supplement IV, pp. 103-4.—**Plate, L., '86.** Untersuchungen einiger an den Kiemenblätten des Gammarus pulex lebenden Ektoparasiten: *Z. w. Z.*, XLIII, pp. 175-241.—**Podwyssozki, W., '94.** Parasitologisches und Bakteriologisches vom V. Pirogow'schen Kongresse der Russischen Aerzte z. St. Petersburg: *Cent. Bakt.*, XV, pp. 481-5.—**Porter, F., '97.** Trichonympha and Other Parasites of Termes flavipes: *Bull. Mus. Comp. Zool.*, XXXI, pp. 47-68.—**Pouchet, F. A., '64.** Embryogénie des Infusoires ciliés: *C. R.*, LIX, pp. 276-83.—**Id., '83.** Contribution à l'histoire des Cilioflagellés: *Journ. d. l'Anat. et de la Phys.*, pp. 379-455.—**Id., '85.** Nouvelle contrib. à l'histoire des Péridiniens marins: *Journ. d. l'Anat. et de la Phys.*, XXI, pp. 28-88.—**Id., '86.** Sur Gymnodinium polyphemus (Pouchet): *C. R.*, CIII, pp. 801-3.—**Id., '92.** Cinquième Contribution à l'histoire des Péridiniens; Peridinium pseudonoctiluca Pouchet: *Journ. de l'Anat. et de la Phys.*, XXVIII, pp. 143-50.—**Pringsheim, N., '69.** Ueber Paarung von Schwärmsporen, *etc.*: *Monatsber. d. Berlin Akad.*, 1869, pp. 721-38.—**Prowazek, S., '00.** Protozoenstudien, II: *Arb. a. Zool. Inst. Wien*, XII, pp. 243-301.—**Przesmycki, M., '94.** Ueber die Zellkörnchen bei den Protozoën: *B. C.*, XIV, pp. 620-6.—**Purkinje and Valentin, '35.** De phenomeno generali et fundamentali motus vibratorii, *etc.*: *Vratislaviæ*.

QUATREFAGES, A. de, '50. Mémoire sur la phosphorescence de quelques invertébrés marins: *Ann. d. Sc. Nat.*, XIV, pp. 236-87.—**Quennerstedt, A., '67.** Bidrag till Sveriges Infusorie-fauna: *Lunds Univ. Arsskrift*, IV (see also Vol. II, 1865, and Vol. VI, 1869).

RABL, C., '89.—Ueber Zelltheilung: *A. A.*, IV, pp. 23-30.—**Racovitza and Labbé, '96.** Pterospora maldaneorum n. g. n. sp.: *Bull. soc. zool. France*, XXII, pp. 92-97.—**Réaumur, R. A. F. de, 1738.** Mém. pour servir à l'histoire des Insectes: *Paris*, 1734 to 1742; Vol. IV, 1838, pp. 430-6.—**Radziszewski, '80.** Ueber das Leuchten des Lophius: *Ber. d. deutsch. chem. Ges.*, X, p. 170.—**Reinhardt, L., '76.** Die Copulation der Zoosporen bei Chlamydomonas pulvisculus und Stigeoclonium sp.: *Arb. der Naturf. Ges. a. d. Univ. z. Charkoff*, X.—**Reinke, F., '94.** Zellstudien, I: *A. m. A.*, XLIII, XLIV.—**Reuss, A. E., '49.** Neue Foraminiferen aus d. Schichten des Osterreichischen Tertiärsbecken: *Denksch. d. math. naturw. Kl. d. k. Akad. Wiss. Wien*, I, p. 365.—**Id., '61.** Entwurf. ein. syst. Zusammenstellung d. Foraminiferen: *Sitz. Ber. d. KK. Akad. d. Wiss. z. Wien*, XLIV, p. 861.—**Rhumbler, L., '88.** Die verschiedenen Cystenbildungen und die Entwicklungsgeschichte der holotrichen Infusorien gattung Colpoda:

Z. w. Z., XLVI, pp. 549-601. — **Id.**, **'93**. Ueber Entstehung und Bedeutung d. in den Kernen vieler Protozoen und in Keimbläschen von Metazoen vorkommenden Binnenkörper (Nucleolen), *etc.*: *Z. w. Z.*, LVI, pp. 328-64. — **Id.**, **'96**. Beiträge zur Kenntniss der Rhizopoden: *Z. w. Z.*, LXI, pp. 38-110. — **Id.**, **'97**. Ueber die phylogenetisch abfallende Schalen-Ontogenie der Foraminiferen und deren Erklärung: *Verh. d. deutsch. Zool. Ges.*, 1897, pp. 162-92. — **Id.**, **'98, 1**. Physikalische Analyse von Lebenserscheinungen der Zelle: *Arch. f. Entwickl.*, VII, pp. 103-350. — **Id.**, **'98, 2**. Zell-leib- Schalen- und Kernverschmelzungen bei den Rhizopoden, *etc.*: *B. C.*, XVIII, pp. 21-33, 69, and 113. — **Risso, J. A.**, **'26**. Histoire naturelle des principales productions de l'Europe, *etc.*; *Paris*, 1826-7. — **Robin, C.**, **'78**. Recherches sur la reproduction gemmipare et fissipare des Noctiluques: *Jour. d. l'Anat. et de la Phys.*, XIV, pp. 563-629. — **Romanes, G. J.**, **'83**. Animal Intelligence: *London*. — **Römpel, J.**, **'94**. Kentrochona nebaliæ n. g. n. sp., *etc.*: *Z. w. Z.*, LVIII, pp. 618-36. — **Rood, O.**, **'53**. On the Paramœcium aurelia: *Silliman's Amer. Journal Sci.* (2), XV, pp. 70-72. — **Ross, R.**, **'97**. On Some Peculiar Pigmented Cells found in Two Mosquitoes fed on Malarial Blood: *Brit. Med. and Surg. Jour.*, Dec. 18, 1897, pp. 1786-88. — **Id.**, **'00**. Diagrams Illustrating the Life-history of the Parasites of Malaria: *Q. J.*, XLIII, pp. 571-81. — **Rossbach, M. J.**, **'72**. Die rhythmischen Bewegungserscheinungen der einfachsten Organismen, *etc.*: *Verh. d. phys. med. Ges. z. Wurzburg* (N. F.), II, pp. 179-242. — **Rostafinski, J.**, **'75**. Quelques mots sur l'Hæmatococcus lacustris et sur les bases d'une classification nat. des Algues chlorosporées: *Mém. soc. nat. d. sci. nat. d. Chérbourg*, XIX, pp. 137-54. — **Rouget, C.**, **'61**. Sur les phénomènes de polarisation qui s'observent dans quelques tissus végétables et des animaux: *Jour. de la Physiol.*, V. — **Roux, W.**, **'83**. Ueber die Bedeutung der Kerntheilungsfiguren: *Leipzig*. — **Id.**, **'94**. Ueber den "Cytotropismus" der Furchungszellen des Grasfrosches (Rana fusca): *Arch. f. Entwickl.*, I, pp. 43-68 and 161-202. — **Rückert, J.**, **'94**. Zur Eireifung bei Copepoden: *Anat. Hefte*, IV, pp. 263-351.

SACHAROFF, N., **'96**. Ueber die Entstehungsmodus der verschiedenen Varietäten der Malaria-parasiten, *etc.*: *Cent. f. Bakt. Paras.*, XIX, pp. 268-73. — **Sandahl**, **'57**. (See On the Genus Astrorhiza of Sandahl, *etc.*, by W. B. Carpenter): *Q. J.*, XVI, p. 221. — **Sars, M.**, **'65**. Bemark. over det dyreske Livsudbredn. in Hav. Dybner.: *Forh. vidensk. selsk.*, *Christiania*, 1865. — **Id.**, **'71**. (See Brady, '84.) — **Sassaki, C.**, **'93**. Untersuchungen über Gymnosphaera albida ein neue Marine Heliozoë: *J. Z.*, XXVIII, pp. 45-51. — **Saussure, H. B.**, **1769**. (See French translation of Spallanzani's work, 1766.) — **Schaaffhausen**, **'68**. Ueber die Organisation der Infusorien: *Verh. d. Naturh. Ver. d. preuss. Rheinl. und Westphal. Correspondenzblatt*, 1868, pp. 52-56. — **Schaudinn, F.**, **'93**. Myxotheca arenilega, nov. gen. nov. spec., *etc.*: *Z. w. Z.*, LVII, pp. 18-32. — **Id.**, **'94, 1**. Die Fortpflanzung der Foraminiferen und eine neue Art der Kernvermehrung: *B. C.*, XIV, pp. 161-6. — **Id.**, **'94, 2**. Ueber Kerntheilung mit nachfolgender Korpertheilung bei Amoeba crystalligera Gruber: *Sitz. Ber. d. Ak. d. Wiss. z Berlin*, 1894, pp. 1029-36. — **Id.**, **'94, 3**. Die systematische Stellung und Fortpflanzung von Hyalopus n. g. (Gromia dujardinii M. Schultze): *Sitz. Ber. Ges. Nat. Freunde, Berlin*, pp. 14-22. — **Id.**, **'95**. Untersuchungen an Foraminiferen, I, Calcituba polymorpha: *Z. w. Z.*, LIX, pp. 191-233. — **Id.**, **'96, 1**. Ueber die Centralkorn der Heliozoën, eine Beitrag zur Centrosomenfrage: *Verh. d. deutsch. Zool. Ges.*, 1896, pp. 113-30. — **Id.**, **'96, 2**. Ueber den Zeugungskreis von Paramoeba Eilhardi n. g. n. sp.: *Sitz. Ber. Akad. Wiss. Berlin*, 1896, I, pp. 31-41. — **Id.**, **'96, 3**. Ueber die Copulation von Actinophrys sol.: *Sitz. Ber. Akad. Wiss. Berlin*, pp. 83-9. — **Id.**, **'96, 4**. Camptonema nutans n. g. n. sp. ein neuer marine Rhizopode: *Sitz. Ber. K. Preuss. Akad. d. Wiss*, LII, pp. 1277-86. — **Schaudinn und Siedlecki**, **'97**. —

Beiträge zur Kenntniss der Coccidien: *Verh. d. Deutsch. Zool. Gesell.*, p. 192.—**Scheel, C., '99.** Beiträge zur Fortpflanzung der Amöben: *Festschrift. zum Siebenzigsten Geburtstag von C. V. Kupffer*, pp. 569-80.—**Schewiakoff, W., '88.** Ueber die karyokinetische Kerntheilung der Euglypha alveolata: *M. J.*, XIII, pp. 193-259.—**Id., '89.** Beiträge zur Kenntniss der Holotrichen Ciliaten: *Bibliotheca Zoologica*, Heft V, pp. 1-77.—**Id., '92.** Ueber die geographische Verbreitung der Süsswasser-Protozoën, *Verh. Nat. Med. Ver. Heidelberg* (2), IV, pp. 544-67.—**Id., '93, 1.** Ueber einen neuen Bakterienähnlichen Organismus: *Habil. Schrift. Heidelberg.*—**Id., '93, 2.** Ueber die Natur der sogenannten Exkretkörner der Infusorien: *Z. w. Z.*, LVII, pp. 32-56.—**Id., '94.** Ueber die Ursache der fortschreitenden Bewegung der Gregarinen: *Z. w. Z.*, LVIII, pp. 340-54.—**Schleiden, M. J., '38.** Beiträge zur Phytogenesis: *Müller's Arch.*, 1838, pp. 137-76. (Tr. in Sydenham Soc., XII, London, 1847.)—**Schlumberger, P., '45.** Observations sur quelques nouvelles espèces d'infusoires de la famille des Rhizopodes: *Ann. d. Sc. Nat.* (3), III, pp. 254-6.—**Id., '83** (See Munier-Chalmas.)—**Schmarda, L. K., '50.** Neue Formien von Infusorien: *Denkschr. der Kaiserlichen Akad. d. Wiss. Wien.*, 1850, I, Abth. 2.—**Id., '54.** Zur Naturgeschichte Ägyptens: *Denk. d. Wiener Akad.*, VII, Abth. 2, p. 1.—**Schmidt, O., '49.** Einige neue Beobachtungen über die Infusorien: *Froriep's Notizen aus dem Gebiete der Natur u. Heilkunde*, 3 Reihe, IX, 1, pp. 5-6.—**Schneider, A., '75, 1.** Contributions à l'histoire des Grégarines des Invertébrés de Paris et de Roscoff (1): *A. Z. E.*, pp. 585-604.—**Id., '75, 2.** Sur un appareil de dissémination des Grégarina et Stylorhynchus: *C. R.*, LIII, pp. 432-5.—**Id., '78, 1.** Monodia confluens, nouvelle monère: *A. Z. E.* (1), VII, pp. 585-8.—**Id., '78, 2.** Beiträge zur Kenntniss der Protozoën: *Z. w. Z.*, XXX, supp., pp. 446-56.—**Id., '81.** Sur les psorospérmies oviformes ou Coccidées, espèces nouvelles ou peu connues: *A. Z. E.* (1), IX, pp. 387-404.—**Id., '82.** Seconde Contribution à l'étude des Grégarines: *A. Z. E.*, X, pp. 423-50.—**Id., '38.** Nouvelles Observations sur la Sporulation du Klossia octopiana. *A. Z. E.*, XI, pp. 77-104.—**Id., '84.** Ophryocystis Bütschlii: *A. Z. E.* (2), II, pp. 111-26.—**Schrank, F. V. P., 1793.** Mikroskopische Wahrnehmungen: *Der Naturforcher*, 27 Stück, pp. 26-37.—**Id., '03.** Fauna boica: Bd. III.—**Schuberg, A., '83.** Die Protozoen des Wiederkauermagens (1): *Z. J.*, III, pp. 365-418.—**Id., '87.** Ueber den Bau der Bursaria truncatella; mit besonderer Berücksichtigung der protoplasmatischen Strukturen: *M. J.*, XII, pp. 333-63.—**Id., '88.** Die Parasitischen Amöben des menschlichen Darmes: *Cent. f. Bakt. und Par.*, XIII, pp. 598-609.—**Id., '96.** Die Coccidien aus dem Darme der Maus: *Verh. Nat. Med. er. Heidelburg* (2), V, pp. 369-98.—**Schütt, F., '90.** Ueber Peridineen-Farbstoffe: *Ber. d. Bot. Ges.*, VIII, pp. 9-32.—**Id., '95.** Die Peridineen der Plankton-Expedition: *Leipzig.*—**Schultze, F. E., '74.** Rhizopodenstudien 1, *etc.*: *A. m. A.*, Bd. X, pp. 328-50.—**Id., '75.** Rhizopodenstudien, V: *A. m. A.*, XL, p. 583.—**Schultze, M., '54.** Ueber den Organismus der Polythalmien (Foraminifera) nebst Bemerk. über den Rhizopoden in allgemeine: *Leipzig.*—**Id., '63.** Das Protoplasma der Rhizopoden und der Pflanzenzellen: *Leipzig.*—**Schwalbe, G., '66.** Ueber die Contractilen Behälter der Infusorien: *A. m. A.*, II, pp. 351-71.—**Schwann, T., '39.** Mikroskopische Untersuchungen über die Uebereinstimmung in der Structur und dem Wachstum der Thiere und Pflanzen: *Berlin* (Tr. in Sydenham Soc. XIII, London, 1847).—**Schweigger, A. F., '20.** Handbuch der Naturgesch. der Skeletlösen umgegliederten Thiere: *Leipzig.*—**Seguenza.** (See Brady, '82.)—**Shaw, R., '94.** Pleodorina, a new genus of the Volvocineæ: *Bot. Gaz.*, XIX, pp. 278-83.—**Siddall, J. D., '80.** On Shepheardella, an Undescribed Type of Marine Rhizopoda, *etc.*: *Q. J.*, XX, pp. 130-45.—**Von Siebold, C. T., '48.** Lehrbuch d. vergleich. Anatomie d. wirbellösen Thiere.—**Siedlecki, M., '98.** Étude cytologique et cycle évolutif de la coccodie de la Seiche: *Ann. d. l'Inst. Pasteur*, XII, pp. 799-836.—

Id., **'99**. Ueber die geschlechtliche Vermehrung der Monocystis ascidiae R. Lank: *Bull. d. l'Acad. d. Sci. d. Cracovie*, 1899, Dec. — **Siedlecki und Schaudinn**. (See Schaudinn and Siedlecki, '96.) — **Simond, P. L.**, **'97, 1**. Recherches sur les formes de reproduction asporulée dans le genre Coccidium: *C. R. S. B.* (10), IV, pp. 425-8. — **Id.**, **'97, 2**. L'Évolution des Sporozoaires der genre Coccidium: *Ann. d. l'Inst. Past.*, II, pp. 545-80. — **Sorokin, N.**, **'78**. Ueber Gloidium quodrifidum, *etc.*: *M. J.*, IV, pp. 399-402. — **Spallanzani, A.**, **1776**. Opuscoli di fisica animale e vegetabile: *Modena*. — **Stache**. (See Brady, '82, '84.) — **Stahl, E.**, **'84**. Zur Biologie der Myxomyceten: *Bot. Zeit.*, XLII, pp. 145, 161, 187. — **Steenstrup, J. J. S.**, **'42**. Am Forplantning og Udvikling gj vexlende generationsrakker, *etc.*: *Trans. Ray Soc.*, '45, Kjöbnh., pp. 52-53, 57. — **Stein, F.**, **'48**. Ueber die Natur der Gregarinen: *Arch. Anat. Phys. Med.*, 1848, pp. 183-223. — **Id.**, **'49**. Untersuch. über die Entwicklung der Infusorien: *A. f. Naturg.*, I, pp. 92-148. — **Id.**, **'51**. Neue Beiträge zur Kenntniss der Entwicklungsgeschichte und des feineren Baues der Infusionsthiere: *Z. w. Z.*, III, pp. 475-509. — **Id.**, **'54**. Die Infusionsthiere auf ihre Entwicklungsgesch. untersucht: *Leipzig*. — **Id.**, **'56**. In Tageblatt der 32 Versammlung deutscher Naturförscher und Ärzte, im Jahr 1856, p. 55. — **Id.**, **'57**. Ueber die aus eigener Untersuchungen bekannt gewordenen Süsswasser Rhizopoden: *Sitz. Ber. d. K. böhm. Ak. W.*, X, pp. 41-3. — **Id.**, **'59**. Der Organismus der Infusionsthiere, 1 Abth.: *Leipzig*. — **Id.**, **'61**. Ueber ein neues paras. infus. aus d. Darmkanal v. Paludina: *Sitz. ber. d. K. böhm. Ges. d. W. in Prag*, p. 85. — **Id.**, **'62**. Neue oder noch nicht genügend bekannte Infusorienformen aus der Ostsee bei Wismar: *Amtl. Ber. d. Vers. deutsch. Naturf. u. Aerzte, Karlsbad*, 1862, p. 165. — **Id.**, **'67**. Der Organismus der Infusionsthiere, Bd. II: *Leipzig*, 1867. — **Id.**, **'78**. Der Organismus der Infusionsthiere, Abth. III, 1878. — **Id.**, **'83**. Der Organismus der Infusionsthiere: *Leipzig*, 1893, Abth. IV. — **Steinhaus, J.**, **'89**. Karyophagus salamandræ. Eine in der Darm epithelzellkern parasitischlebende Coccidie: *Arch. Path. An. u. Phys.*, CXV, pp. 176-85. — **Sterki, V.**, **'78**. Beiträge zur Morphologie der Oxytrichinen: *Z. w. Z.*, XXXI, pp. 29-58. — **Stöhr, P.**, **'80**. Die Radiolarienfauna der Tripoli von Grotte, *etc.*: *Palæontographica*, XXVI, p. 121. — **Stokes, A. C.**, **'84**. Notices of Some New Parasitic Infusoria: *Amer. Nat.*, XVIII, pp. 1081-6. — **Id.**, **'86**. Some New Infusoria from American Fresh Waters: *A. M. N. H.* (5), XVII. — **Id.**, **'93**. The Contractile Vacuole: *Amer. Mon. Mic. Jour.*, XVI, pp. 182-8. — **Stolc, A.**, **'00**. Beobachtungen und Versuche über die Verdauung und Bildung der Kohlenhydrate bei einem Amöbenartigen Organismus *Pelomyxa palustris* Greeff: *Z. w. Z.*, LXVIII, pp. 625-68. — **Strasburger, E.**, **'78**. Wirkung des Lichtes und der Wärme auf Schwärmsporen: *J. Z.*, XII, pp. 551-625. — **Id.**, **'92**. Histologische Beiträge, IV: *Jena*. — **Suriray**, **'36**. Recherches sur la cause ordinaire de la phosphorescence marine et description du Noctiluca miliaris: *Guérin, Mag. d. Zoologie*, VI.

THÉLOHAN, P., **'91**. Sur deux sporozoaires nouveaux, parasites des muscles des poissons: *C. R. S. B.* (9), III, pp. 27-9. — **Id.**, **'92**. Observations sur les Myxosporides et essai de classification de ces organismes: *Bull. Soc. Phil.* (8), IV, pp. 165-78. — **Id.**, **'95**. Recherches sur les Myxosporidies: *Bull. Sci. France et de la Belgique*, XXVI, pp. 101-394. — **Thompson, Wyv.**, **'76**. Reports from the "Challenger": *Proc. Roy. Soc.*, XXIV, 1876. — **Thuret**, **'43**. Recherches sur les organes locomoteurs des spores des algues: *Ann. d. Sc. Nat.* (Botan.) (2), XIX, p. 266. — **Topsent, E.**, **'92**. Sur un nouveau Rhizopode marin (Pontomyxa flava): *C. R.*, CXIV, pp. 774-5. — **Trembley, A.**, **1744**. Translation of a Letter from A. Trembley to the President, with Observations upon Several newly discovered Species of Fresh-water Polypi: *Phil. Trans.*, *London*, XLIII, pp. 169-83. — **Id.**, **1747**. Observations upon Several Species of Small Water Insects of the Polypus Kind: *Phil.*

Trans., XLIV, pp. 627–55. — **Treviranus, G. R.**, **'03.** Biologie oder Philosophie d. lebenden Natur: Bd. II, *Göttingen*.

VEJDOVSKY, F., **'92.** Sur la Monodontophrya, nouvelle espèce d'Opalinide: *Cong. Intern. d. Zool.*, 1892, pp. 24–31. — **Verworn, M.**, **'88.** Biologische Protistenstudien: *Z. w. Z.*, XLVI, pp. 455–70. — **Id.**, **'89.** Psychophysiologische Protistenstudien. Experimentale Untersuchungen: *Jena*, 1889. — **Id.**, **'91.** Die physiologische Bedeutung des Zellkerns: *Arch. f. de Ges. Phys.*, LI, pp. 1–118. — **Id.**, **'94.** Allgemeine Physiologie (translated by F. C. Lee, 1898). — **Id.**, **'97.** Die polare Erregung der lebendigen Substanz durch den constanten Strom: *Arch. f. d. Ges. Phys.*, LXV, pp. 47–62. — **Villot, A.**, **'91.** La classification zoologique dans l'état actuel de la science: *Rev. Biol. d. Nord d. l. France, Lille*, III, pp. 245–61. — **Virchow, R.**, **'58.** Die Cellularpathologie in ihrer Begründung auf Physiologische und pathologische Gewebelehre: *Berlin*.

WALKER and BOYS, 1784. Testacea minuta rariora, *etc.*: *London*. — **Wallich, G. C.**, **'63.** Further Observations on an Undescribed Indigenous Amœba, *etc.*: *A. M. N. H.* (3), XI, pp. 365–71. — **Id.**, **'64.** On the Process of Mineral Deposit in the Rhizopods and Sponges as affording a Distinctive Character: *A. M. N. H.* (3), XIII, pp. 72–82. — **Wasielewsky, von**, **'96.** Sporozoenkunde, ein Leitfaden für Aerzte, Tierärzte und Zoologen: *Jena*. — **Watasé, S.**, **'92.** On the Phenomena of Sex-differentiation: *J. M.*, VI, 481–93. — **Id.**, **'98.** Protoplasmic Contractility and Phosphorescence: *Wood's Hole Biol. Lect.*, 1898, pp. 177–92. — **Weber, E. H.**, **'55.** Mikroskopische Beobachtungen sehr gesetzmässiger Bewegungen, welche die Bildung von Niederschlagen harziger Körper aus Weingeist begleiten: *Ann. f. Phys. u. Chemie*, XCIV, pp. 447–59. — **Whitman, C. O.**, **'93.** The Inadequacy of the Cell Theory of Development: *Wood's Hole Biol. Lect.*, 1893, p. 105. — **Wiegmann, A. F. A.**, **'35.** Bericht über d. Fortschritte d. Zoologie in Jahre 1834: *Arch. f. Naturg.*, I, p. 12, Anmerk. — **Weismann, A.**, **'84.** Ueber Leben und Tod. Lectures and Essays, 1884. — **Id.**, **'91.** Essays upon Heredity. First Series: *Oxford*. — **Williamson, W. C.**, **'58.** On the Recent Foraminifera of Great Britain: *Ray Soc.* — **Wilson, E. B.**, **'00.** The Cell in Development and Inheritance. *2d Ed.*: *N. Y.* — **Id.**, **'00.** Some Aspects of recent Biological Research: *The Internat. Monthly*, July, 1900. — **Wolters, M.**, **'91.** Die Conjugation und Sporenbildung bei Gregarinen: *A. m. A.*, XXXVII, pp. 99–138. — **Wright, S.**, **'58.** Description of New Protozoa: *Edin. New Phil. Jour.*, VII, pp. 276–81. — **Wrzésniowsky, A.**, **'69.** Ein Beitrag zur Anatomie der Infusorien: *A. m. A.*, V, pp. 25–49. — **Id.**, **'70.** Beobachtungen über Infusorien a. d. Umgebung von Warschau: *Z. w. Z.*, XX, pp. 467–511. — **Id.**, **'77.** Beiträge zur Naturgeschichte der Infusorien: *Z. w. Z.*, XXIX, pp. 267–323.

ZACHARIAS, O., **'93.** Faunistische und Biologische Beobachtungen am Gr. Plöner See: *Forsch. Ber. a. d. Biol. Stat. z. Plön.*, I. — **Id.**, **'95.** Faunistische Mittheilungen: *Forch. Ber. a. d. Biol. St. z. Plön.* Ab. III, pp. 78–83. — **Zenker, W.**, **'66.** Beiträge zur Naturgeschichte der Infusorien: *A. m. A.*, II, pp. 332–48. — **Zschokke, F.**, **'98, 1.** Die Myxosporidien in der Musculatur der Gattung Coregonus: *Z. A.*, XXI, pp. 213–14. — **Id.**, **'98, 2.** Die Myxosporidien der Gattung Coregonus: *Cent. Bakt.*, XXIII, pp. 602–76, 46–55, and 699–703.

INDEX OF AUTHORS

Agardh, 139.
Agassiz, L., 12.
Alder, 12.
Allman, *Noctiluca*, 136.
Altmann, structure of protoplasm, 35.
Archer, classification, 106, 109.
Aristotle, animals and plants, 23, 25.
Auerbach, excretory granules, 286.

Balbiani, 2, 13; conjugation, 15; spore-formation in Sporozoa, 154; classification, 167; nucleus of *Loxophyllum*, 189, 190; nucleus, 250, 252; mitosis in micronuclei, 261; in *Spirochona*, 262; function of vacuole, 289; merotomy, 302.
Barbagello, dysentery, 64.
Barry, nucleus, 2; Protozoa as single cells, 11.
De Bary, 18; pseudopodia in mycetozoa, 84; irritability, 299.
Van Beneden, maturation, 233.
Bergh, 21; animals and plants, 25; cuticula, 38; coloring matter in Dinoflagellidia, 117; feeding in Dinoflagellidia, 126; inter-relations, 135; classification, 140, 206; nuclei, 250, 251.
Berthold, explanation of pseudopodia-formation, 84.
Bichat, 2.
Blainville, spontaneous generation, 28.
Blochmann, conjugation in *Euglypha*, 219, 234; division of *Euglena*, 256, 259.
Boeck, function of vacuole, 291.
Bosc, Sporozoa, 159.
Bourne, bacteria in Protozoa, 83; nuclei of *Pelomyxa*, 87.
Boveri, origin of the mitotic spindle, 248; of the centrosome, 276.
Brady, classification, 107, 108, 109; shell, 305.
Brandt, symbiosis in Radiolaria, 71; conjugation, 95; nucleus, 249; function of vacuole, 289; merotomy, 301.
Brauer, encystment of *Actinosphærium*, 90; maturation, 235; mitosis, 264.
Bruch, Sporozoa, relationships, 166.
Brücke, 2; pseudopodia-formation, 83.
Buck, swarm-spores in *Arcella*, 95.
Buffon, 13; animals and plants, 23; spontaneous generation, 26; theory of generation, 27.
Bühler, origin of spindle, 274.
Burnet, 12.
Bütschli, 2, 12, 14; animals and plants, 25; structure of protoplasm, 35, 79; odors, 62; pseudopodia-formation, 84; swarm-spores in *Arcella*, 95; conjugation, 60, 96, 193, 228; classification, 20–22, 99, 101, 169, 206, 208, 209, 210; shells of Dinoflagellidia, 117; flagella-motion, 121; nutrition in Mastigophora, 126; catenation, 131; inter-relationships of Mastigophora, 135; spore-formation in Sporozoa, 152, 154; Sporozoa, inter-relationships, 166; membrane of ciliates, 176, 178; contraction of the vacuole, 188; tentacles of suctoria, 195; inter-relationships of the Infusoria, 199–200; mouth-shifting in ciliata, 185; rejuvenescence, 213, 220; conjugation in Mastigophora, 228, 241; nucleus, 249, 250; mitosis, 264; division-centre, 266; digestion of fat, 282; function of vacuole, 289; gas vacuoles, 290.

Calendruccio, dysentery, 64.
Calkins, odors due to Protozoa, 62; *Uroglena*, 130; *Lymphosporidium*, 169; mitosis in Mastigophora, 250, 252; historic origin of mitotic figure, 274.
Carpenter, classification, 18, 107, 108, 109; types of shells, 73; food-selection, 305.
Carter, 109; conjugation in *Eudorina*, 134; classification, 138; conjugation in Mastigophora, 224, 228, 232.
Carus, C. G., 10; spontaneous generation, 25.
Carus, J. V., classification, Sporozoa, 168.
Cassagrandi, dysentery, 64.
Cattaneo, swarm-spores in *Arcella*, 95.

Celli, malaria, 163.
Celli and Fiocca, dysentery, 64.
Certes, respiration, 289.
Cienkowsky, 13; animals and plants, 23; symbiosis in Radiolaria, 71; pseudopodia-formation, 83; budding *Clathrulina*, 96; classification, 106, 109, 137, 138, 139; conjugation of monads, 132; *Multicilia*, 200; conjugation in *Noctiluca*, 219; food-selection, 305.
Claparède, 12; classification, 18, 21, 22; gas vacuoles in rhizopods, 89.
Claparède and Lachmann, classification, 106, 140; myonemes, 180; classification, 206, 207, 208, 209, 210; conjugation, 217.
Clark-James, feeding in Choanoflagellida, 127; classification, 137.
Claude-Bernard, 29.
Cohn, F., 23; conjugation of *Volvox*, 134; of *Gonium*, 226; irritability, 296.
Cohn, L., dysentery, 64; mesoplasm, 144; myonemes of *Vorticella*, 180, 207, 209.
Corti, 7; digestion, 281; respiration, 288.
Councilman, dysentery, 64.
Cunningham, dysentery, 64.

Dallinger and Drysdale, spontaneous generation, life-history of monads, 28, 132; flagellum-formation, 120; conjugation of flagellates, 219, 221.
Dangeard, feeding in Dinoflagellidia, 126; in Suctoria, 196; conjugation, 219.
Danilewsky, *Polymitus* form, 162; classification, 168.
Le Dantec, symbiosis in Radiolaria, 71; in Infusoria, 174; digestion, 282, 284.
Dawson, Eozoön, 29.
Defrance, classification, 107, 108.
Delage and Hérouard, 18; animals and plants, 22, 25; flagella-motion, 121; contractile vacuole of ciliates, 188; tentacles of Suctoria, 195; nervous organ of *Vorticella*, 183; function of the vacuole, 290.
Dewitz, thigmotaxis, 300.
Diesing, 20; animals and plants, 23; classification, 139, 140.
Doflein, structure of Sporozoa, 142, 144; classification, 209; centrosomes in Ciliata, 272.
Donné, classification, 138.
Dreyer, skeleton-formation, Radiolaria, 77; secretion, 292, 308.
Drüner, origin of the spindle, 248.
Dufour, classification, 21, 167.
Dujardin, classification, 6, 10, 15, 16, 18, 20, 23, 33, 106, 107, 137, 138, 139, 206, 207, 208, 209; *sarcode*, 34; immortality of Protozoa, 60; pseudopodia and flagella, 44; gastric vacuoles in rhizopods, 91; consciousness in Protozoa, 280, 303; function of the vacuole, 289.
Dunal, odors due to Protozoa, 62.

Ecker, pseudopodia-formation, 83.
Ehrenberg, 9, 13, 15; classification, 15, 18, 20, 33, 106, 137, 138–140, 206, 207, 208, 210; immortality of Protozoa, 60; odors due to Protozoa, 62; *Uroglena*, 130; markings on ciliates, 177; conjugation, 222; function of vacuole, 292.
Eimer, consciousness in Protozoa, 304.
Eismond, suction in the tentacles of Suctoria, 196.
Ellis, discovery of trichocysts, 11.
Engelmann, 12; conjugation, 14, 61, 193, 212, 213, 220, 222, 241; stigmata, 37, 118; inotogmata and pseudopodia formation, 83; gas vacuoles in rhizopods, 89; chlorophyl in *Vorticella*, 174; nematocysts in *Vorticella*, 176; myonemes, 180; classification, 207, 209; digestion, 283; gas vacuoles, 290; irritability, 296, 297.
Entz, *Actinobolus*, 50; symbiosis in Radiolaria, 71; gas vacuoles, 89; classification, 106, 109, 206, 207; mouth of Choanoflagellida, 123; nutrition in Mastigophora, 126, 127; myonemes in ciliates, 178, 180; inter-relations of Infusoria, 201; tentacles of *Actinobolus*, 183; gas vacuoles, 290; function of vacuole, 291.

Fabre-Dumergue, contractile vacuole, 54, 187; classification, 209; digestion, 281, 284.
Feletti, malaria, 163.
Fisch, classification, 138.
Fischer, classification, 109; structure of flagella, 121.
Fiszer, function of the vacuole, 288.
Flemming, origin of the mitotic spindle, 248.
Focke, 11.
Fœttinger, classification, 207.
Fol, classification, 208.
Foulke, *Actinosphærium*, 97.
Francé, collar of Choanoflagellida, 123, 127.
Frenzel, classification, 106, 137, 138, 167, 210; pyxinine granules, 143; nucleus, 250.
Fresenius, classification, 138, 207.
Fromentel, classification, 139.

Gabriel, Sporozoa as plants, 166.

Garry, irritability, 299.
Gasser, dysentery, 64.
Geddes and Thompson, origin of sex, 212.
Giard, classification Sporozoa, 168.
Gleichen, 7; function of the vacuole, 288.
Goeze, digestion, 281.
Goldfuss, spontaneous generation, 28.
Goroschankin, conjugation of Mastigophora, 224, 228.
Gould, Bacteria in Rhizopoda, 83.
Grassi, dysentery, 64; classification, 138, 168; malaria, 163, 164.
Gray, classification, 109.
Greeff, 33; contractile fibres in *Amœba*, 83; budding in *Clathrulina*, 96; membrane in *Actinosphærium*, 105; classification, 106, 109; conjugation in *Zoothamnium*, 222; division-centre in Heliozoa, 266.
Greenwood, digestion in rhizopods, 92, 280; in Ciliata, 285.
Grenacher, division-centre in Heliozoa, 266.
Griffiths, excretion of urea, 291.
Gruber, 2; conjugation, 15, 97; *Actinophrys*, 69; classification, 106, 206, 208, 210; inter-relations of Sarcodina, 101; maturation, 235, 241; nucleus, 250; mitosis, *Actinosphærium*, 264; merotomy, 302, 305.
Gruby, classification, 137.
Gruithuisen, 11.
Gurley, Myxosporidiida, reproduction, 154; classification, 169.

Haeckel, 12; classification, 17, 21, 106, 109, 110, 208; animals and plants, 23; structure of Radiolaria, 69, 76; conjugation in Radiolaria, 95; development of Radiolaria, 95; origin of Sporozoa, 167; phylogeny of Infusoria, 205, irritability, 300, 304, 306.
Hallez, classification, 106.
Hammerschmidt, classification, 168.
Harrington and Leaming; see Leaming.
Harting, shell-formation, 293.
Hartog, polar bodies, 238.
Harvey, 5.
Heidenhain, origin of spindle, 248.
Heitzmann, pseudopodia-formation, 83.
Henle, Sporozoa as worms, 166.
Henneguy, classification, 169.
Hermann, origin of spindle, 247, 248.
Hertwig, O., rejuvenation, 212.
Hertwig, R., 2, 12; conjugation, 15, 211; classification, 21, 106, 109, 140; structure of Radiolaria, 69; encystment of *Actinosphærium*, 90; swarm-spores in *Arcella*, 95; budding in *Clathrulina*, 96; origin of tentacles, 197; sex, 211; maturation, 235, 237, 238; nucleus, 249, 261; mitosis in *Actinosphærium*, 261, 266; origin of centrosome, 276.
Hertwig and Lesser, 109; excretory granules, 286.
Hieronymous, conjugation in *Gonium*, 227.
Hill, classification, 206.
Hofer, 2; enucleated *Amœba*, 86, 302; secretion, 292.
Hofmeister, pseudopodia-formation, 84.
Holman, conjugation in *Amœba*, 97, 217.
Hooke, 6.
Howell, starch, 281; calcium in man, 288.
Hoyer, conjugation, 194.
Huxley, classification, 18, 207; animals and plants, 23; spontaneous generation, 29; epidemic among silkworms, 165.

Ishikawa, tentacles of Suctoria, 197; conjugation in *Noctiluca*, 219; nucleus, 252; mitosis in *Noctiluca*, 267.
Israel, bacteria in Rhizopoda, 83.

Jackson, classification, 209.
Jennings, irritability in Ciliata, 298.
Jensen, lifting power of *Paramœcium*, 181.
Jickeli, 15; conjugation in Rhizopoda, 97, 219.
Johlot, 7; spontaneous generation, 26.
Johnson, division of *Stentor*, 192; pseudopodia of *Stentor*, 183; plastogamy in *Actinosphærium*, 218.
Joly, odors due to Protozoa, 62.
Jones, Rupert, classification, 108.
Joukowsky, conjugation in Ciliata, 220.
Julin, function of the macronucleus, 188.

Kartulis, dysentery, 64.
Kent, 21; nutrition in Mastigophora, 126; *Uroglena*, 130; classification, 137, 138, 139, 208, 209, 210; conjugation, 217; secretion, 292.
Keuten, division of *Euglena*, 256.
King and Rowney, Eozoön, 29.
Kirschner, conjugation in *Volvox*, 232.
Klebs, inter-relations of the Sarcodina, 99; of Mastigophora, 135; ectoplasmic modifications, 113, 117; classification, 137, 138, 139; irritability, 296.
Kofoid, classification, 140.
Kölliker, 12; classification, 21; Sporozoa, 166; function of vacuole, 291.
Korotneff, budding in Heliozoa, 95.

Kostanecki, origin of spindle, 248, 274.
Kowalewsky, classification, 209.
Krassilstschik, conjugation in *Polytoma*, 221.
Kruse, classification, 168.
Kühne, conjugation of *Amœba*, 218.

Labbé, 22; protoplasmic structure of Sporozoa, 142, 144; division in Sporozoa, 149; spore-formation, 151; malaria, 163; classification, 167–169; nucleus, 250, 254.
Lachmann, myonemes in *Vorticella*, 178; classification, 209. See Claparède and Lachmann.
Lamarck, 9; spontaneous generation, 28; classification, 107, 108, 109, 208.
Lambl, dysentery, 64.
Lameere, 12.
Langmann, Myxosporidiida in reptiles and amphibia, 165.
Lankester, classification, 18, 169; flagella-motion, 121.
Lauterborn, classification, 137, 139; division of *Ceratium*, 260; historic origin of centrosome, 276.
Laveran, dysentery, 64; Sporozoa, 159; malaria, 162–165.
Leaming and Harrington, effect of light on *Amœba*, 297.
Le Blanc, swarm-spores in *Difflugia*, 95.
Le Clerc, classification, 106.
Ledermüller, 8, 22.
Leeuwenhoek, 5; spontaneous generation, 26; classification, 140.
Léger, structure of gregarines, 144; microsporozoites, 159; classification, 167.
Leidy, classification, 106, 138, 208.
Leuckart, 21; animals and plants, 23; life-cycle of *Coccidium*, 158; Sporozoa relationships, 166.
Leydig, 12; Sporozoa inter-relationships, 166; respiration, 289; function of vacuole, 291.
Limbach, function of the vacuole, 288.
Linnæus, 5.
Lister, alternation of generations in Reticulariida, 74, 95.
Loeb, irritability, 299.
Loeb and Hardesty, macronucleus, 188.
Loesch, dysentery, 64.
Logan, Eozoön, 29.
Lutz, classification, 169.

Maccallum, malaria, 164.
Macfarland, origin of spindle, 248.
Mannaberg, dysentery, 64.
Manson, malaria, 163–165.
Marchiafava and Celli, *Plasmodium malariæ*, 168.
Mark, polar bodies, 238.
Massart, irritability, 298.
Maupas, conjugation, 15, 193, 207, 234, 237; food-taking, 48; senescence, 60; predatory ciliates, 175, 199; inter-relationships of Infusoria, 201, 204; mitosis in micronuclei, 261; digestion, 283; excretory granules, 286; function of the vacuole, 288, 291; rejuvenescence, 212, 220.
Meissner, digestion in Rhizopoda, 92, 280, 284.
Mendelssohn, irritability, 295.
Mereschowsky, classification, 109, 138; phylogeny of Ciliata, 205.
Mesnil, conjugation in Sporozoa, 225.
Metschnikoff, 13; gastric vacuoles in Rhizopoda, 91; malaria, 163; myonemes of *Vorticella*, 180; digestion, 283, 284.
Meyen, 11.
Meyer, nutrition in Mastigophora, 126.
Milne-Edwards, spontaneous generation, 29.
Mingazzini, life-cycle of Coccidiida, 158; classification, 168, 169; nuclei, 254.
Minot, rejuvenescence, 212.
Möller, classification, 109.
Montfort, classification, 107.
Moore, nerve-organ of *Vorticella*, 183.
Mottier, origin of spindle, 248.
Müller, Johannes, 13; classification, 18, 21; conjugation of *Gonium*, 226.
Müller, O. F., 8, 13; classification, 15, 208; spontaneous generation, 28.
Munier-Chalmas et Schlumberger, dimorphism in Reticulariida, 95.

Nägeli, spontaneous generation, 29, 169.
Needham, theory of generation, 27.
Nitsch, 25; classification, 138, 206, 207.
Nussbaum, merotomy, 302.

Oersted, animals and plants, 24.
Oken, 9, 12; spontaneous generation, 28; classification, 208.
d'Orbigny, classification, 18, 107, 108, 109.
Osterhout, origin of spindle, 248.
Owen, 11.

Parker, classification, 107, 108.
Pasteur, spontaneous generation, 29.
Pearl, irritability, 300.
Pénard, structure of *Pelomyxa*, 38; contractile vacuole, 54; structure of *Euglypha*,

68; food-taking in Heliozoa, 91; conjugation in Heliozoa, 97, 217, 219; classification, 106, 109; phylogeny of Ciliata, 204.
Perty, 12, 13, 22; animals and plants, 25; nutrition in Mastigophora, 126; classification, 138, 139, 207, 208; digestion, 282; gas vacuoles, 290; consciousness, 303.
Petridis, dysentery, 64.
Pfeffer, chemotaxis in reproductive elements, 297.
Pfeiffer, L., classification, 169.
Pfeiffer, R., life-cycle of *Coccidium*, 158; *Serumsporidium*, 165.
Phillips, classification, 207.
Pineau, 12.
Plate, mitosis in *Spirochona*, 262.
Podwyssozki, microsporozoite and macrosporozoite, 158.
Pouchet, stigmata, 119; catenation, 131; respiration, 288.
Pringsheim, conjugation of *Pandorina*, 132, 228.
Prowazek, conjugation in *Euglypha*, 234.
Przesmycki, excretory granules, 287.
Purkinje and Valentin, classification, 207.

Quatrefages, spontaneous generation, 29; phosphorescence in *Noctiluca*, 293.
Quennerstedt, classification, 206.

Rabl, origin of the spindle, 274.
Racovitza and Labbé, classification, 168.
Radziszewski, phosphorescence, 294.
Reaumur, spontaneous generation, 26.
Redi, Sporozoa, 21.
Reinhardt, conjugation in flagellates, 224.
Reuss, classification, 107, 108.
Rhumbler, structure of *Euglypha*, 68; pseudopodia-formation, 84; ecto- and endoplasmic interchange, 85; gas vacuoles, 89; axial filaments, 101; contractile vacuoles in ciliates, 188; conjugation in Rhizopoda, 217; action of vacuole, 289; function, 291; secretion, 292; irritability, 297; experiments with inorganic fluids, 306–309.
Risso, classification, 109.
Robin, classification, 136.
Roboz, classification, 107.
Römer, classification, 109.
Römpel, classification, 209.
Rood, function of the vacuole, 291.
Rosenhof, irritability, 299.
Ross, malaria, 164.
Rossbach, function of the vacuole, 288, 289.
Rostafinski, conjugation in *Gonium*, 227.
Rouget, *Vorticella* myonemes, 180.
Roux, cytotropy, 218.
Rückert, origin of the centrosome, 276.

Sacharoff, malaria, 163.
St. Vincent, classification, 137, 140.
Sandahl, classification, 107.
Sanfelice, malaria, 163.
Sars, classification, 107.
Sassaki, division-centre in Heliozoa, 266.
Schaffhausen, *Vorticella* movement, 180.
Schaudinn, 2; axial filaments, 79, 82, 101; *Paramœba*, 93; conjugation in Reticulariida, 95, 99; in *Actinophrys*, 97; in Sporozoa, 156, 158; classification, 109; conjugation of Rhizopoda, 218, 219; nucleus, 250; division of *Amœba crystalligera*, 260; mitosis in Heliozoa, 266; in *Paramœba*, 268; origin of centrosome, 276.
Schewiakoff, structure of *Euglypha*, 68; membrane of gregarine, 145; movement of gregarines, 149; protoplasm of ciliates, 174, 178; myonemes, 180; classification, 208, 209; nucleus, 250; division of *Euglypha*, 265; excretory granules, 286; secretion, 292.
Schilling, feeding in Dinoflagellidia, 126.
Schleiden, 10.
Schlumberger, dimorphism in Reticulariida, 95; classification, 106.
Schmarda, feeding in Dinoflagellidia, 126; classification, 139.
Schmidt, 11; function of vacuole, 289.
Schneider, classification, 106, 109, 167–169; carminophilous granules, 143; nuclei of Sporozoa, 146; spore-formation in Sporozoa, 152; nucleus, 254.
Schrank, classification, 140, 206.
Schuberg, dysentery, 64; life-cycle of Coccidiida, 158; membrane of ciliates, 176; classification, 206, 207, 208.
Schultze, F. E., inter-relations of the Sarcodina, 103; classification, 106, 109, 137; conjugation of Rhizopoda, 217; division-centre in Heliozoa, 266; excretory granules, 286; irritability, 295.
Schultze, M., protoplasm and sarcode, 12; axial filaments, 79, 101; classification, 107.
Schütt, coloring matter in Dinoflagellidia, 117; classification, 140.
Schwalbe, function of the vacuole, 288.
Schwann, 2.

Schweigger, 9.
Shaw, classification, 140.
Siddall, classification, 106.
Siebold, V., 11; classification, 16; animals and plants, 22; function of vacuole, 289.
Siedlecki, movement of gregarines, 149; conjugation, 156, 159, 225, 232; nucleus, 250; division-centre in Sporozoa, 258; secretion, 292.
Simond, conjugation in Coccidiida, 159.
Sjöbring, Sporozoa, 159.
Sorokin, classification, 106.
Spallanzani, 7, 11; spontaneous generation, 27; respiration, 288; irritability, 294.
Spencer, growth and reproduction, 212.
Stache, classification, 107.
Stahl, irritability, 295.
Steenstrup, alternation of generations, 12.
Stein, 12, 13, 22; animals and plants, 24; classification, 15, 21, 106, 107, 109, 137, 138, 139, 167, 168, 206–210; shells of Dinoflagellidia, 117; nutrition in Mastigophora, 126; inter-relationships of Mastigophora, 135; *Symphyta*, 166; myonemes, 177, 178; phylogeny of Ciliata, 204; conjugation of *Codosiga*, 222; *Chlorogonium*, 225; *Gonium*, 226; *Volvox*, 232; excretory granules, 286; function of vacuole, 291.
Steinhaus, classification, Sporozoa, 168.
Steinmann, shell-formation, 293.
Sterki, classification, 209.
Stöhr, classification, 140.
Stokes, contractile vacuole of *Amœba*, 88; classification, 206, 207.
Stolc, starch digestion in Rhizopoda, 92, 284.
Strasburger, origin of the spindle, 248; irritability, 295, 297.
Suriray, classification, 140.
Thélohan, protoplasmic structure of Sporozoa, 142; reproduction in Sporozoa, 154; classification, 169.
Thompson, classification, 108.
Topsent, classification, 106.
Trembley, 7; conjugation of *Zoothamnium*, 222.
Treviranus, 9, 28.
Tyndall, spontaneous generation, 29.

Vejdowsky, classification, 207.
Verworn, 2, 32; isolated central capsule, 69; enucleated Protozoa, 86; cytotropy, 218; secretion, 292; irritability, 295, 297, 300, 302, 305, 306, 308, 309.
Villot, 12.

Walker and Boys, classification, 108.
Wallich, pseudopodia-formation, 84; chitosarc, 293.
Wasielewsky, granules in Sporozoa, 143; conjugation, 159; origin of Sporozoa, 167.
Watasé, sex, 243.
Weber, pseudopodia-formation, 84.
Weismann, origin of death, 60; maturation, 233.
Wiegmann, function of the vacuole, 288, 289.
Williamson, classification, 109.
Wilson, maturation, 233; origin of the spindle, 248; vital phenomena, 279.
Wolters, nuclei of Sporozoa, 146; conjugation, 156, 225; maturation, 235; nucleus, 250.
Wright, classification, 107, 210.
Wrzesniowsky, classification, 206, 209.
Wysotzki, classification, 139.

Zacharias, function of pseudopodia, 90; *Uroglena*, 130, 131; classification, 210.
Zenker, classification, 210; function of the vacuole, 288.
Zschokke, spore-cysts of *Myxobolus*, 155.

INDEX OF SUBJECTS

Acantharia, 70, 110.
Acanthocystis, skeleton, 75, 82; budding, 95; axial filaments, 101; classification, 109; centrosomes, 266.
Acanthonida, 110.
Acanthospora, 168.
Acanthosporidæ, 168.
Acephalina, 168.
Acineta, budding, 199; classification, 210.
Acineta-theory, Stein, 13.
Acinetidæ, 210.
Actinelida, 110.
Actinobolus, food-taking, 50; fig. p. 51; tentacles, 183; classification, 206.
Actinocephalidæ, 167.
Actinocephalus, 168.
Actinolophus, skeleton, 74; pseudopodia, 82; classification, 109; phylogeny, 204.
Actinomonas, 103; fig. p. 103; classification, 137.
Actinophrys, 109; fig. p. 16; pseudopodia, 82; contractile vacuole, 87; conjugation, 97, 218; axial filaments, 101.
Actinosphærium, alveoli in protoplasm, 35; skeleton, 36; encystment, 90; artificial reproduction, 97; axial filaments, 101; membrane, 105; classification, 109; conjugation, 218; mitosis, 264; maturation, 236; irritability, 295, 301.
Actinotricha, 209.
Actipylea, 70, 110.
Actissa, fig. p. 17.
Adelea, conjugation, 158, 229; classification, 168; maturation, 237.
Adinida, 5, 140.
adoral zone, 6, 171.
Ægyria, 207.
Aggregata, 167.
Aggregatidæ, 167.
Allomorphina, 108.
alternation of generations, 12; in Reticulariida, 74, 95.
alveolar structure of protoplasm, 35.
Alveolina, 107.
Alveolininæ, 107.
Ammodiscus, 107.
Amœba coli, fig. p. 63.
Amœbæa, 18.
Amœba proteus, ectoplasm, fig. p. 36; pseudopodia, 81; nucleus, 87; contractile vacuole, 54, 88; fig. p. 89; conjugation, 97, 105; classification, 105; conjugation, 218; nucleus, 252, 276; secretion, 291, 292; irritability, 295, 297, 301.
Amœbida, classification, 105.
Amœbidæ, classification, 105.
Amœbosporidia, 169.
Amphidinium, 140.
Amphidoma, 140.
Amphileptus, 206; irritability, 295; food-selection, 305.
Amphilonchidæ, 110.
Amphimonas, 137.
Amphisia, 209.
Amphisolenia, 140.
Amphistegina, 109.
Amphistominæ, 106.
Amphitholus, 140.
Amphitrema, 106.
Amphizonella, 106.
Amphoroides, 168.
anal tube, in Infusoria, 186.
Ancistrum, 207.
Ancyromonas, 137.
Ancyrophora, 168.
androgonidia, in *Volvox*, 134.
Androspyridæ, 110.
Anentera, 9.
Angiosporea, 167.
Animalcula, 7.
animals and plants, 22.
anisogamy, 224.
Anisonema, 139.
anisospore, 95.
Anomalina, 109.
Anoplophrya, 207.
Anthocyrtidæ, 111.
Anthophysa, 137; secretion, 292.

Anthorhynchus, 168.
Aphrosina, 109.
Aphrothoracida, 109.
arboroid colony, 57.
Arcella, pseudopodia, 86; contractile vacuole, 87; budding, 95; classification, 106; conjugation, 218; gas vacuoles, 290.
Arcellidæ, 106.
Archeodiscus, 109.
archigony, 29.
archispore, 151.
archoplasm, 274.
Arcuothrix, 106.
Articulina, 107.
Artiscidæ, 110.
Artodiscus, cause of motion, 101.
Aschemonella, 107.
Ascoglena, 138.
asexual reproduction; see Reproduction.
Aspidisca, 209.
Astasia, membrane, 114; classification, 138.
Astasiida, 138.
Asterophora, 168.
Astomea, 138.
astral granule (Centralkorn), 18, 82.
Astrodisculus, 109.
Astrolonchidæ, 110.
Astrolophidæ, 110.
Astropyle, 70.
Astrorhiza, 107.
Astrorhizidæ, 107.
Astrorhizinæ, 107.
Astrosphæridæ, 110.
Astylozoön, 209.
Atractonema, 138.
attraction sphere, 246.
Aulocanthidæ, 111.
Aulosphæridæ, 111.
auto-infection, 158.
axial filament, 81, 265.
axoneme, in Vorticella, 179.
axopodia (pseudopodia with axial filaments), *Dimorpha*, 101.

Balantidiopsis, 208.
Balantidium, fig. p. 63; classification, 208.
Balladina, 209.
barotaxis, 300.
Barrouxia, 168.
Bathysiphon, 107.
Bdelloidina, 107.
Belloides, 168.
Benedenia, 168.
Bicœcidæ, 137.
Bicosœca, 137.
Bifarina, 108.
Bigenerina, 108.
Biloculina, fig. p. 74; classification, 1[illegible]
Blanchardina, 169.
Blepharacysta, 140.
Blepharisma, 208.
Bodo, 138; conjugation, 217; irrit[illegible] 297.
Bodonidæ, 138.
Bolivina, 108.
Botellina, 107.
Bothriopsis, 168.
Bradyina, 108.
budding (see Reproduction), in Sar[illegible] 95.
Bulimina, 108.
Bulimininæ, 108.
Bullaria, obsolete, 15.
Bursaria, sphincter, 181; classificatio[illegible]
Bursariidæ, 208.
Bütschlia, 206.

Cænomorpha, 208.
Calcarina, 73, 109.
Calcituba, classification, 107; nucleus chromosomes, 253, 254.
calymma, 23, 69.
Calyptotricha, 207.
Campanella, adoral zone, 202.
Campascus, 106.
Camptonema, pseudopodia, 82.
Candeina, 108.
Cannobotryidæ, 110.
Cannopylea, 70, 111.
Cannoraphidæ, 111.
Cannosphæridæ, 111.
capitate tentacles, fig. p. 196; functio[illegible]
carbohydrates in digestion, 282.
carbon dioxid in contractile vacuo[illegible] 288, 289.
Carchesium, 209; digestion in, 285.
carminophilous granules, 143.
Carpenteria, 109.
Carteria, 139.
Carterina, 107.
Caryolysus, 168.
Caryophagus, 168.
Cassidulina, 108.
Cassidulinæ, 108.
Castanellidæ, 111.
catenoid colony, 57.
cell-structure, 246.
cellulose shells, 39.
Cenodiscidæ, 110.
central capsule, 17, 67.

Centralkorn; see central granule; division-centre.
centrosome; see division-centre.
centrosphere, 246.
Cephalina, 167.
cephalont, 145.
Cephalothamnium, 137.
Ceratium, 41; shell, 116; catenation, 131; classification, 140; mitosis, 260.
Ceratocorys, 136, 140.
Ceratomyxa, 169.
Ceratospora, 168.
Cercomonadidæ, 137.
Cercomonas, reproduction, 132; classification, 137; conjugation, 214.
Cercoporidæ, 111.
Certesia, 209.
Chætoproteus, 106.
Chalarothoracida, 109.
Challengeridæ, 111.
Chasmostoma, 207.
chemotaxis, 297.
Chiastolidæ, 110.
Chiliferidæ, 207.
Chilodon, 207; nucleus, 263.
Chilodontinæ, 207.
Chilomonas, fig. p. 36; protoplasmic structure, 113; starch, 118; nucleus, 124; classification, 139; irritability, 296, 300.
Chilostomella, 108.
Chilostomellidæ, 108.
chitin-shell, 39.
chitosarc, 293.
Chlamydodon, 207.
Chlamydodontidæ, 207.
Chlamydomonadidæ, 139.
Chlamydomonadina, 139.
Chlamydomonas, odor in water, 62; classification, 139.
Chlamydophorida, 109.
Chlorangium, 139.
Chloraster, 139.
Chlorogonium, odor in water, 62; vacuoles, 127; reproduction, 131; classification, 139; conjugation, 224.
Chloromonadina, 139.
Chloromyxidæ, 169.
Chloromyxum, spore-capsules, 156; classification, 169.
chlorophyl, distinguishing animals and plants, 23; in *Paramæcium*, 282.
chlorophyllin, in Dinoflagellidia, 118.
Choanoflagellida, 137.
Chænia, 206.
Chromatella, 106.
chromatin, 40, 253.
chromatoid granules, 144.
chromatophore, 37.
chromatospherite, 254.
Chromomonadina, 139.
chromosomes, 247.
Chromulina, fig. p. 19; nutrition, double nature, 25, 126; classification, 139.
Chrysalidina, 108.
Chrysamœba, 139.
Chrysococcus, 139.
Chrysomonadidæ, 139.
Chrysopyxis, 139.
Chrysosphærella, 139.
chyle, 92.
cilia, structure, 181.
Ciliata, classification, 206.
Cilio-flagellata, 121.
Ciliophrys, 103, 137.
Cinetochilum, 207.
cirri, structure, 182.
Citharistes, 139.
Cladomonas, 137.
classification, history of, 15.
Clathrulina, skeleton, 75; swarm-spores, 96; classification, 109.
Clavulina, 108.
Climacostomum, vacuole, 187; classification, 208.
Cnemidiophora, 167.
Coccidiida, 21, 168.
Coccidium, spores, 153; life-cycle, 158; classification, 168.
Coccodiscidæ, 110.
Coccomonas, 139.
Cochliopodium, 106.
Codonella, 208.
Codonocladium, stalk, 131; classification, 137.
Codonæca, gelatinous cup, 114; classification, 137.
Codonœcidæ, 137.
Codosiga, fig. p. 57; colony-form, 131; classification, 137; conjugation, 222.
Cœlodendridæ, 111.
Cœlographidæ, 111.
Cœlomonas, 139.
Colacium, 138.
Coleorhynchus, 168.
Coleps, covering, 176; fig. p. 176; classification, 206.
collar, in Choanoflagellida, 122.
Collodictyon, 138.
Colloidida, 110.
Collosphæridæ, 110.

Collozoida, 110.
colony-formation, varieties of, 57.
color and odors in drinking water, 62.
Colpidium, 207.
Colpoda, 27; cyclosis, 173; classification, 207.
Colponema, 138.
Cometoides, 168.
Concharidæ, 111.
Conchopthirus, 208.
Condylostomum, 208.
conjugation, 41, 60; in ciliata, 193; general, 214.
consciousness in Protozoa, 280, 304.
contractile vacuole, 52; function, 187.
Cornuspira, 107.
Coronidæ, 110.
cortical plasm, 174.
Coskinolina, 107.
Cothurnia, 209.
Craspedomonadidæ, 137.
Cryptoglena, 138.
Cryptomonadidæ, 139.
Cryptomonas, 139; irritability, 297.
Crystallispora, 168.
Cubosphæridæ, 110.
Cuneolina, 108.
Cyathomonas, 139.
Cyclammina, 108.
Cyclidium, 207.
Cyclochæta, 209.
Cycloclypeina, 109.
Cycloclypeus, 109.
Cyclospora, 168.
Cyphinidæ, 110.
Cyphoderia, protoplasmic structure, 68, 106.
Cyrtoidida, 111.
cyst, varieties, fig. p. 47.
Cystobia, 168.
Cystocalpidæ, 111.
Cystocephalus, 168.
Cystodiscus, 169.
Cystoflagellidia, 140.
Cytamœba, 168.
cytolymph, 246.
cytophan, 179.
cytoplasm, 246.
Cytosporidia, 142.
cytotropy, 217.

Dactylophoridæ, 167.
Dactylophorus, 167.
Dactylosphæra, 106.
Dallasia, 207.
Dallingeria, 138.
Dasytricha, cortical plasm, 175; c tion, 207; cilia, 177.
Dendrocometes, 210.
Dendrometidæ, 210.
Dendromonas, stalk, 131; classificati
Dendrophrya, 107.
Dendrosoma, 197, 210.
Dendrosomidæ, 210.
Desmothoracida, 109.
deutomerite, 144.
Diaspora, 168.
diastole, 288.
diatomin, 37.
Dictyocysta, 208.
dictyotic moment, 51, 76.
Didinium, phylogeny, 206; classi 206; food-selection, 305.
Didymophyes, protoplasmic structur classification, 167.
Didymophyidæ, 167.
Difflugia, pseudopodia, 86; buddi conjugation, 97, 106, 218.
digestion, 280.
Dileptus, form, 172; trichocysts, 175 fication, 206; food-selection, 305.
Dimorpha, pseudopodia, 82; fig. classification, 137.
dimorphism, in Reticulariida, 94.
Dinema, 138.
Diniferidæ, 140.
Dinobryon, 57; fig. p. 115; classi 139.
Dinoflagellidia, 140.
Dinophrya, 206.
Diophrys, 209.
Dinophysidæ, 140.
Dinophysis, 140.
Diploconidæ, 110.
Diplocystis, skeleton, 75; classificati
Diplomita, 137.
Diplophrys, 106.
Diplopsalis, 140.
Diplosiga, collar, 123; classification,
Diplospora, 168.
Discophrya, vacuole, 187; classi 207.
Discorbina, 109; conjugation, 219.
Discorhynchus, 168.
Disporocystidæ, 168.
Distephanus, skeleton, 114; classi 140.
Distomea, 138.
Ditrema, 106.
division-centre, 261, 266, 268, 272.
dizoic cyst or spore, 153.

Doliocystidæ, 168.
Doliocystis, 168.
Dorataspidæ, 110.
Drepanomonas, 207.
Druppulidæ, 110.
dysentery, 64.
Dysteria, 207.

Echinomera, 167.
Ectoplasm, 34; interchange with endoplasm, 85.
Ehrenbergina, 108.
Eiermocystis, fig. p. 58.
Eikenia, 106.
Eimeria, conjugation, 229.
Ellipsidæ, 110.
Ellipsoidina, 108.
Enchelinidæ, 206.
Enchelys, 206; irritability, 295; food selection, 305.
encystment, 46, 193.
endogenous budding, 198.
endogenous sporozoite-formation, 158.
endoplasm, 34; interchange with ectoplasm, 85.
Endosphæra, 210.
endospore, 151.
Endothyra, 108.
Endothyrinæ, 108.
Enterodela, 9.
Entodinium, 208.
Entosiphon, 139.
Eozoön, 29.
Ephelota, fig. p. 58; classification, 210.
Epiclintes, 209.
epimerite, 144.
Epipyxis, division, 128; classification, 137.
epispore, 151.
Epistylis, nematocysts, 37; classification, 209.
Erviliinæ, 207.
Eudorina, colony-formation, 130; classification, 140; conjugation, 227.
Euglena, fig. p. 37, 62; membrane, 114; pyrenoids, 118; stigmata, 118; classification, 138; chromosomes, 256; secretion, 292; irritability, 296.
Euglenida, 138.
Euglenidæ, 138.
Euglenopsis, 138.
Euglypha, protoplasmic structure, 38, 68; shell, 40; encystment, fig. p. 41; budding, fig. p. 55, 93; contractile vacuole, 87; encystment, 90; classification, 106; conjugation, 219; mitosis, 264.
Euglyphidæ, 106.
Euplotes, fig. p. 54; classification, 209.
Euplotidæ, 209.
Euspora, 167.
Eutreptia, 138.
excretion, 52.
excretory granules, 286.
exogenous budding, 198.
extra-capsular protoplasm, 69.
extra-nuclear division-centre, 265.
Exuviælla, inter-relations, 136; classification, 140.
eye-spot; see stigma.

Fabularia, 107.
fertilization, in Coccidiida, 159, 229.
Flagellidia, 20; classification, 137.
flagellum, 60; structure, absorption, etc., 119.
flagellum fissure, 117.
Folliculina, covering, 176; classification, 208.
food-selection, 305, 309.
food-taking, in Mastigophora, 125; in Sporozoa, 146.
Foraminifera, 18.
Frondicularia, 108.
Frontonia, trichocysts, 175; vacuoles, 187; classification, 207.
Fusulina, 109.
Fusulininæ, 109.

galvanotaxis, 300.
Gamocystis, 167.
gastric vacuole, in Sarcodina, 91; in Infusoria, 185.
Gastrostyla, 50, 209.
gas vacuole, 290.
gemmation; see reproduction.
generation *de novo*, 26.
Gerda, degeneration of peristome, 204; classification, 209.
Glaucoma, 207.
Glenodinium, color in water, 62; shell, 116; food-taking, 126; classification, 140.
Globigerina, 108.
Globigerina ooze, 32.
Globigerinidæ, 108.
Gloidium, 106.
Glossatella, 140, 209.
Glugea, spore-capsules, 156; silk-worm epidemic, 165; classification, 169.
Goniodoma, 140.
Gonium, colony-formation, 129; classification, 140; conjugation, 226.

Gonospora, 168.
Gonyaulax, 140.
Gonyostomum, trichocysts, 119; classification, 209.
Goussia, 168.
Grassia, 138; relation to Ciliata, 200.
gregaloid colony, 57.
Gregarina, 167.
Gregarinida, 167.
Gregarinidæ, 167.
Gringa, 105.
Gromia, food-taking, fig. p. 91; classification, 106; digestion, 285.
Gromidæ, 106.
Gymnamœbina, classification, 105.
Gymnodinium furrow, 116; stigma, 119; food-taking, 126; primitive, 136.
Gymnophrys, 106.
Gymnosphæra, pseudopodia, 82.
Gymnostomata, 22.
Gymnostomina, 206.
Gymnospora, 167.
gynogonidia, in *Volvox*, 134.
Gypsina, 109.
Gyrocorys, 208.

hæmatochrome, in Mastigophora, 118.
Hæmatococcus, odor in water, 62; food-taking, 125; classification, 139; irritability, 295.
Hæmogregarina, 168.
Hæmoproteus, 168.
Hæmosporidia, 22.
Hæmosporidiida, 168.
Haliphysema, 107.
Halteria, food of *Actinobolus*, 50; classification, 208.
Halteridium, 168.
Halteriidæ, 208.
Haplophragmium, 107.
Haplostiche, 107.
Hastigerina, 108.
Hauerina, 107.
Haueríninæ, 107.
Heliozoa, 17, 109.
Hemidinium, shell and furrows, 116; classification, 140.
Henneguya, 169.
Herpetomonas, 137.
Heteromastigida, 137.
Heteromita, conjugation, 221.
Heteromonadidæ, 137.
Heteronema, 138.
Heterophrys, 93.
Heterostegina, 109.
Heterotrichida, 171, 208.
Hexalaspidæ, 110.
Hexamitus, 138.
Hippocrepina, 107.
Hirmidium, 137.
Hirmocystis, 167.
Histioneis, 140.
Holophrya, cilia, 177; classification, 206.
holophytic nutrition, 24.
Holosticha, 209.
Holotrichida, 171, 206.
holozoic nutrition, 24.
Hoplitophrya, vacuole, 187; classification, 207.
Hoplorhynchus, 168.
Hormosina, 107.
Hyalobryon, 139.
Hyalodiscus, 106.
Hyaloklossia, 168.
Hyalopus, conjugation, 99, 225.
Hyalosphænia, 106.
Hyalospora, 167.
Hymenomonas, 139.
Hypocoma, cilia, 195; classification, 210.
Hypocomidæ, 210.
Hypotrichida, 171; peristomial structures, 181; classification, 208.

Ileonema, tentacles, 183; classification, 206.
Imperforina, 106.
Infusoria, 15; protoplasmic structure, 173; mouth-shifting, 185; vacuoles, 186; nucleus, 188; reproduction, 192; inter-relations, 199; classification, 206.
inorganic fluids simulating Protozoa, 306.
Inotogmata, 83.
inter-alveolar substance, 35.
intermediate skeleton, 73.
intra-capsular protoplasm, 69.
irritability, 294.
isogamy, 214.
Isospora, 168.
isospore, 95.
Isotricha, 207.
Isotrichidæ, 207.

Jaculella, 107.
Jenia, 138.

karyogamy, Sarcodina, 90, 97; general, 220.
karyolymph, 246.
karyophan, 179.
karyosome, 42, 87, 146.

Kentrochona, 209.
Kentrochonopsis, 209.
Keramosphæra, 107.
Keramosphærinæ, 107.
Kerona, 209.
kinetic centre; see division-centre.
kinoplasm, 248, 274.
Klossia, 168; conjugation, 229; nucleus, 250, 254.

Lacrymaria, cilia, 177; phylogeny, 206; classification, 206.
Lagena, 108.
Lagenidæ, 108.
Lageninæ, 108.
Lagenophrys, 209.
Lankesterella, 168.
Larcaridæ, 110.
Larcoidida, 110.
Larnacidæ, 110.
Laverania, 168.
Lecquereusia, 106.
Légeria, 168.
Leidyonella, 138; relation to Ciliata, 200.
Lembadion, cilia, 177; vacuole, 187; classification, 207.
Lembus, 207.
Leptocinclis, 138.
Leptodiscus, 21, 112; origin, 136; classification, 140.
Leptotheca, 169.
Leucophrys, 207; conditions of conjugation, 220.
Lichnaspis, fig. p. 78.
Lichnophora, phylogeny, 203; classification, 209.
Lichnophoridæ, 209.
Lieberkühnia, 106; digestion, 284.
Lieberkühnidæ, 208.
Lingulina, 108.
linin-reticulum, 246, 252.
Lionotus, form, 172; trichocysts, 175; classification, 206; food-selection, 305.
Liosphæridæ, 110.
Lithelidæ, 110.
Lithobotryidæ, 110.
Lithocampidæ, 110.
Lithocystis, 168.
Litholophidæ, 110.
Lituola, 107.
Lituolidæ, 107.
Lituolinæ, 107.
Loftusia, 108.
Loftusinæ, 108.
Lophocephalus, 168.
Lophomonas, 138; relation to Ciliata, 200.
lorication moment, 76.
Loxocephalus, 207.
Loxodes, 206.
Loxophyllum, trichocysts, 175; nucleus, 189, 250, 257; classification, 206.
Lymphosporidium, fig. p. 36; life-cycle, 147; classification, 169.

macrogamete, *Vorticella*, 195.
macronucleus, 14, 188.
macrospore, 151.
macrosporozoite, 158.
Magosphæra, 26, 57, 130.
Malaria, 65, 160.
Mallomonas, 139.
Marginulina, 108.
Marsipella, 107.
Maryna, 208.
Mastigamœba, 99; protoplasm, 113; classification, 137.
Mastigophora, 15; structure, 113; flagella, 119; nucleus, 124; vacuoles, 127; interrelations, 135; classification, 136.
Mastigophrys, 103, 137.
maturation, 233.
megalosphæric chamber, 74, 95.
Megastoma, fig. p. 63, 126; classification, 138.
melanin granules, 144.
membrane, structure, 182.
membranelle, 171; structure, 182.
Menoidium, 138.
Menospora, 168.
Menosporidæ, 168.
merozoite, 157, 229.
Mesodinium, tentacles, 183; phylogeny, 206; classification, 206.
mesoplasm, 144.
Mesostigma, 139.
Metacineta, 210.
Metacinetidæ, 210.
metasitism, 30.
Metopus, 208.
micellæ, 29.
microgamete, 195.
Microglena, 139.
Microgromia, fig. p. 56; in division, 93; classification, 106.
micronucleus, 14, 188.
micropyle, 229.
microsphæric chamber, 74.
microspore, 151.
Microsporidiina, 169.
microsporozoite, 158.

Microthoracidæ, 207.
Microthorax, 207.
Miliolidæ, 107.
Miliolina, 107.
Miliolinæ, 107.
mitosis, 246.
mitotic figure, 246.
Monadida, 137.
Monas, 104, 137; irritability, 297.
Monaster, 140.
monaxonic forms, 34.
Monera, 30.
Monobia, 109.
Monocercomonas, 138.
Monocystis, fig. p. 63; pseudopodia, 144; life-cycle, 151; conjugation, 157, 225; classification, 168; nucleus, 250.
Monodontophrya, vacuole, 187; classification, 208.
monophagous Sporozoa, 142.
Monopylea, 70, 110.
Monosiga, 137.
Monostomea, 138.
Monostominæ, 106.
monozoic spore, 153.
morulit, 254.
mosquitoes and malaria, 164.
motor response, 298.
mouth-shifting, 178.
movement, Sporozoa, 147.
Müllerian law, 77.
Multicilia, flagella, 121; classification, 138; relation to Ciliata, 200.
Mycetozoa, 18.
Mycetozoida, 18.
myoneme, 38; in Sporozoa, 145; in Infusoria, 178.
Myxastrum, 109.
Myxidiidæ, 169.
Myxidium, 169.
Myxinia, fig. p. 20.
Myxobolidæ, 169.
Myxobolus, fig. p. 39; spore-formation, 154; classification, 169.
Myxodictyum, contractile vacuole, 87; classification, 106.
Myxomycetes, 18.
Myxosoma, 169.
Myxosporidia, 21.
Myxosporidiida, 169.

Nadinella, 106.
Nasselaria, 70; classification, 110.
Nasselidæ, 109.
Nassoidida, 109.
Nassula, trichocysts, 175; classification, 207; irritability, 295.
Nassulinæ, 207.
Nebenkern (micronucleus), 14, 188.
Nebenkörper, 93.
nematocysts, 37; in Mastigophora, 119; in *Vorticella*, 176.
Neosporidia, 168.
Nephroselmis, 139.
Noctiluca, 21, 112; alveoli, 35; feeding, 50, 127; budding, 59; color in water, 62; origin, 136; classification, 140; conjugation, 219; nucleus, 249, 252; mitosis, 256, 257; phosphorescence, 293.
Nodosaria, fig. p. 72; classification, 108.
Nodosarina, fig. p. 73.
Nodosinella, 108.
Nonionina, 109.
Nosema, 169.
Nosematidæ, 169.
Nubecularia, 107.
Nubecularinæ, 107.
Nuclearia, shell, 74; pseudopodia, 82; division, 93; fig. p. 102; classification, 109.
nuclein, 246.
nucleolus, 253.
nucleus, types in Protozoa, 40; in Sarcodina, 86; in Mastigophora, 124; in Sporozoa, 146; in Infusoria, 188; general, 246, 301.
Nuda, 106.
Nummulinidæ, 109.
Nummulites 109.
Nummulitinæ, 109.
nutrition, 24, 280.
Nyctotherus, 208.

Ochromonas, food-taking, 126; classification, 139.
odors caused by Protozoa, 62.
Oikomonas, fig. p. 49; food-taking, 125; classification, 137.
old age, 60.
Oligotrichina, 208.
Onychodactylus, 207.
Onychodromus, degeneration, 60; classification, 209; conjugation, 220, 234.
Oöcephalus, 168.
Opalina, contractile vacuole, 187; nucleus, 190; classification, 207.
Opalinidæ, 207.
Opalinopsis, 207.
Opercularia, fig. p. 73; classification, 209.
Ophrydium, covering, 176; vacuole, 187; classification, 209.
Ophryocystis, 169.

Ophryodendridæ, 210.
Ophryodendron, budding, 199; classification, 210.
Ophryoglena, 207.
Ophryoscolecidæ, 208.
Ophryoscolex, 208.
Opisthodon, 207.
Orbiculina, 107.
Orbitoides, 109.
Orbitolites, 73, 107.
Orbulinella, 109.
Ornithocercus, 140.
Orosphæridæ, 111.
Orthodon, 207.
Oxyrrhis, division, 128; classification, 138; nucleus and division-centre, 271.
Oxytoxum, 140.
Oxytricha, phylogeny, 202; classification, 209.
Oxytrichidæ, 209.

Pachymyxa, 106.
Panartidæ, 110.
Pandorina, fig. p. 37; colony-formation, 129; reproduction, 132; classification, 140; conjugation, 227.
pan-sporoblast, 155.
paraglycogen, 143.
Paramœba, reproduction, 59; spore-formation, 93; classification, 105; conjugation, 218; chromosomes, 233; reduction, 234; division-centre, 276.
Paramœcidæ, 207.
Paramœcium, trichocysts, 37, 175; cyclosis, 173; strength of, 181; classification, 207; digestion, 282; urea, 291; motor response. 298; irritability, 301.
paramylum, in Mastigophora, 117.
parasitic Protozoa, 63.
Parkeria, 108.
parthenogonidia, in *Volvox*, 134, 232.
Patellina, 109; conjugation, 219.
Pavonina, 108.
pellicula, 38, 114, 176.
Pelomyxa, 106; protoplasmic structure, 35, 38; nuclei, 87; vacuole, 87; budding, 95; starch digestion, 281; irritability, 297.
Pelosina, 107.
Peneroplidinæ, 107.
Peneroplis, 107.
Peranema, motion, 121; classification, 138.
Peranemidæ, 138.
Perforina, 108.
Peridinidæ, 140.
peridinin, in Dinoflagellidia, 118.
Peridinium, fig. p. 20; color, 62; shell, 116; *pseudonoctiluca*, 136; classification, 140.
periplast, 113.
Peripylea, 69, 110.
peristome, 184.
Peritrichida, 171, 209.
Peritromidæ, 208.
Peritromus, cilia, 181; phylogeny, 202; classification, 208.
Petalomonas, 138.
Phacodiscidæ, 110.
Phacotidæ, 139.
Phacotus, shell, 114; classification, 139; conjugation, 224.
Phacus, fig. p. 19; classification, 138.
Phænocystina, 169.
Phæoconchida, 111.
Phæocystinida, 111.
Phæodaria, 111.
Phæodinidæ, 111.
phæodium, 111.
Phæogromida, 111.
Phæospherida, 111.
Phalacroma, 140.
Phalansteridæ, 137.
Phalansterium, fig. p. 122; classification, 137.
Phascolodon, 207.
Phialoides, 168.
Phormocampidæ, 111.
Phormocyrtidæ, 111.
Phormospyridæ, 110.
Phorticidæ, 110.
phosphorescence, 62, 293.
phototaxis, 296.
Phractopeltidæ, 110.
phycopyrrin, in Dinoflagellidia, 118.
Phyllomitus, 138.
Phyllomonas, 137.
Phytoflagellida, 19, 25, 139.
Phytomastigoda, 25.
Pileocephalus, 168.
Pilulina, 107.
Pilulininæ, 107.
Pinaciophora, skeleton, 75; classification, 109.
Pinacocystis, 109.
Plæodorina, 140.
Plagiopogon, 206.
Plagiotoma, 208.
Plagiotomidæ, 208.
Plagonidæ, 110.
Plakopus, 106.
Planorbulina, 109.
Plasmodium malariæ, 65; conjugation and life-cycle, 160.

plasmosome, 246.
plastic granules in Sporozoa, 144.
plastogamy, in Sarcodina, 90, 97; general, 218.
Platoum, 106.
Platydorina, 140.
Platytheca, 137.
Plectanidæ, 110.
Plectoidida, 110.
Plectophrys, 106.
Pleuronema, 207; irritability, 296.
Pleuronemidæ, 207.
Pleurotricha, 209.
Plistophora, 169.
Podocampidæ, 111.
podoconus, 70.
Podocyathus, 210.
Podocyrtidæ, 111.
Podolampas, 140.
Podophrya, reproduction, 198; classification, 210.
Podophryidæ, 210.
Podostoma, 106.
polar bodies, 237.
polar capsule, 154.
pole-plates, 262.
Polydinida, 140.
Polygastrica, 9.
Polykrikos, 37, 41; trichocysts, 119; food-taking, 126.
Polymastigida, 138.
Polymitus form, 162.
Polymorphina, 108.
Polymorphininæ, 108.
Polyœca, 137.
polyphagous Sporozoa, 142.
Polyphragma, 108.
Polysporocystidæ, 168.
Polystomella, 72, 109; regeneration, 302; irritability, 306.
Polystomellinæ, 109.
Polytoma, starch, 118; reproduction, 131; classification, 139; irritability, 297, 301.
Polytrema, 109.
Polytrichina, 208.
polyzoic spore, 153.
Pompholyxophrys, 109.
Pontomyxa, 106.
Porodiscidæ, 110.
Porospora, 167.
Porosporidæ, 167.
Poteriodendron, 137.
primite, 156.
Prorocentridæ, 140.
Prorocentrum, color, 62; flagella, 121; inter-relations, 136; classification, 140.
Prorodon, trichocysts, 175; classification, 206; myonemes, 177.
Protamœba, contractile vacuole, 87, 105.
protein, Sporozoa, 143.
Proterospongia, fig. p. 113; classification, 137.
Protista, 25.
Protoceratium, 140.
Protogenes, contractile vacuole, 87, 106.
protomerite, 144.
Protomonas, 103.
Protomyxa, 106.
protoplasm, of Sarcodina, 67; Mastigophora, 113; Sporozoa, 142; Ciliata, 173.
Protozoa, name, 28; mode of life, 32; old age and immortality, 60; economic aspects, 62; cancer, malaria, etc., 65; sex, 211; physiology, 279.
Prunoidida, 110.
Prunophractida, 110.
Psammosphæra, 107.
Pseudochlamys, 40, 106.
pseudoconjugation, 156.
pseudocyst, 151.
Pseudodifflugia, 106.
Pseudonavicella, 151.
Pseudopodia, 17, 79–86; explanation of movement, 83.
Pseudospora, 103.
Pseudosporæ, 103.
Psilotricha, 209.
psorosperm, 21.
Pterocephalus, 167.
Pterospora, 168.
Ptychodiscus, 140.
Ptychostomum, 207.
Pullenia, 108.
Pylobotryidæ, 110.
Pylodiscidæ, 110.
Pylonidæ, 110.
Pyramimonas, 139.
pyrenoid, 117.
Pyrophacus, 140.
Pyrsonympha, 138.
Pyxidicula, 106.
Pyxinia, fig. p. 3; 143; classification, 168.
pyxinine, 143.

Quadrilonchidæ, 110.
Quadrula, budding, 93; classification, 106.

Radiolaria, 17; central capsule, 67, 70; protoplasmic structure, 69; modifications of skeleton, 77; spore-formation, 95; conjugation, 96; development, 105; classification, 110.

Radiolarian ooze, 32.
Rainey's corpuscles, 156.
Rainey's tubes, 145.
Ramulina, 108.
Ramulininæ, 108.
Raphidiophrys, fig. p. 49; skeleton, 75, 93.
réjeunissement; see rejuvenescence.
rejuvenescence, 14, 61, 213.
reorganization; see rejuvenescence.
reproduction, 54; division, 55; spore-formation, 58; budding, 59; in Sporozoa, 149; in Infusoria, 192.
respiration, 288.
Reticulariida, 18, 67, 106.
Rhabdammina, 107.
Rhabdammininæ, 107.
Rhabdomonas, 138.
Rhabdostyla, 209.
Rhipidodendron, shell, 114; classification, 137.
Rhizammina, 109.
Rhizomastigidæ, 99, 137.
Rhizopoda, 105.
Rhopalonia, 167.
Rhynchæta, 210.
Rimulina, 108.
Rotalia, 109.
Rotalidæ, 108.
Rotalinæ, 109.

Saccammina, 107.
Saccammininæ, 107.
Sagenella, 107.
Sagosphæridæ, 111.
Sagrina, 108.
Salpingœca, 137.
Saltonella, 106.
saprophytic nutrition, 24.
Sarcocystis, 169.
sarcodictyum, 69.
Sarcodina, 15, 67; shells and tests, 71; dimorphism, 74; pseudopodia, 82; nuclei, 86; contractile vacuole, 87; nutrition, 90; encystment, 90; reproduction, 92; classification, 105.
sarcomatrix, 69.
Sarcosporidiida, 169.
satellite, 156.
Scaphidiodon, 207.
Schaumplasma, 35.
schizogony, 159; in malaria organism, 161; general, 229.
schizont, 159.
Schneideria, 168.
Schwagerina, 109.
Sciadiophora, 168.
Scyphidia, 209.
Scytomonas, 138.
secretion and excretion, 290.
selection of food, 48.
Semantidæ, 110.
senile degeneration, 220.
sensory organs, 61.
Serumsporidia, 169.
Serumsporidium, 165; classification, 169.
Sethocyrtidæ, 111.
sex-differentiation, definition, 213.
shells and tests, Reticulariida, 71.
Shepheardella, 106.
Silicoflagellida, 140.
Solenophrya, 210.
Soreumidæ, 110.
Sorosphæra, 107.
Sparotricha, 209.
spasmoneme, in *Vorticella*, 179.
Spathidium, 206.
Sphærastrum, pseudopodia, 82, 93; classification, 109; division-centre, 266.
Sphærocapsida, 110.
Sphærocystis, 167.
Sphæræca, 137.
sphæroid colony, 57.
Sphæroidida, 110.
Sphæroidina, 108.
Sphæromyxa, 169.
Sphærophractida, 110.
Sphærophrya, 210.
Sphærorhynchus, 168.
Sphærospora, 169.
Sphærozoidæ, 110.
Sphenomonas, 138.
Spirillina, 109.
Spirillinæ, 109.
Spirochona, mitosis, 191, 261; classification, 209.
Spirochonidæ, 209.
Spiroloculina, 107.
Spironema, 138.
spironeme, 179.
Spirostomum, vacuole, 187; nuclei, 190; classification, 208; irritability, 301.
Spondylomorum, 139.
Spongodiscidæ, 110.
Spongomonas, 137.
Sponguridæ, 110.
spontaneous division, 59.
spontaneous generation, 26.
spore-formation, 58.
sporoblast, 151, 225.
sporocyst, 151.

sporoduct, 151.
sporont, 145.
Sporozoa, 15; structure, 142, 146; food-taking, 146; motion, 148; reproduction, 149; inter-relationships, 166; classification, 167.
sporozoite, 141.
Spumellaria, 70, 110.
Spyroidida, 110.
Squamulina, 107.
Stacheia, 108.
Staurophrya, 210.
Staurophryidæ, 210.
Staurosphæridæ, 110.
Steiniella, 140.
Stenophora, 167.
Stentor, pseudopodia, 175, 183; nuclei, 190; division, 192; classification, 208.
Stentoridæ, 208.
Stephanidæ, 110.
Stephanophora, 168.
Stephanopogon, phylogeny, 201; classification, 206.
Stephanosphæra, 140.
Stephoidida, 110.
Stichotricha, 209.
Stictospora, 168.
stigma (eye-spot), fig. p. 37; 61; in Infusoria, 174.
Streblonidæ, 110.
Strombidium, trichocysts, 175; classification, 208.
Strongylidium, 209.
Stylamœba, 106.
styliform tentacles, 195.
Stylochrysalis, division, 128; classification, 139.
Stylocometes, 210.
Stylonychia, 209; conjugation, 220; nucleus, 257; irritability, 295.
Stylorhynchidæ, 168.
Stylorhynchus, 168.
Stylosphæridæ, 110.
Succto-ciliata, 205.
Suctorella, 210.
Suctoria, 18, 22, 195; food-taking, 52; tentacles, fig. p. 52; reproduction, 198; classification, 209.
symbiosis, in Radiolaria, 70; in *Paramæcium*, 282.
Symphyta, 166.
Syncrypta, 139.
Syncystis, 168.
Synura, odor, 62; classification, 139.
Syringammina, 107.
Telosporidia, 167.
tentacles, 52, 196.
Testacea, 18.
Tetramitus, flagella, 121; distributed nucleus, 124, 251; classification, 138; conjugation, 214.
Tetrasporocystidæ, 168.
Tetrastyla, 209.
Tetratoma, 139.
Textularia, 108.
Textularidæ, 108.
Textularinæ, 108.
Thalamophora, 18.
Thalamopora, 109.
Thalassicolidæ, 18, 110.
Thalassicolla, 110; enucleated central capsule, 302.
Thalassosphærida, 110.
Thaumatomastix, 139.
Thecamœbina, 106.
Thélohania, 169.
Theocyrtidæ, 111.
thermotaxis, 295.
thigmotaxis, 300.
Tholonidæ, 110.
Tholospyridæ, 110.
Thurammina, 107.
Thylakidium, 208.
Tiarina, 206.
Tinoporinæ, 109.
Tinoporus, 109.
Tintinnidæ, 208.
Tintinnidium, 208.
Tintinnopsis, 208.
Tintinnus, 208.
Tokophrya, 210.
Trachelinidæ, 206.
Trachelius, trichocysts, 175; classification, 206.
Trachelocerca, 206.
Trachelomonas, shell, 114; division, 128; classification, 138.
Trepomonas, 138; irritability, 297.
trichocysts, in Mastigophora, 119; in Infusoria, 175.
Trichodina, phylogeny, 203; classification, 209.
Trichodinopsis, 209.
Trichogaster, 209.
Trichomonas, 138.
Trichonympha, 138; relation to ciliata, 200.
Trichonymphinea, 138.
Trichophrya, 210.
Trichorhynchus, 167.
Trichosphærium, 106.

Trichostomata, 22.
Trichostomina, 207.
Trigonomonas, 138.
Trimastigidæ, 138.
Trimastix, 138.
Trinema, 106.
Tripocalpidæ, 111.
Tripocyrtidæ, 111.
Tritaxia, 108.
Trochammina, 107.
Trochammininæ, 107.
Trochilia, 207.
trophoplasm, 274.
Tropidoscyphus, 139.
Truncatulina, 109; shell, 305.
Trypanosoma, 137.
Tubulata, 18.
Tuscaroridæ, 111.
Tympanidæ, 110.

Urceolarinæ, 209.
Urceolus, 138.
urea, 54, 291.
Urnula, 210.
Urnulidæ, 210.
Urocentridæ, 207.
Urocentrum, caudal cilia, 182; vacuoles, 187; classification, 207.
Uroglena, 25; odor, 62; fig. p. 57; colony-formation, 129; division of the colony, 131; classification, 139.
Uroleptus, 209.
Uronema, 207.
Uronychia, 209.
Urophagus, 138.
Urosoma, 209.
Urospora, 168.
Urostyla, 209; nucleus, 251.
Urotricha, 206.
Urozona, 207.
Urthiere, 28.
Uvigerina, 108.

Vacuolaria, 139.
vacuoles, in Sarcodina, 87; in Mastigophora, 127; in Infusoria, function, 291.
Vaginicola, 209.
Vaginulina, 108.
Valvulina, 108.
Vampyrella, pseudopodia, 82; encystment, 90; division, 92; conjugation, 98, 104; classification, 109; food-selection, 305.
Verjüngung; see rejuvenescence.
Verneuilina, 108.
Virgulina, 108.
Volvocina, 139.
Volvox, 26, 57; colony-formation, 129; reproduction, 134; classification, 140; conjugation, 232.
Vorticella, chlorophyl, 174; origin of, 202; nematocysts, 176; myonemes, 178; classification, 209; urea, 291.
Vorticellidæ, origin of, 202; classification, 209.
Vorticellidinæ, 209.

Wagnerella, 109.
Webbina, 107.
Wimperring, 179.

Zonaridæ, 110.
Zoöglæa, 28.
zoöphyte, 25.
Zoothamnium, conjugation, 195, 222; classification, 209.
Zygartidæ, 110.
Zygocystis, 168.
Zygoselmis, 138.
zygospore, in Mastigophora, 133.
Zygospyridæ, 110.

www.ingramcontent.com/pod-product-compliance
Ingram Content Group UK Ltd.
Pitfield, Milton Keynes, MK11 3LW, UK
UKHW021430280726
14060UKWH00001BA/6